A Collection of Problems on the Equations of Mathematical Physics

Edited by V. S. Vladimirov

Translated from the Russian by Eugene Yankovsky

Springer-Verlag Berlin Heidelberg GmbH

Edited by Academician V. S. Vladimirov
Contributors: V. S. Vladimirov, V. P. Mikhailov,
A. A. Vasharin, Kh. Kh. Karimova, Yu. V. Sidorov,
M. I. Shabunin

First published 1986
Revised from the 1982 Russian edition

With 4 Figures

ISBN 978-3-662-05560-1 ISBN 978-3-662-05558-8 (eBook)
DOI 10.1007/978-3-662-05558-8

Originally published by English translation, Mir Publishers, in 1986.
Softcover reprint of the hardcover 1st edition 1986

Preface

The extensive application of modern mathematical techniques to theoretical and mathematical physics requires a fresh approach to the course of equations of mathematical physics. This is especially true with regards to such a fundamental concept as the solution of a boundary value problem. The concept of a generalized solution considerably broadens the field of problems and enables solving from a unified position the most interesting problems that cannot be solved by applying classical methods. To this end two new courses have been written at the Department of Higher Mathematics at the Moscow Physics and Technology Institute, namely, "Equations of Mathematical Physics" by V.S. Vladimirov and "Partial Differential Equations" by V.P. Mikhailov (both books have been translated into English by Mir Publishers, the first in 1984 and the second in 1978).

The present collection of problems is based on these courses and amplifies them considerably. Besides the classical boundary value problems, we have included a large number of boundary value problems that have only generalized solutions. Solution of these requires using the methods and results of various branches of modern analysis. For this reason we have included problems in Lebesgue integration, problems involving function spaces (especially spaces of generalized differentiable functions) and generalized functions (with Fourier and Laplace transforms), and integral equations.

The book is aimed at undergraduate and graduate students in the physical sciences, engineering, and applied mathematics who have taken the typical "methods" course that includes vector analysis, elementary complex variables, and an introduction to Fourier series and boundary value problems. Asterisks denote the more difficult problems.

We would like to express our gratitude to all who helped with constructive comment to improve this book, our colleagues at the Department of Higher Mathematics at the Moscow Physics and Technology Institute, and especially T.F. Volkov, Yu.N. Drozhzhinov, A.F. Nikiforov, and V.I. Chekhlov.

V.S. Vladimirov
V.P. Mikhailov
A.A. Vasharin
Kh.Kh. Karimova
Yu.V. Sidorov
M.I. Shabunin

Contents

Symbols and Definitions

1. We denote a real n-dimensional Euclidean space by R^n and its points by $x = (x_1, x_2, \ldots, x_n)$, $y = (y_1, y_2, \ldots, y_n)$, ξ and the like.

2. $dx = dx_1\, dx_2 \ldots dx_n$,

$$\int f(x)\, dx = \int_{R^n} f(x_1, \ldots, x_n)\, dx_1 \ldots dx_n.$$

3. $\alpha = (\alpha_1, \alpha_2, \ldots, \alpha_n)$ is a multi-index, with the α_j nonnegative integers. We will also use the abbreviations

$$\alpha! = \alpha_1!\, \alpha_2! \ldots \alpha_n!, \quad x^\alpha = x_1^{\alpha_1} x_2^{\alpha_2} \ldots x_n^{\alpha_n}.$$

4. $(x, y) = x_1 y_1 + x_2 y_2 + \ldots + x_n y_n$

$$r = |x| = \sqrt{(x, x)} = \sqrt{x_1^2 + x_2^2 + \ldots + x_n^2}.$$

5. $U(x_0; R) = \{x: |x - x_0| < R\}$ is an *open ball* of radius R centered at point x_0, and $S(x_0; R) = \{x: |x - x_0| = R\}$ is a *sphere* of radius R centered at x_0; $U_R = U(0, R)$ and $S_R = S(0; R)$.

6. A set A will be said to be *lying strictly* in a region $G \subset R^n$ (this is denoted by $A \Subset G$) if it is bounded and $\bar{A} \subset G$.

7. A function $f(x)$ is said to be *locally integrable in a region* G if it is absolutely integrable in every subregion $G' \Subset G$. Functions that are locally integrable in R^n will be said to be simply *locally integrable.*

8. $$D^\alpha f(x) = \frac{\partial^{|\alpha|} f(x_1, x_2, \ldots, x_n)}{\partial x_1^{\alpha_1}\, \partial x_2^{\alpha_2} \ldots \partial x_n^{\alpha_n}}.$$

9. $C^p(G)$ is a class of functions f that are continuous together with the derivatives $D^\alpha f$, $|\alpha| \leqslant p$ $(0 \leqslant p < \infty)$ in the region $G \subset R^n$. The functions f of class $C^p(G)$ for which all the derivatives $D^\alpha f$, $|\alpha| \leqslant p$, allow continuous continuation into the closure $\bar{G}$ form the class $C^p(\bar{G})$; $C(G) = C^0(G)$, $C(\bar{G}) = C^0(\bar{G})$. We denote

the class of functions belonging to $C^p(G)$ for all p's by $C^\infty(G)$; the class $C^\infty(\bar{G})$ is defined in a similar manner.

10. The uniform convergence of a sequence of functions $\{f_k\}$ to a function f on a set A is denoted by

$$f_k(x) \overset{x \in A}{\Rightarrow} f(x), \quad k \to \infty$$

11. $A \cup B$ is the *union* of sets A and B, $A \cap B$ is the *intersection* of A and B, $A \setminus B$ is the *complement* of B with respect to A, and $A \times B$ is the *direct product* (or simply product) of A and B (the set of pairs (a, b) with $a \in A$ and $b \in B$).

12. The *support* of a continuous function f is denoted by supp f and is the closure of the set of all points x for which $f(x)$ is nonzero. If a function $f(x)$ that is measurable on a region G vanishes almost everywhere in $G \setminus G'$, where $G' \Subset G$, then it is *finite in* G; a function that is finite in R^n is said to be simply *finite*.

13. $\nabla^2 = \frac{\partial^2}{\partial x_1^2} + \frac{\partial^2}{\partial x_2^2} + \ldots + \frac{\partial^2}{\partial x_n^2}$ is Laplace's operator;

$\Box = \frac{\partial^2}{\partial t^2} - a^2\nabla^2$ is the wave operator; $\Box_1 = \Box$;

$\frac{\partial}{\partial t} - a^2\nabla^2$ is the heat conduction operator.

14. $\Gamma^+ = \{x, t\colon at > |x|\}$ is a future cone.

15. $\Phi(\xi) = \frac{1}{\sqrt{2\pi}} \int_{-\infty}^{\xi} e^{-z^2/2}\, dz$

16. $\omega_\varepsilon(x) = \begin{cases} C_\varepsilon e^{-\varepsilon^2/(\varepsilon^2 - |x|^2)} & |x| \leqslant \varepsilon, \\ 0, & |x| > \varepsilon, \end{cases}$

where $C_\varepsilon = \varepsilon^{-n}\varkappa, \quad \frac{1}{\varkappa} = \int_0^1 e^{-1/(1-x^2)}\, dx,$

(ω_ε is the averaging kernel, or "cap").

17. $\mathbb{C}$ is the complex plane.

18. $\theta(x)$ is the Heaviside unit function:

$$\theta(x) = \begin{cases} 1 \text{ if } x \geqslant 0, \\ 0 \text{ if } x < 0. \end{cases}$$

19. $\sigma_n = \int_{S_1} ds = \frac{2\pi^{n/2}}{\Gamma(n/2)}$ is the surface area of the unit sphere S_1 in R^n.

20. In $C^p(\bar{G})$ the norm is

$$\| f \|_{C^p(\bar{G})} = \sum_{|\alpha| \leqslant p} \max_{x \in \bar{G}} |D^\alpha f(x)|$$

21. The totality of (measurable) functions $f(x)$ for which $|f|^p$ is integrable on G is denoted by $L_p(G)$. In $L_p(G)$ the norm is

$$\|f\|_{L_p(G)} = \Big[\int_G |f|^p\,dx\Big]^{1/p}, \quad 1 \leqslant p < \infty,$$

$$\|f\|_{L_\infty(G)} = \operatorname*{vrai\,sup}_{x \in G} |f(x)|, \quad p = \infty$$

The scalar product in $L_2(G)$ is introduced thus:

$$(f, g) = \int_G f\bar{g}\,dx, \quad f, g \in L_2(G).$$

22. Let $\rho(x)$ be a continuous positive-valued function in a region G. The totality of (measurable) functions $f(x)$ for which $\rho(x)\,|f(x)|^2$ is integrable on G is denoted by $L_{2,\rho}(G)$ and constitutes a Hilbert space with the scalar product

$$(f, g)_{L_{2,\rho}}(G) = \int_G \rho f \bar{g}\,dx.$$

23. Cylinder functions.

(a) Bessel functions:

$$J_\nu(x) = \sum_{k=0}^{\infty} \frac{(-1)^k}{\Gamma(k+\nu+1)\,\Gamma(k+1)} \left(\frac{x}{2}\right)^{2k+\nu}, \quad -\infty < x < \infty;$$

(b) Neumann's Bessel functions:

$$N_\nu(x) = \frac{1}{\sin \pi\nu}\,[J_\nu(x)\cos \pi\nu - J_{-\nu}(x)], \quad \nu \neq n,$$

$$N_n(x) = \frac{1}{\pi}\Big[\frac{\partial J_\nu(x)}{\partial \nu} - (-1)^n \frac{\partial J_{-\nu}(x)}{\partial \nu}, \quad \nu = n;$$

(c) Hankel functions:

$$H_\nu^{(1)}(x) = J_\nu(x) + iN_\nu(x), \quad H_\nu^{(2)}(x) = J_\nu(x) - iN_\nu(x);$$

(d) modified Bessel and Hankel functions:

$$I_\nu(x) = e^{-\frac{\pi}{2}\nu i} J_\nu(ix), \quad K_\nu(x) = \frac{\pi i}{2} e^{\frac{\pi}{2}\nu i} H_\nu^{(1)}(ix).$$

Chapter I

Statement of Boundary Value Problems in Mathematical Physics

1 Deriving Equations of Mathematical Physics

We start by introducing the following notation:

$\rho(x) = \rho$ is the density of a material (linear, surface, or volume);

T_0 is the tensile force acting on a string or membrane;

E is Young's modulus;

k is the coefficient of elasticity of a string or rod with elastically fixed ends or of a membrane;

S is the cross-sectional area of a rod, shaft, and the like;

$\gamma = c_p/c_v$ is the specific heat ratio;

p and p_0 are pressures in a fluid;

m and m_0 denote mass;

g is the gravitational acceleration near the earth's surface;

ω is angular velocity;

k, $k(x)$ and $k(x, u)$ are internal thermal conductivities;

α is the external thermal conductivity (the heat-exchange coefficient);

D is the diffusion coefficient.

Here are some examples dealing with deriving equations.

Example 1: *Transverse vibrations of strings.* A string of length l with a tensile force T_0 is in the horizontal state of equilibrium. At time $t = 0$ the string is displaced with an initial velocity. We wish to state the problem of determining the small transverse vibrations of the string at $t > 0$ if the ends of the string are (a) rigidly fixed, (b) free (i.e. can freely move along straight lines parallel to the direction of displacement u), (c) fixed elastically (i.e. each end is under a resistance force from the fixture, and this force is proportional to the displacement and is directed opposite to the displacement), or (d) move in the transverse direction according to specified laws. We ignore the force of gravity and the resistance of the medium in which the string moves.

Solution. We send the x axis along the string in the equilibrium position. We assume that the string is a thin cord that does not resist bending not related to changes in length. This means that if we mentally cut the string at a point x, the action of one portion of the string on the other (or simply the tensile force T) is directed along the tangent to the string at that point. To derive the equation of motion (vibrations) of the string we examine the portion of the string between x and $x + \Delta x$ and project all the forces acting on this portion (including the forces of inertia) on the axes of coordinates. According to D'Alembert's principle, the sum of the projections of all forces on each axis must be zero. Since we are studying only transverse vibrations, we can assume that the external forces and the force of inertia act along the u axis. We will consider only small

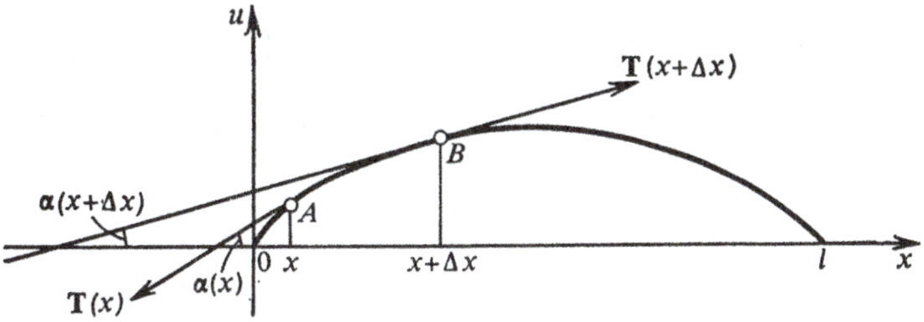

Fig. 1

vibrations, which means that in deriving our equation we will neglect squares of $u_x(x, t)$. The length S of the arc AB is expressed by the integral $S = \int_x^{x+\Delta x} (1 + u_x^2)^{1/2}\, dx \cong \Delta x$. This means that no portion of the string changes its length during vibrations and, hence, by Hooke's law the magnitude of the tensile force $T_0 = |\mathbf{T}|$ depends neither on time nor on x. Let us find the projections of all the forces at time t onto the u axis. Up to first-order infinitesimals, the projection of the tensile force is (see Fig. 1)

$$
\begin{aligned}
&T_0[\sin\alpha(x+\Delta x) - \sin\alpha(x)] \\
&= T_0\left[\frac{\tan\alpha(x+\Delta x)}{\sqrt{1+\tan^2\alpha(x+\Delta x)}} - \frac{\tan\alpha(x)}{\sqrt{1+\tan^2\alpha(x)}}\right] \\
&= T_0\left[\frac{u_x(x+\Delta x, t)}{\sqrt{1+u_x^2(x+\Delta x, t)}} - \frac{u_x(x, t)}{\sqrt{1+u_x^2(x, t)}}\right] \\
&\cong T_0[u_x(x+\Delta x, t) - u_x(x, t)] \\
&\cong T_0 u_{xx}(x, t)\,\Delta x.
\end{aligned}
$$

Let $p(x, t)$ be the continuous linear density of the external forces acting on the string. Then the force that acts on the portion AB along the u axis is $p(x, t)\,\Delta x$. To find the force of inertia acting on AB, we recall the expression $-mu_{tt}$, with m the mass of that portion of the string. If $\rho(x)$ is the continuous linear density of the material of the string, then $m = \rho\,\Delta x$. Thus the projection of the force of inertia on the u axis is $-\rho u_{tt}\,\Delta x$, and the projection of all the forces on the same axis obeys the equation

$$[T_0 u_{xx} + p(x, t) - \rho(x) u_{tt}]\,\Delta x = 0. \tag{1.1}$$

Hence,

$$T_0 u_{xx} - \rho(x) u_{tt} + p(x, t) = 0,$$

which is the *equation of forced vibrations of a string*. If ρ is constant all along the string, the equation takes the form

$$u_{tt} = a^2 u_{xx} + g(x, t),$$

where $a^2 = T_0/\rho$ and $g(x, t) = p(x, t)/\rho$. The function $u(x, t)$ satisfies the initial conditions $u|_{t=0} = \varphi(x)$ and $u_t|_{t=0} = \psi(x)$, where $\varphi(x)$ and $\psi(x)$ are given functions. We now turn to deriving the boundary conditions:

(a) If ends of the string are rigidly fixed, $u|_{x=0} = 0$ and $u|_{x=l} = 0$.

(b) If the ends are free, then to obtain the boundary condition at $x = 0$ we project the forces that act on the portion KM onto the u

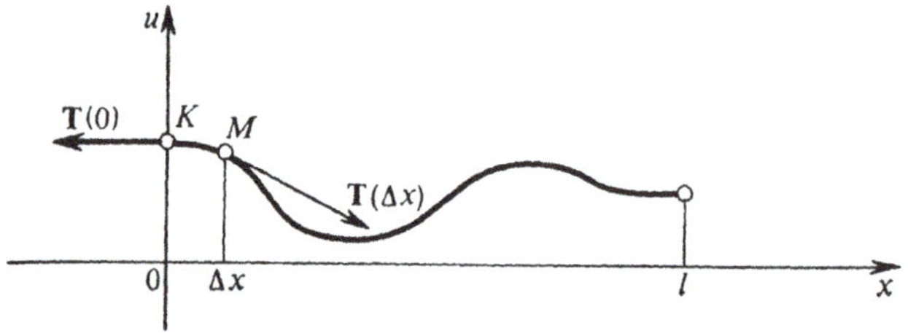

Fig. 2

axis (Fig. 2). Since the tensile force at point $x = 0$ can act only parallel to the x axis, the projection of this force on KM is $T_0 u_x(\Delta x, t)$. The projection of the external force on this axis is $p(0, t)\Delta x$, while the projection of the force of inertia is $-\rho u_{tt}(0, t)\Delta x$. Nullifying the sum of these projections yields the following equation:

$$T_0 u_x(\Delta x, t) + p(0, t)\Delta x - \rho u_{tt}(0, t)\Delta x = 0. \qquad (1.2)$$

Let us send Δx to zero. In view of the continuity and boundedness of the functions in this equation we arrive at the boundary condition $u_x|_{x=0} = 0$. Similarly, $u_x|_{x=l} = 0$.

(c) We take the case where the left fixture acts on the left end of the string with an elastic force $-ku(0, t)$. We nullify the sum of the projections of all the forces acting on the portion KM onto the u axis. The left-hand side of Eq. (1.2) has an additional term equal to $-ku(0, t)$. The resulting equation is

$$T_0 u_x(\Delta x, t) - ku(0, t) + p(0, t)\Delta x - \rho u_{tt}(0, t)\Delta x = 0.$$

which transforms into the following equation as $\Delta x \to 0$:

$$(u_x - hu)|_{x=0} = 0, \quad h = k/T_0$$

At the right end of the string (see Fig. 3) the sum of the projections of all the forces must obey the equation

$$-T_0 u_x(l - \Delta x, t) - ku(l, t) + p(l, t)\Delta x - \rho u_{tt}(l, t)\Delta x = 0,$$

since $\sin\alpha(l - \Delta x) \cong u_x|_{x=l-\Delta x}$. If we take the limit as $\Delta x \to 0$, we obtain $(u_x + hu)|_{x=l} = 0$.

(d) $u|_{x=0} = \mu_1(t)$ and $u|_{x=l} = \mu_2(t)$, where the functions $\mu_1(t)$ and $\mu_2(t)$ determine the law by which the two ends of the string move ($\mu_1(0) = \varphi(0)$ and $\mu_2(0) = \varphi(l)$).

Example 2: *Vibrations of a rod.* An elastic straight rod of length l is taken out of the state of rest by imparting small longitudinal

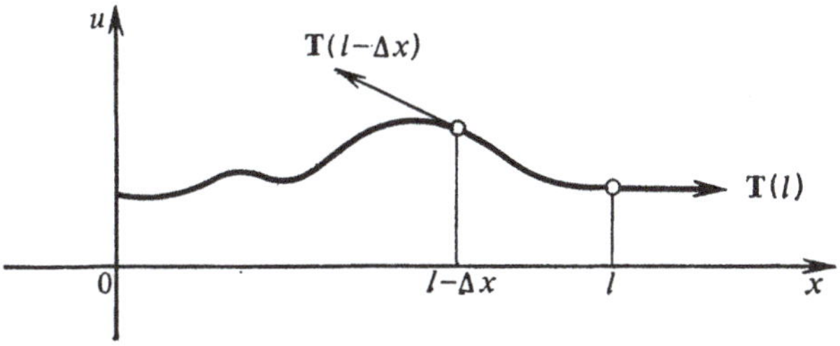

Fig. 3

displacements and velocities to its cross sections at time $t = 0$. Assuming that during the motion all cross sections remain parallel to a plane that is normal to the rod's axis, we wish to state the problem of determining the small longitudinal vibrations of the rod at $t > 0$. The ends of the rod may (a) be fixed rigidly, (b) move

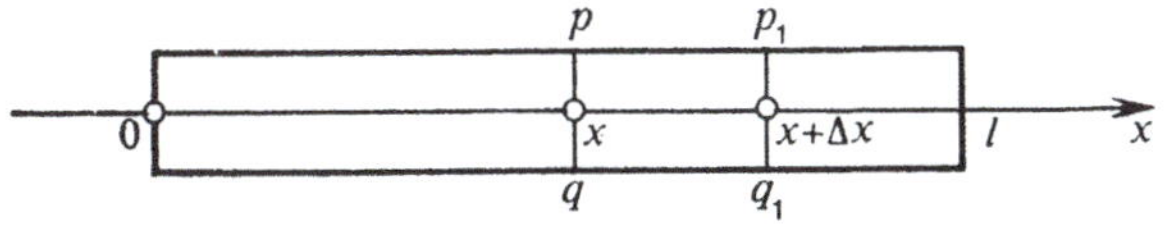

Fig. 4

in the longitudinal directions according to specified laws, (c) move freely, or (d) be fixed elastically (i.e. each end is under a longitudinal force from the fixture, and this force is proportional to the displacement and is directed opposite to the displacement).

Solution. We send the x axis along the axis of the rod (see Fig. 4), with x the coordinate of the cross section pq in the state of rest. We are interested only in small longitudinal vibrations, which means that all external forces and the forces of inertia are directed along the x axis. If $u(x, t)$ denotes the displacement of this cross section at time t, then (under our assumptions) the displacement of the section at point $x + \Delta x$ will be

$$u(x + \Delta x, t) \cong u(x, t) + u_x(x, t)\,\Delta x.$$

For this reason the elongation (per unit length of the rod) at section x will be $u_x(x, t)$. Hooke's law then states that the tensile force in this cross section is $T = ESu_x(x, t)$, with S the cross-sectional area and E the elastic modulus of the material of the rod. We can obtain the vibration equation if we nullify the sum of all forces acting on the portion from section pq to section p_1q_1, including the forces of inertia.

The sum of the tensile forces is

$$T(x + \Delta x) - T(x) = ES\,[u_x(x + \Delta x, t) - u_x(x, t)] \cong ESu_{xx}(x, t)\,\Delta x.$$

Let $p(x, t)$ be the volume density of the external forces acting on the rod. There are two other forces acting on the chosen section: the external force $Sp(x, t)\,\Delta x$ and the force of inertia $-\rho(x)Su_{tt} \times \times (x, t)\,\Delta x$. The sum of all three forces must be zero, according to D'Alembert's principle, that is,

$$[ESu_{xx}(x, t) + p(x, t)\,S - \rho(x)\,Su_{tt}(x, t)]\,\Delta x = 0. \quad (1.3)$$

Whence

$$\rho(x)\,u_{tt}(x, t) = Eu_{xx}(x, t) + p(x, t); \quad (1.4)$$

in addition, $u(x, t)$ satisfies the initial conditions $u|_{t=0} = \varphi(x)$ and $u_t|_{t=0} = \psi(x)$, with $\varphi(x)$ and $\psi(x)$ given functions. If $\rho(x) = \rho$ is constant (i.e. the rod is homogeneous), the equation takes the form

$$u_{tt} = a^2 u_{xx} + g(x, t)$$

where

$$a^2 = E/\rho, \quad g(x, t) = p(x, t)/\rho \quad (1.5)$$

Now we must derive the boundary conditions:

(a) In the case of rigidly fixed ends we can write $u|_{x=0} = u|_{x=l} = 0$.

(b) $u|_{x=0} = \mu_1(t)$ and $u|_{x=l} = \mu_2(t)$, where $\mu_1(t)$ and $\mu_2(t)$ are the functions that define the law of the motion of the rod's ends ($\mu_1(0) = \varphi(0)$ and $\mu_2(0) = \varphi(l)$).

(c) When both ends are free, we must construct the balance equation for the forces acting at the ends. At the left end the resultant of the elastic tensile forces is $T(\Delta x) = ESu_x(\Delta x, t)$, the external force is $Sp(0, t)\,\Delta x$, and the force of inertia is $-\rho Su_{tt}(0, t)\,\Delta x$. The sum of all forces acting on the chosen portion must be zero, that is,

$$ESu_x(\Delta x, t) + p(0, t)\,S\,\Delta x - \rho Su_{tt}(0, t)\,\Delta x = 0, \quad (1.6)$$

and the limit as $\Delta x \to 0$ is $u_x|_{x=0} = 0$. Similarly, at the right end $u_x|_{x=l} = 0$.

(d) To the left-hand side of Eq. (1.6) we must add the force $-ku(0, t)$. If we now take the limit as $\Delta x \to 0$, we obtain

$$ESu_x(0, t) - ku(0, t) = 0, \quad \text{or} \quad (u_x - hu)|_{x=0} = 0,$$

where $h = k/ES$. At the right end we have

$$-T(l - \Delta x) = -ESu_x(l - \Delta x, t),$$

the external force $Sp(l, t)\,\Delta x$, and the force of inertia $-\rho(x)\,Su_{tt}(l, t)\,\Delta x$. The resultant must be zero, or

$$-ESu_x(l - \Delta x, t) - ku(l, t) + Sp(l, t)\,\Delta x - u_{tt}(l, t)\,S\rho(x)\,\Delta x = 0,$$

and the limit as $\Delta x \to 0$ is $(u_x + hu)\,|_{x=l} = 0$, which constitutes the second boundary condition.

Example 3: *Vibrations of a membrane.* A membrane is a stretched film that resists stretching but does not resist twisting. The work produced by an external force that is applied to the membrane and changes the surface area of a selected portion is proportional to the change in area. The positive proportionality factor T depends neither on the shape of the portion nor on its position and is called a tensile force.

Let us derive the equation that determines the equilibrium of the membrane, assuming that at time $t = 0$ the membrane coincided with a region G of the x_1, x_2 plane bounded by a sufficiently smooth curve L. The work performed by internal elastic forces is equal in magnitude to the work of the external forces on the membrane but has an opposite sign. Let $f(x)$ be the density of external forces at point x acting perpendicularly to the x_1, x_2 plane. As a result the membrane occupies a new position given by the equation $u = u(x)$. We will assume that the membrane is not bent too strongly, so that we can ignore $u_{x_1}^4$ and $u_{x_2}^4$ when deriving the equation. An additional assumption is that under external forces the points of the membrane move along normals to the x_1, x_2 plane, which means that the coordinates (x_1, x_2) of an arbitrary point of the membrane do not change.

The work performed by an external force that takes the membrane out of its initial position ($u \equiv 0$, $x \in G$) into a position given by the equation $u = u(x)$, $x \in G$, is equal to the integral $\int_G f(x)\, u(x)\, dx$. Under this displacement the area of the membrane changes by $\int_G (\sqrt{1 + u_{x_1}^2 + u_{x_2}^2} - 1)\, dx$, while the work performed by the internal elastic forces is

$$-T \int_G [\sqrt{1 + u_{x_1}^2 + u_{x_2}^2} - 1]\, dx \cong -\frac{T}{2} \int_G (u_{x_1}^2 + u_{x_2}^2)\, dx.$$

Hence, the sum of all works is

$$A(u) = \int_G \left[-\frac{T}{2} (u_{x_1}^2 + u_{x_2}^2) + fu \right] dx. \tag{1.7}$$

The variation of functional (1.7) is given by the formula

$$\delta A(u) = \int_G [-T(u_{x_1} \delta u_{x_1} + u_{x_2} \delta u_{x_2}) + f \delta u]\, dx.$$

According to the virtual work principle, in the state of equilibrium we have ($u = u(x)$) $\delta A(u) = 0$ for all admissible $\delta u(x)$. Since

$$\int_G (u_{x_1} \delta u_{x_1} + u_{x_2} \delta u_{x_2})\, dx = \int_L \frac{\partial u}{\partial n} \delta u\, dl - \int_G \nabla^2 u \delta u\, dx,$$

where **n** is the outward unit normal vector to contour L, we have

$$\delta A(u) = -T\int_L \frac{\partial u}{\partial \mathbf{n}}\,\delta u\,dl + \int_G (T\nabla^2 u + f)\,\delta u\,dx = 0. \qquad (1.8)$$

Since every function that is continuously differentiable in $\overline{G}$ and vanishes at the boundary is admissible and the functions $u(x)$ and $f(x)$ can be assumed to be sufficiently smooth, Eq. (1.8) yields

$$T\nabla^2 u = -f(x), \quad x \in G. \qquad (1.9)$$

Now we go over to the boundary conditions:

(a) A fixed membrane. If the edge of the membrane is rigidly fixed, the boundary L does not experience any deflections and $u|_L = 0$.

(b) If the membrane's edge is free, it can move along the vertical lateral surface of a cylinder with base L. In this case δu can be arbitrary both on G and on L, and condition (1.8) yields $(\partial u/\partial \mathbf{n})|_L = 0$.

(c) If a force with a linear density f_1 is applied to the edge of the membrane, the line integral in (1.8) must be replaced by $\int_L [-T(\partial u/\partial \mathbf{n}) + f_1]\,\delta u\,dl$. Since δu is arbitrary on L, we obtain $[-T(\partial u/\partial \mathbf{n}) + f_1]|_L = 0$.

(d) If the edge of the membrane is elastically fixed, the force that acts on the edge has a density $-ku$, where k characterizes the rigidity with which the edge is fixed. To obtain the required boundary condition, we must substitute $-ku$ for f_1 in the boundary condition $[-T(\partial u/\partial \mathbf{n}) + f_1]|_L = 0$. We then have

$$\left(\frac{\partial u}{\partial \mathbf{n}} + hu\right)\Big|_L = 0, \text{ where } h = k/T$$

Now we wish to derive the equation that describes the vibrations of the membrane. Let $u = u(x, t)$ be the equation that describes the position of the membrane at time t. According to D'Alembert's principle, the function $u(x, t)$ satisfies the equation $T\nabla^2 u = -(f - \rho u_{tt})$, with $f = f(x, t)$ the density of the external force and $-\rho(x)\,u_{tt}$ the density of the force of inertia. Thus, the equation of vibrations of the membrane is

$$a^2\nabla^2 u - u_{tt} = F(x, t), \text{ where } a^2 = T/\rho \text{ and } F = -f(x, t)/\rho. \qquad (1.10)$$

From the physics of the problem it is clear that to give an unambiguous description of the vibration process we must, besides using Eq. (1.10) and one of the boundary conditions (a)-(d), specify the initial position of the membrane (i.e. its shape at $t = 0$) and the initial velocities of its points.

We have thus arrived at the following problem involving Eq. (1.10): to find a twice differentiable solution $u(x, t)$, $x \in G$, $t > 0$, that is continuously differentiable in $\overline{G}$ at $t \geqslant 0$ and satisfies the following boundary value problem:

$$a^2\nabla^2 u - u_{tt} = F(x, t), \quad u|_{t=0} = \varphi(x), \quad u_t|_{t=0} = \psi(x),$$

where $\varphi(x)$ and $\psi(x)$ are given functions. In addition, depending on the mode in which the edge of the membrane is fixed, we must select one of the boundary conditions (a)-(d).

Example 4: The continuity equation, the flow problem, and the equations of acoustics. We start by examining the motion of an ideal fluid, that is, a fluid in which viscosity does not manifest itself. (The motion of the fluid is studied in Eulerian coordinates). Suppose $\mathbf{v} = (v_1, v_2, v_3)$ is the velocity of the fluid, $\rho(x, t)$ its density, and $f(x, t)$ the source strength. We isolate a volume Ω of the fluid bounded by a surface S. Then the rate at which the mass of fluid inside Ω changes is

$$\frac{\partial}{\partial t}\int_{\Omega}\rho\,dx=\int_{\Omega}\frac{\partial\rho}{\partial t}\,dx.$$

On the other hand, the rate must be equal to the amount Q_1 of fluid produced by the sources inside Ω minus the amount Q_2 flowing out of Ω through S. Obviously,

$$Q_1=\int_{\Omega}f(x,\ t)\,dx,\qquad Q_2=\int_{S}\rho\,(\mathbf{v},\ \mathbf{n})\,ds=\int_{\Omega}\operatorname{div}(\rho\mathbf{v})\,dx,$$

where $\mathbf{n}$ is the outward unit normal vector. Thus,

$$\int_{\Omega}[\rho_t+\operatorname{div}(\rho\mathbf{v})-f]\,dx=0.$$

Since Ω is taken arbitrary and the integrand is continuous, we can write

$$\rho_t+\operatorname{div}(\rho\mathbf{v})=f(x,\ t). \tag{1.11}$$

This equation is known as the *equation of continuity* of an ideal fluid.

Now let us turn to the problem of studying the flow of a homogeneous incompressible liquid around a solid Ω with a boundary S. The liquid has a given velocity $\mathbf{v}_0$ at infinity and no sources inside it. Since $\rho\equiv\text{const}$ and $f\equiv 0$, the problem results in solving the equation

$$\operatorname{div}\mathbf{v}=0 \tag{1.12}$$

with the condition that

$$v_n|_S=0, \tag{1.13}$$

where $v_n=(\mathbf{v},\ \mathbf{n})$ and $\mathbf{n}$ is the outward unit normal vector. Let u be the velocity potential, that is, $\mathbf{v}=\operatorname{grad}u$. Then Eq. (1.12) takes the form $\operatorname{div}\operatorname{grad}u=\nabla^2u=0$ and the boundary condition becomes $(\partial u/\partial\mathbf{n})|_S=0$, since

$$v_n=(\mathbf{v},\ \mathbf{n})=(\operatorname{grad}u,\ \mathbf{n})=\frac{\partial u}{\partial\mathbf{n}}.$$

The physics of the problem implies that $\mathbf{v}(x)$ must tend to $\mathbf{v}_0$ as $|x|\to\infty$, where $\mathbf{v}_0$ is the flow velocity at infinity.

Thus the problem has reduced itself to the following boundary value problem:

$$\nabla^2 u = 0, \quad x \notin \overline{\Omega},$$

$$\left.\frac{\partial u}{\partial \mathbf{n}}\right|_S = 0, \quad \lim_{|x| \to \infty} \operatorname{grad} u = \mathbf{v}_0.$$

Finally, we derive the *equations of acoustics*. Suppose that a portion of ideal gas within a certain volume performes small vibrations about the position of equilibrium due to certain external forces of a density $\mathbf{F}(x, t)$ and that these vibrations are adiabatic, that is, the pressure $p(x, t)$ of the gas and its density $\rho(x, t)$ are related by the following formula (the equation of state):

$$\frac{p}{p_0} = \left(\frac{\rho}{\rho_0}\right)^{\gamma}, \tag{1.14}$$

where p_0 and ρ_0 are the initial pressure and density, and the specific heat ratio γ is positive.

Let us denote the vector by which the gas is displaced with respect to the equilibrium position by $\mathbf{u}(x, t) = (u_1(x, t), u_2(x, t), u_3(x, t))$ and the velocity vector by $\mathbf{v}(x, t) = (v_1(x, t), v_2(x, t), v_3(x, t))$, with

$$\frac{\partial \mathbf{u}}{\partial t} = \mathbf{v}. \tag{1.15}$$

Under the assumptions we have made, that is, that $\rho - \rho_0$, $\mathbf{u}$, $\mathbf{v}$, and their derivatives are small, we can rewrite Eq. (1.14) thus:

$$p = p_0\left(1 + \gamma \frac{\rho - \rho_0}{\rho_0}\right), \tag{1.16}$$

while the continuity equation (1.11) can be rewritten in the form

$$\rho_t + \rho_0 \operatorname{div} \mathbf{v} = 0 \tag{1.17}$$

(we assume that the source strength is zero).

According to the first law of Newton the balance of the forces acting on a small volume ΔV of the gas is zero, i.e. $\rho(\partial \mathbf{v}/\partial t)\Delta V + \operatorname{grad} p\, \Delta V = \mathbf{F}\, \Delta V$, whence after we substitute ρ_0 for ρ (within the framework of our assumption) we have

$$\rho_0 \frac{\partial \mathbf{v}}{\partial t} = \mathbf{F} - \operatorname{grad} p. \tag{1.18}$$

Differentiating (1.18) with respect to t and employing Eqs. (1.16) and (1.17), we find an equation for the velocity vector $\mathbf{v}$, namely,

$$\frac{\partial^2 \mathbf{v}}{\partial t^2} = a^2 \operatorname{grad} \operatorname{div} \mathbf{v} + \frac{1}{\rho_0} \frac{\partial \mathbf{F}}{\partial t}, \tag{1.19}$$

with $a^2 = p_0\gamma/\rho_0$.

If we assume that at $t = 0$ $\operatorname{div} \mathbf{u} = -1$, then from Eqs. (1.17) and (1.15) we find that in all subsequent moments of time $\rho + \rho_0 \operatorname{div} \mathbf{u} = 0$. This together with (1.15), (1.16), and (1.18) yields an equation for the displacement vector **u**, namely,

$$\frac{\partial^2 \mathbf{u}}{\partial t^2} = a^2 \operatorname{grad} \operatorname{div} \mathbf{u} + \frac{1}{\rho_0} \mathbf{F}. \tag{1.20}$$

Finally, if we differentiate Eq. (1.17) with respect to time and employ (1.16) and (1.18), we obtain the equations for p and ρ:

$$\rho_{tt} = a^2 \nabla^2 \rho - \operatorname{div} \mathbf{F}, \quad p_{tt} = a^2 \nabla^2 p - a^2 \operatorname{div} \mathbf{F}. \tag{1.21}$$

Equations (1.19)-(1.21) are known as the *equations of acoustics.*

Example 5: Propagation of heat. The derivation of the heat conduction equation is based on Fourier's law, according to which the amount of heat passing during a time interval Δt through a small area ΔS lying inside the object being studied is given by the formula

$$\Delta Q = -k(x, u) \frac{\partial u(x, t)}{\partial \mathbf{n}} \Delta S \, \Delta t, \tag{1.22}$$

where **n** is the unit normal vector pointing in the direction of the heat flow, $k(x, u)$ the internal thermal conductivity, and $u(x, t)$ the temperature of the object at point $x = (x_1, x_2, x_3)$ and at time t. Let us assume that the object is isotropic in relation to heat conductivity, which means that $k(x, u)$ does not depend on the orientation of the area. To derive an equation that $u(x, t)$ would obey we isolate inside the object a volume Ω bounded by a surface S. According to Fourier's law, the amount of heat flowing into Ω through S during an interval of time from t_1 to t_2 is

$$\int_{t_1}^{t_2} dt \int_S k \frac{\partial u}{\partial \mathbf{n}} \, ds = \int_{t_1}^{t_2} dt \int_\Omega \operatorname{div}(k \operatorname{grad} u) \, dx.$$

If $F(x, t)$ is the density of the heat sources, then the amount of heat produced by them in Ω over the specified time interval is

$$\int_{t}^{t_2} dt \int_\Omega F(x, t) \, dx.$$

The total amount of heat flowing into Ω during the time from t_1 to t_2 can be calculated from the temperature difference:

$$\int_\Omega c\rho [u(x, t_2) - u(x, t_1)] \, dx = \int_{t_1}^{t_2} dt \int_\Omega c\rho \frac{\partial u}{\partial t} \, dx,$$

with $c(x)$ and $\rho(x)$ the specific heat and density of the substance. Hence,

$$\int_{t_1}^{t_2} dt \int_\Omega \left(c\rho \frac{\partial u}{\partial t} - \operatorname{div}(k \operatorname{grad} u) - F(x, t) \right) dx = 0 \tag{1.23}$$

(here we assume that the integrand is a continuous function). Since both Ω and the time interval from t_1 to t_2 are arbitrary, Eq. (1.23) yields

$$c\rho u_t - \operatorname{div}(k \operatorname{grad} u) = F(x, t), \tag{1.24}$$

which is known as the *heat conduction equation.*

If the thermal conductivity k is independent of temperature u, or $k(x, u) = k(x)$, then Eq. (1.24) is linear. If the object is homogeneous, then $c(x) \equiv \text{const}$, $\rho \equiv \text{const}$, $k \equiv \text{const}$, and Eq. (1.24) takes the form

$$u_t = a^2\nabla^2 u + f(x, t), \tag{1.25}$$

where $a^2 = k/c\rho$ and $f(x, t) = F(x, t)/c\rho$. The physics of the problem implies that to describe the heat propagation process unambiguously we must, besides writing Eq. (1.24) or (1.25), state the initial temperature $u|_{t=0} = \varphi(x)$ and the temperature (or heat) condition at the boundary. If the boundary Γ of the object is kept at a given temperature, the boundary condition is

$$u|_\Gamma = \psi.$$

If a heat flux q is specified at the boundary, the boundary condition is

$$\left.\frac{\partial u}{\partial \mathbf{n}}\right|_\Gamma = h,$$

where $h = q/k$ and $\mathbf{n}$ is an outward unit normal vector. In particular, if the object is thermally isolated at its boundary, then

$$\left.\frac{\partial u}{\partial \mathbf{n}}\right|_\Gamma = 0.$$

When the temperature of the surrounding medium is known, we may assume that heat exchange occurs according to Newton's law, that is, $q|_\Gamma = \alpha(u_1 - u)_\Gamma$, where q is the heat flux, α the external thermal conductivity (the heat-exchange coefficient), and u_1 the temperature of the surrounding medium. On the other hand, Fourier's law states that a unit area of the boundary Γ produces every second a heat flux $q_1 = k(\partial u/\partial \mathbf{n})$ pointed within the object. The two fluxes must be equal, that is,

$$k\left.\frac{\partial u}{\partial \mathbf{n}}\right|_\Gamma = \alpha(u_1 - u)|_\Gamma, \text{ or } \left.\left(\frac{\partial u}{\partial \mathbf{n}} + hu\right)\right|_\Gamma = \varphi_1(s).$$

Example 6: Diffusion. We wish to derive the equation that describes the diffusion of a substance in a medium at rest that occupies a limited region Ω with a boundary Γ. We assume that we know the density $F(x, t)$ of the sources and that diffusion takes place along with absorption (for instance, particles of the diffusing substance react chemically with the medium). Also, let us assume that the absorption rate at each point of space x that belongs to Ω is proportional to the density $u(x, t)$ of the diffusing substance. Besides finding this equation, we must state the boundary conditions, which may be as follows:

(a) A given density u is maintained at the boundary.

(b) The boundary is impenetrable.

(c) The boundary is semipermeable, with diffusion of particles through the boundary obeying a law that is similar to Newton's law for heat exchange by convection.

We derive the diffusion equation starting from Nernst's law, according to which the amount of substance passing through a small area ΔS during a time interval Δt is

$$\Delta Q = -D(x)\frac{\partial u}{\partial \mathbf{n}}\Delta S\,\Delta t,$$

where $D(x)$ is the diffusion coefficient, and $\mathbf{n}$ the unit vector normal to ΔS and pointing in the direction of the flow of the substance. Let $\rho(x)$ be the density of the medium. Just as in the case with the heat conduction equation, we isolate a volume Ω with a boundary S and calculate the balance of the amount of substance that is transferred into Ω during a time interval from t_1 to t_2.

The amount of substance transferred into Ω during this time interval is, according to Nernst's law,

$$\int_{t_1}^{t_2} dt \int_S D(x)\frac{\partial u}{\partial \mathbf{n}}\,ds = \int_{t_1}^{t_2} dt \int_\Omega \operatorname{div}(D \operatorname{grad} u)\,dx.$$

The amount of substance that is produced within Ω by sources is

$$\int_{t_1}^{t_2} dt \int_\Omega F(x, t)\,dx.$$

The amount of substance dissipated in Ω due to absorption by the medium is

$$\int_{t_1}^{t_2} dt \int_\Omega q(x)\,u(x, t)\,dx,$$

where $q(x)$ is the absorption coefficient. Since, on the other hand, the amount of substance in Ω changes during the same time interval by

$$\int_\Omega \rho(x)\,[u(x, t_2) - u(x, t_1)]\,dx = \int_{t_1}^{t_2} dt \int_\Omega \rho\frac{\partial u}{\partial t}\,dx,$$

we can write

$$\int_{t_1}^{t_2} dt \int_\Omega (\rho u_t - \operatorname{div}(D \operatorname{grad} u) - F + qu)\,dx = 0 \qquad (1.26)$$

(the integrand is assumed to be continuous). Since both Ω and the time interval are arbitrary, Eq. (1.26) yields

$$\rho u_t + qu = \operatorname{div}(D \operatorname{grad} u) + F. \qquad (1.27)$$

This is known as the *diffusion equation*. The physics of the problem implies that to describe the diffusion process unambiguously we must know the initial distribution of the density $u|_{t=0} = \varphi(x)$, $x \in \Omega$, and the diffusion mode at the boundary. Just as in Example 5, the boundary conditions are

(a) $u|_{\Gamma} = u_0$,

(b) $\left.\frac{\partial u}{\partial n}\right|_{\Gamma} = 0$,

(c) $\left. D\frac{\partial u}{\partial n}\right|_{\Gamma} = \alpha(u_1 - u)|_{\Gamma}$,

where u_0 and u_1 are given functions, and α the penetrability of the boundary Γ.

1.1. Find the static sag of a string with its ends fastened and under a continuously distributed load (per unit length).

1.2. Derive the equation of small transverse vibrations of a string that has a small ball of mass m attached to it at an inner point x_0.

1.3. Derive the equation of vibrations of a string in an elastic medium.

1.4. The periodic motion of a rod or shaft in which the cross sections are rotated with respect to each other and about the axis is known as *torsional vibrations*. Derive the equation of small torsional vibrations for a homogeneous cylindrical rod. Consider the cases when (a) the ends of the rod are free, (b) the ends are rigidly fixed, and (c) the ends are elastically fixed.

1.5. An elastic homogeneous rectangular bar has one of its ends rigidly fixed and the other free. At time $t = 0$ small transverse deflections and velocities parallel to the vertical lateral symmetry plane of the bar are imparted to the points of the bar. State the boundary value problem of determining the small transverse deviations of the points of the bar at $t > 0$ assuming that the bar performs small transverse vibrations.

1.6. A pipe filled with an ideal gas and open at one end is in translational motion in the direction of its axis with a constant velocity v. At time $t = 0$ the pipe suddenly stops. State the boundary value problem of determining the displacement of the gas inside the pipe at a distance x from the closed end.

1.7. An ideal gas occupying a cylindrical pipe performs small lateral vibrations. The planar cross sections, consisting of gas particles, do not deform and all gas particles move parallel to the axis of the cylinder. State the boundary value problem of determining the displacement $u(x, t)$ of the gas particles when the ends of the pipe are (a) closed with impenetrable partitions, (b) open, and (c) closed by small pistons of infinitesimal masses, the pistons being mounted on springs with a stiffness ν and moving without friction inside the pipe.

1.8. At time $t = 0$ one end of a straight elastic homogeneous rod begins to perform longitudinal vibrations according to a certain law, while the other end is under a force $\Phi(t)$ directed along the axis of the rod. Before $t = 0$ the cross sections of the rod were at rest and in nondeflected positions. State the boundary value problem of determining the small longitudinal deflections of the points of the rod at $t > 0$.

1.9. State the boundary value problem of the small transverse vibrations of a string in a medium with a resistance proportional to the first power of velocity. Both ends of the string are fixed.

1.10. Construct the equation of longitudinal vibrations of a rod whose cross-sectional area is a given function of x, assuming that the material of the rod is homogeneous.

1.11. State the boundary value problem of the lateral vibrations of an elastic rod in the form of a frustum of a cone. Both ends of the rod are fixed, and the rod is taken out of the state of rest by imparting initial velocities and longitudinal deflections at time $t = 0$. The length of the rod is l, the radii of the bases are R and r $(R > r)$, the material of the rod is homogeneous, and all deformations of the cross sections are ignored.

1.12. A light string is rotating in the horizontal plane with a constant angular velocity ω around a vertical axis, with one end of the string being fixed to a point on the axis and the other being free. At time $t = 0$ small deflections and velocities normal to the plane are imparted to the points of the string. State the boundary value problem of determining the deflections of the points of the string from the plane of equilibrium motion.

1.13. Suppose a ball of mass m_0 is fixed at point $x = 0$ of an infinite homogeneous string. The initial velocities and deflections of the string's points are zero. State the boundary value problem of determining the deflections of the points of the string from the position of equilibrium when (a) a force $F = F_0 \sin \Omega t$ begins to act on the ball at time $t = 0$, (b) momentum p_0 is imparted to the ball transversely at time $t = 0$, and (c) the ball is elastically fixed with an effective rigidity k^2 with case (b) otherwise being effective.

1.14. State the boundary value problem of small lateral vibrations of an elastic homogeneous rod one end of which is rigidly fixed and the other moves with resistance proportional to the first power of velocity. The resistance of the medium can be ignored.

1.15. Masses m_i, $i = 1, 2, \ldots, n$, are concentrated at the internal points $x = x_i$, $i = 1, 2, \ldots, n$, of a string with fixed ends. State the boundary value problem of small transverse vibrations of the string with arbitrary initial conditions.

1.16. Two semi-infinite elastic homogeneous rods with the same cross sections are connected rigidly at their ends and constitute an infinite rod. Let ρ_1 and E_1 be the density and elasticity modulus of one rod and ρ_2 and E_2 the respective quantities of the other. State the boundary value problem of the deflections of the cross sections of the infinite rod from their positions of equilibrium if at time $t = 0$ certain longitudinal deflections and velocities are imparted to the cross sections.

1.17. A heavy homogeneous cord of length l is fixed by its upper end ($x = l$) to a vertical axis and rotates about the axis with a constant angular velocity ω. Prove that the equation of small vibrations of the cord about its vertical position of equilibrium is

$$\frac{\partial^2 u}{\partial t^2} = g \frac{\partial}{\partial x}\left(x \frac{\partial u}{\partial x}\right) + \omega^2 u.$$

1.18. State the boundary value problem of transverse vibrations of a heavy homogeneous string about its vertical position of equilibrium if the upper end of the string is rigidly fixed and the lower end is free.

1.19. State the problem of finding the magnetic field inside and outside of a cylindrical conductor with a current J travelling along the surface of the conductor.

1.20. An electric cable with a potential v_0 is grounded at one of its ends through a lumped capacitance (or inductance), while the other end is insulated. State the problem of determining the electric current flowing through the cable.

1.21. The end $x = 0$ of a round homogeneous shaft is fixed, while the end at $x = l$ is rigidly attached to a disk with a moment of inertia J_0. The disk is twisted through an angle α and at time $t = 0$ is let go without any initial velocity imparted to it. State the boundary value problem of determining the angles of rotation of the cross sections of the shaft at $t > 0$.

1.22. A heavy rod is hung up and is jammed in such a manner that the deflection of each of its points is zero. At time $t = 0$ the rod is released. State the boundary value problem of the forced vibrations of the rod.

1.23. Let all conditions of the previous problem be fulfilled except that at the lower end, namely, a load Q is attached to the lower end of the rod, with the unstressed state of the rod being the equilibrium position (e.g. at time $t = 0$ a support is taken out from under the load and the rod begins to stretch).

1.24. State the problem of the motion of a semi-infinite string ($0 \leqslant x < \infty$) at $t > 0$ if at $t < 0$ a wave $u(x, t) = f(x + at)$ travels along the string; the end $x = 0$ is rigidly fixed.

1.25. State the boundary value problem of small radial vibrations of a homogeneous ideal gas occupying a cylindrical pipe of radius R; the pipe is so long that it can be considered to extend infinitely in both directions. The initial deflections and initial velocities are given functions of r.

1.26. State the problem of the steady state flow of an ideal fluid around a sphere (potential flow). Give an electrostatic analog of this process.

1.27. State the boundary value problem of small radial vibrations of a homogeneous ideal gas occupying a spherical vessel of radius R if the initial velocities and deflections are given functions of r.

1.28. State the boundary value problem of transverse vibrations of a membrane to which a pressure P is applied (per unit surface area and normal to the surface of the membrane) if in the unperturbed state the membrane is flat and the surrounding medium does not resist the membrane vibrations. Consider the cases where (a) the membrane is rigidly fixed at the edge L, (b) the membrane is free at L, and (c) a part L_1 of the edge is rigidly fixed while the remaining part L_2 is free.

1.29. State the boundary value problem of the vibrations of a round homogeneous membrane fixed at its edge and placed in a medium whose resistance is proportional to the first power of velocity. At time $t = 0$ an external force of density $f(r, \varphi, t)$ is applied to the surface of the membrane in the direction normal to the plane of the unperturbed membrane. The initial velocities and deflections of the points of the membrane are zero.

1.30. A homogeneous rectangular membrane is fixed at its edge. At time $t = 0$ it receives a shock in the vicinity of its central point, so that

$$\lim_{\varepsilon \to 0} \int_{U_\varepsilon} v_0(x)\, dx = A, \quad x = (x_1, x_2),$$

where A is a constant, and $v_0(x)$ is the initial velocity. State the boundary value problem of free vibrations of this membrane.

1.31. Suppose an electric circuit consists of a resistance R, a self-inductance L, and a capacitance C. At time $t = 0$ an emf E_0 is introduced into the circuit. Show that the current $i(t)$ in the circuit obeys the following equation:

$$Li'(t) + Ri(t) + \frac{1}{C} \int_0^t i(\tau)\, d\tau = E_0, \quad t > 0.$$

1.32. Let us consider an electromagnetic field in a medium. Starting with Maxwell's equations, derive equations whose solutions are the various components of the electric and magnetic fields for the cases where (a) the charge density $\rho = 0$, $\varepsilon = \text{const}$, $\lambda = \text{const}$, $\mu = \text{const}$, $\mathbf{J} = \lambda \mathbf{E}$ (Ohm's law), and (b) the medium is a vacuum and no currents exist.

1.33. State the problem of the penetration by a magnetic field into the right half-space filled by a medium with conductivity σ if starting from time $t = 0$ a magnetic field $H = H_0 \sin \Omega t$ is maintained at the surface $x = 0$ (the magnetic field is directed parallel to the surface).

1.34. State the boundary value problem of determining the temperature of a rod $0 \leqslant x \leqslant l$ with heat-insulated lateral surface. Consider the cases where (a) the ends of the rod are kept at a given temperature, (b) at the ends of the rod a given heat flux is maintained, and (c) at the ends of the rod there is convective heat exchange by Newton's law with a medium whose temperature is given.

1.35. Derive the equation of diffusion in a medium at rest assuming that the planes perpendicular to the x axis are at each moment of time t the surfaces of constant density. Write the boundary conditions assuming that diffusion occurs in the flat layer $0 \leqslant x \leqslant l$ and consider the cases where (a) the concentration of the diffusing substance is kept equal to zero at the boundary planes, (b) the boundary planes are impenetrable, and (c) the boundary surfaces are semipermeable, with diffusion through these planes governed by a law similar to Newton's law for convective heat exchange.

1.36. Derive the equation of diffusion for a disintegrating gas (the number of molecules decaying every second at a given point is proportional to the density with a positive proportionality factor α).

1.37. Suppose we have a thin homogeneous rod of length l whose initial temperature is $f(x)$. State the boundary value problem of determining the temperature of the rod if a constant temperature u_0 is maintained at the end $x = 0$, while the lateral surface and the end $x = l$ are involved in convective heat exchange (governed by Newton's law) with the surrounding medium whose temperature is zero.

1.38. State the problem of determining the temperature distribution in a thin heat-insulated rod of infinite length along which a point heat source begins to move at time $t = 0$. The source moves in the positive direction, its velocity is v_0, and it generates q units of heat every second.

1.39. State the boundary value problem of the cooling off of a thin homogeneous ring of radius R whose surface is involved in convective heat exchange with the surrounding medium whose temperature is known. The nonuniformity of the temperature distribution in the ring can be ignored.

1.40. Derive the equation that describes the diffusion of suspended particles and allows for precipitation. Assume that the velocity of the particles caused by gravity is constant and the density of the particles depends only on the height z and the time t. Write the boundary condition that corresponds to an impenetrable partition.

1.41. State the boundary value problem of the cooling off of a uniformly heated rod in the form of a frustum of a cone (the warpage of isothermal surfaces can be ignored) if the ends of the rod are thermally insulated and the lateral surface is involved in convective heat exchange with a medium whose temperature is zero.

1.42. A solute with an initial concentration c_0 = const diffuses from the solution that occupies the space between two planes, $x = 0$ and $x = h$, into the solvent that occupies the space between the planes $x = h$ and $x = l$. State the boundary value problem of the process of equalization of the density assuming that the boundaries at $x = 0$ and $x = l$ are impenetrable for the solute.

1.43. Heat sources that are distributed with a constant density Q within a homogeneous ball begin to operate at time $t = 0$. State the boundary value problem of determining the temperature distribution within the ball at $t > 0$ if the initial temperature of every point of the ball depends only on the distance between this point and the ball's center. Consider the cases where (a) the temperature of ball's surface is maintained at zero, and (b) the ball's surface is involved in convective heat exchange (obeying Newton's law) with the surrounding medium whose temperature is zero.

1.44. Let us take a uniform ball of radius R whose initial temperature is zero. State the boundary value problem of determining the temperature distribution inside the ball at $t > 0$. Suppose that (a) the ball is heated uniformly over its entire surface by a constant heat flux q, and (b) the surface of the ball is involved in convective heat exchange with the surrounding medium whose temperature depends only on time

1.45. Suppose that the initial temperature of an unlimited plate of thickness $2h$ is zero. State the boundary value problem of finding the temperature distribution at $t > 0$ over the thickness if (a) the plate is heated from both sides by equal heat fluxes q, and (b) a heat source with a constant density Q begins to operate inside the plate at $t = 0$ while the temperature of both sides of the plate is kept constant and equal to zero.

1.46. An infinitely long cylinder of radius R has an initial temperature $f(r)$. State the boundary value problem of the radial propagation of heat in the cylinder if (a) the temperature of the lateral surface is maintained constant, and (b) the lateral surface emits heat into a medium whose temperature is zero.

1.47. Given a thin rectangular plate with edges l and m and with a known initial temperature distribution, state the boundary value problem of heat propagation in the plate if the lateral sides of the plate are heated to the following temperatures:

$$u|_{y=0} = \varphi_1(x), \quad u|_{y=m} = \varphi_2(x),$$
$$u|_{x=0} = \psi_1(x), \quad u|_{x=l} = \psi_2(y).$$

1.48. The initial temperature distribution in a homogeneous ball is given by the function $f(r, \theta, \varphi)$. State the boundary value problem of heat distribution in the ball if the temperature of the ball's surface is maintained constant and equal to u_0.

1.49. Two semi-infinite straight rods made out of different materials are brought into contact at time $t = 0$ through their ends. State the boundary value problem of heat distribution in the new infinitely long rod if the initial temperatures of both rods are known.

1.50. State the boundary value problem of the time-independent temperature distribution in a thin rectangular plate $OACB$ with edges $OA = a$ and $OB = b$ if (a) the temperatures of the lateral surfaces of the plate are known, and (b) known heat fluxes cross the edges OA and OB while the edges BC and AC are thermally insulated.

1.51. A time-independent transverse load with a density $f(x, y)$ is applied to a flat membrane bounded by a closed curve L. State the boundary value problem of determining the deviations of the points of the membrane from the plane if (a) the edge of the membrane is fixed, (b) the edge is free, and (c) the edge is elastically fixed.

1.52. Given a cylinder with a base of radius R and altitude h, state the boundary value problem of the time-independent temperature distribution inside the cylinder if the temperature of the upper and lower bases is a given function of r, while (a) the lateral surface is thermally insulated, (b) a temperature that depends only on z is maintained at the surface, and (c) the surface freely gives off heat to a medium whose temperature is zero.

1.53. State the boundary value problem of finding the time-independent temperature distribution inside a hemisphere if the spherical surface is kept at a given temperature $f(\varphi, \theta)$, while the temperature of the base of the hemisphere is maintained constant and equal to zero.

1.54. A ball of radius R is heated by a plane-parallel heat flux of density q that falls on its surface, and gives its heat off to the surrounding medium in accordance with Newton's law. State the boundary value problem of the temperature distribution inside the ball.

1.55. Let $n(x, \mathbf{s}, t)$ be the time-dependent density at point x of particles that fly with a constant velocity v in the direction specified by the vector $\mathbf{s} = (s_1, s_2, s_3)$, $\alpha(x)$ the absorption coefficient at point x, and $h(x)$ the multiplication factor at point x. Assuming that the scattering at each point x is isotropic, show that $n(x, \mathbf{s}, t)$ satisfies the integro-differential transport equation

$$\frac{1}{v}\frac{\partial n}{\partial t} + (\mathbf{s}\,\mathrm{grad}\, n) + \alpha(x)\, n = \frac{\beta(x)}{4\pi} \int\limits_{|\mathbf{s}'|=1} n(x, \mathbf{s}', t)\, d\mathbf{s}' + F,$$

where $F(x, \mathbf{s}, t)$ is the source density, and $\beta(x) = \alpha(x)\, h(x)$.

1.56. State the boundary value problem for the equation of Problem 1.55 with a given initial density distribution and a given incident flux of particles onto the boundary S of a region G.

1.57. Show that for the solution $n(x, \mathbf{s})$ of the time-independent boundary value problem

$$(\mathbf{s}, \operatorname{grad} n) + \alpha(x) n = \frac{\beta(x)}{4\pi} \int\limits_{|\mathbf{s}'|=1} n(x, \mathbf{s}')\, d\mathbf{s}' + F(x),$$

with $n|_S = 0$ if $(\mathbf{s}, \mathbf{n}) < 0$, where $\mathbf{n}$ is the outward unit vector normal to S, the average density $n_0(x) = \frac{1}{4\pi} \int\limits_{|\mathbf{s}|=1} n(x, \mathbf{s})\, d\mathbf{s}$ satisfies Peierls's integral equation

$$n_0(x) = \frac{1}{4\pi} \int \frac{e^{-|x-x'|\int_0^1 \alpha[tx+(1-t)x']\,dt}}{|x-x'|^2} [\beta(x') n_0(x') + F(x')]\, dx'.$$

1.58. Expanding the solution $n(x, \mathbf{s})$ of the time-independent boundary value problem 1.57 in a series in spherical functions of $\mathbf{s}$ and retaining only terms with the zeroth and first harmonics, show that the function

$$n_0(x) = \frac{1}{4\pi} \int\limits_{|\mathbf{s}|=1} n(x, \mathbf{s})\, d\mathbf{s}$$

is the solution of the boundary value problem (the diffusion approximation)

$$-\frac{1}{3} \operatorname{div}\left(\frac{1}{\alpha} \operatorname{grad} h_0\right) + (1-h) h_0 = \frac{f}{\alpha},$$

$$\left(n_0 + \frac{2}{3\alpha} \frac{\partial n_0}{\partial \mathbf{n}}\right)\Big|_S = 0.$$

Answers to Problems of Sec. 1

1.1. $Tu_{xx} + f(x) = 0$, $0 < x < l$, $u|_{x=0} = u|_{x=l} = 0$, where $f(x)$ is the load density.

1.2. $\rho u_{tt} = T_0 u_{xx}$, $0 < x < l$, $x \neq x_0$, $t > 0$, $u|_{x=0} = u|_{x=l} = 0$, $u(x_0 + 0, t) = u(x_0 - 0, t)$, $u_x(x_0 + 0, t) - u_x(x_0 - 0, t) = \frac{m}{T_0} u_{tt}(x_0, t)$.

1.3. $\rho u_{tt} = Tu_{xx} - \alpha u$, $0 < x < l$, $t > 0$, where α is the elasticity of the medium.

1.4. $\theta_{tt} = a^2 \theta_{xx}$, $0 < x < l$, $0 < t < \infty$, $\theta(x, 0) = f(x)$, $\theta_t(x, 0) = F(x)$, $0 \leqslant x \leqslant l$, where $\theta(x, t)$ is the angle through which the cross section with coordinate x has turned at time t, $a^2 = GJ/\Phi$, G is the shear modulus, J the polar moment of inertia of the cross section in relation to the point at which the axis intersects the cross section, and Φ is the axial moment of inertia per unit length of rod. The boundary conditions are: (a) $\theta_x(0, t) = \theta_x(l, t) = 0$, (b) $\theta(0, t) = \theta(l, t) = 0$, (c) $(\theta_x - h\theta)|_{x=0} = 0$ and $(\theta_x + h\theta)|_{x=l} = 0$, where $h = k/GJ$, and k is the rigidity of the elastic fixture.

1.5. $u_{tt} + a^2 u_{xxxx} = 0$, $0 < x < l$, $t > 0$, $u(x, 0) = f(x)$, $u_t(x, 0) = F(x)$, $0 \leqslant x \leqslant l$, $u(0, t) = u_x(0, t) = u_{xx}(l, t) = u_{xxx}(l, t) = 0$, where $a^2 = EJ/\rho S$, and J is the geometric moment of inertia of the cross section in relation to its center line perpendicular to the plane of vibrations.

1.6. $u_{tt} = a^2 u_{xx}$, $0 < x < l$, $t > 0$, where $a^2 = \gamma p_0/\rho_0$ is the speed of sound, $u(x, 0) = 0$, $u_t(x, 0) = v$, $0 \leqslant x \leqslant l$, $u(0, t) = 0$, $u_x(l, t) = 0$, $t > 0$.

1.7. $u_{tt} = a^2 u_{xx}$, $a^2 = \gamma p_0/\rho_0$, $0 < x < l$, $t > 0$, $u(x, 0) = f(x)$, $u_t(x, 0) = F(x)$, $0 \leqslant x \leqslant l$. The boundary conditions are (a) $u(0, t) = u(l, t) = 0$, (b) $u_x(0, t) = u_x(l, t) = 0$, (c) $(u_x - hu)|_{x=0} = 0$ and $(u_x + hu)_{x=l} = 0$, where $h = v/s\gamma p_0$, and S is the pipe's cross-sectional area.

1.8. $u_{tt} = a^2 u_{xx}$, $0 < x < l$, $t > 0$, $u(0, t) = \varphi(t)$, $u_x(l, t) = \Phi(t)/ES$, $t > 0$, $u(x, 0) = 0$, $u_t(x, 0) = 0$, $0 \leqslant x \leqslant l$, $a^2 = E/\rho$.

1.9. $u_{tt} = a^2 u_{xx} - 2\nu^2 u_t$, $0 < x < l$, $t > 0$, $u(x, 0) = \varphi(x)$, $u_t(x, 0) = \psi(x)$, $0 \leqslant x \leqslant l$, $u(0, t) = u(l, t) = 0$, $t > 0$, where $2\nu^2 = k/\rho$, and k is the friction coefficient.

1.10. $\dfrac{\partial}{\partial x}\left[S(x)\dfrac{\partial u}{\partial x}\right] = a^2 \dfrac{\partial^2 u}{\partial t^2}$, $a^2 = \rho S/E$.

1.12. $\dfrac{\partial^2 u}{\partial t^2} = a^2 \dfrac{\partial}{\partial x}\left(x^2 \dfrac{\partial u}{\partial x}\right)$, $0 < x < l$, $t > 0$, $a^2 = \omega^2/2$, $|u(0, t)| < \infty$, $u(l, t) = 0$, $t > 0$, $u(x, 0) = f(x)$, $u_t(x, 0) = F(x)$, $0 \leqslant x \leqslant l$.

1.13. $u_{tt} = a^2 u_{xx}$, $x \neq 0$, $t > 0$, $a^2 = T_0/\rho$, $u(x, 0) = 0$, $u_t(x, 0) = 0$, $x \neq 0$; the condition at point $x = 0$ is (a) $-m_0 u_{tt}(0, t) + T_0[u_x(+0, t) - u_x(-0, t)] + F_0 \sin \Omega t = 0$, $t > 0$, (b) $u(-0, t) = u(+0, t)$, $-m_0 u_{tt}(0, t) + T_0[u_x(+0, t) - u_x(-0, t)] = 0$, $t > 0$, $u(-0, 0) = u(+0, 0) = 0$, $m_0 u_t(-0, 0) = m_0 u_t(+0, 0) = p_0$, (c) $u(-0, t) = u(+0, t)$, $t > 0$, $m_0 u_{tt}(0, t) + T_0[u_x(+0, t) - u_x(-0, t)] - k^2 u(0, t) = 0$, $m_0 u_t(-0, 0) = m_0 u_t(+0, 0) = p_0$, $u(-0, 0) = u(+0, 0) = 0$.

1.14. $u_{tt} = a^2 u_{xx}$, $0 < x < l$, $t > 0$, $a_2 = E/\rho$, $u(x, 0) = f(x)$, $u_t(x, 0) = g(x)$, $0 \leqslant x \leqslant l$, $u(0, t) = 0$, $(ESu_x - ku_t)|_{x=l} = 0$, $t > 0$, where k is the friction coefficient for the end of the rod at $x = l$.

1.15. $u_{tt} = a^2 u_{xx}$, $x \neq x_i$, $i = 1, \ldots, n$, $0 < x < l$, $t > 0$, $u(0, t) = u(l, t) = 0$, $u(x_i - 0, t) = u(x_i + 0, t)$, $u_x(x_i + 0, t) - u_x(x_i - 0, t) = \dfrac{m_i}{T} u_{tt}(x_i, t)$, $t > 0$, $i = 1, \ldots, n$; $u|_{t=0} = f(x)$, $u_t|_{t=0} = F(x)$, $0 \leqslant x \leqslant l$.

1.16. $\left.\begin{array}{l} u^1_{tt} = a_1^2 u^1_{xx}, \quad -\infty < x < 0 \\ u^2_{tt} = a_2^2 u^2_{xx}, \quad 0 < x < +\infty \end{array}\right\} t > 0$, $u^1(0, t) = u^2(0, t)$, $E_1 u^1_x(0, t) = E_2 u^2_x(0, t)$, $t > 0$, $u^1(x, 0) = f(x)$, $u^1_t(x, 0) = F(x)$, $-\infty < x < 0$, $u^2(x, 0) = f(x)$, $u^2_t(x, 0) = F(x)$, $x > 0$, where u^1 and u^2 are the deflections of the points of the left and right rods, and $a_i^2 = E_i/\rho_i$, $i = 1, 2$.

1.18. $\frac{\partial^2 u}{\partial t^2} = g \frac{\partial}{\partial x}\left(x \frac{\partial u}{\partial x}\right)$, $0 < x < l$, $t > 0$, $|u(0, t)| < \infty$, $u(l, t) = 0$, $t > 0$, $u|_{t=0} = f(x)$, $u_t|_{t=0} = F(x)$, $0 \leqslant x \leqslant l$.

1.19. $\nabla^2 \Phi^{(l)} = 0$, $r > R$, $\nabla^2 \Phi^{(i)} = 0$, $0 \leqslant r < R$, $\operatorname{grad} \Phi = H$, $\Phi_r^{(l)}|_{r=R} = \Phi_r^{(i)}|_{r=R}$, $\Phi_\varphi^{(l)}|_{r=R} = \left(\Phi_\varphi^{(i)} + \frac{4\pi}{c} j_{\text{sur}}\right)\Big|_{r=R}$, $|\Phi^{(i)}(0, t)| < \infty$, $j_{\text{sur}} = J/2\pi R$ is the surface current density, and $\Phi^{(i)}$ and $\Phi^{(l)}$ are the magnetic field potentials inside and outside the conductor, respectively.

1.20. $J_x = -c v_t$, $v_x = -L J_t$, $0 < x < l$, $t > 0$, $v|_{t=0} = v_0$, $v(0, t) = -\frac{1}{c}\int_0^1 J\,dt$ at the grounded end and $v_x(l, t) = 0$ at the insulated end.

1.21. $\theta_{tt} = a^2 \theta_{xx}$, $0 < x < l$, $t > 0$, $\theta|_{t=0} = \alpha x/l$, $\theta_t|_{t=0} = 0$, $0 \leqslant x \leqslant l$, $\theta|_{x=0} = 0$, $\theta_x|_{x=l} = -\frac{\Phi}{JG}\theta_{tt}$, where the constants, a^2, Φ, J, and G have the same meaning as in Problem 1.4.

1.22. $u_{tt} = a^2 u_{xx} + g$, $0 < x < l$, $t > 0$, $u(x, 0) = u_t(x, 0) = 0$, $0 \leqslant x \leqslant l$, $u(0, t) = 0$, $u_x(l, t) = 0$, $t > 0$, $a^2 = E/\rho$.

1.23. $u_{tt} = a^2 u_{xx} + g$, $0 < x < l$, $t > 0$, $u|_{t=0} = u_t|_{t=0} = 0$, $0 \leqslant x \leqslant l$, $u|_{x=0} = 0$, $\frac{Q}{g} u_{tt}|_{x=l} = -ESu_x|_{x=l} + Q$.

1.25. $u_{tt} = a^2\left(u_{rr} + \frac{1}{r} u_r\right)$, $0 < r < R$, $t > 0$, $u(r, 0) = f(r)$, $u_t(r, 0) = F(r)$, $0 \leqslant r \leqslant R$, $|u(0, t)| < \infty$, $u_r|_{r=R} = 0$.

1.26. $\nabla^2 \varphi = 0$, $r > R$, $t > 0$, $\frac{\partial \varphi}{\partial r}\Big|_{r=R} = 0$, $t > 0$, $\lim_{r\to\infty} v = \lim_{r\to\infty} \operatorname{grad} \varphi = v_0$, where v_0 is the flow velocity at infinity.

1.27. $u_{tt} = a^2\left(u_{rr} + \frac{2}{r} u_r\right)$, $0 \leqslant r < R$, $t > 0$, $u(r, 0) = f(r)$, $u_t|_{t=0} = F(r)$, $0 \leqslant r \leqslant R$, $|u(0, t)| < \infty$, $u_r|_{r=R} = 0$, where $a^2 = \gamma p_0/\rho_0$.

1.29. $u_{tt} + k u_t = a^2 \nabla^2 u + f(r, \varphi, t)/\rho$, $0 \leqslant r < R$, $0 \leqslant \varphi < 2\pi$, $t > 0$, $u|_{t=0} = u_t|_{t=0} = 0$, $|u(0, \varphi, t)| < \infty$, $u(R, \varphi, t) = 0$, where $a^2 = T/\rho$, $k = \alpha/\rho$, and α is the coefficient of the elastic resistance of the medium.

1.32. (a) $u_{tt}-a^2\nabla^2 u+\frac{4\pi\lambda}{\varepsilon}u_t=0$, $a^2=\frac{c^2}{\varepsilon\mu}$, (b) $\left(\frac{\partial^2}{\partial t^2}-a^2\nabla^2\right)\varphi_0=-\frac{4\pi c^2}{\varepsilon^2\mu}\rho$, $\left(\frac{\partial^2}{\partial t^2}-a^2\nabla^2\right)\boldsymbol{\varphi}=0$, $\frac{\mu\varepsilon}{c}\frac{\partial\varphi_0}{\partial t}-\operatorname{div}\boldsymbol{\varphi}=0$, where $\mathbf{E}=(E_1, E_2, E_3)$ is the electric field strength, $\mathbf{H}=(H_1, H_2, H_3)$ the magnetic field strength, $\rho(x)$ the charge density, ε the dielectric constant of the medium, μ the magnetic permeability of the medium, and $\mathbf{I}(x, t)=(I_1, I_2, I_3)$ the conduction current. In the case (a) the components of $\mathbf{E}$ and of $\mathbf{H}$ obey the same telegrapher's equation. In the case (b) one must introduce the four-component electromagnetic potential $(\varphi_0, \boldsymbol{\varphi})$, with $\boldsymbol{\varphi}=(\varphi_1, \varphi_2, \varphi_3)$. Then the solution to Maxwell's equations can be sought in the form $\mathbf{E}=\operatorname{grad}\varphi_0-\frac{1}{c}\frac{\partial\boldsymbol{\varphi}}{\partial t}$ and $\mathbf{H}=\frac{1}{\mu}\operatorname{curl}\boldsymbol{\varphi}$.

1.33. $H_{xx}=\frac{4\pi\sigma}{c^2}H_t+\frac{1}{c^2}H_{tt}$, $x>0$, $t>0$, $H|_{t=0}=0$, $H_t|_{t=0}$, $x>0$, $H|_{x=0}=H_0\sin\Omega t$, $t>0$, where c is the speed of light.

1.34. $u_t=a^2u_{xx}$, $0<x<l$, $t>0$, $u(x, 0)=f(x)$, $0\leqslant x\leqslant l$; the boundary conditions are (a) $u|_{x=0}=\varphi_1(t)$, $u|_{x=l}=\varphi_2(t)$, $t>0$, (b) $-kSu_x|_{x=0}=q_1(t)$, $kSu_x|_{x=l}=\varphi_2(t)$, $t>0$, (c) $u_x|_{x=0}=h[u(0, t)-\varphi_1(t)]$, $u_x|_{x=l}=-h[u(l, t)-\varphi_2(t)]$, $a^2=k/c\rho$ is the specific heat, $\varphi_1(t)$ and $\varphi_2(t)$ are the temperatures of the ends of the rod in the case (a) or the temperatures of the surrounding medium at the two ends in the case (b), and the q_i are the heat fluxes at the ends of the rod.

1.35. $u_t=Du_{xx}$, $0<x<l$, $t>0$, $u(x, 0)=f(x)$, $0\leqslant x\leqslant l$; the boundary conditions are (a) $u(0, t)=u(l, t)=0$, $t>0$, (b) $u_x(0, t)=u_x(l, t)=0$, $t>0$, (c) $u_x|_{x=0}=h[u(0, t)-\varphi_1(t)]$, $t>0$, $u_x|_{x=l}=-h[u(l, t)-\varphi_2(t)]$, where $\alpha/D=h$, and α is the penetrability at the ends.

1.36. $u_t=D\nabla^2u-\alpha u$, $t>0$, $x=(x_1, x_2, x_3)\in R^3$.

1.37. $u_t=a^2u_{xx}-\frac{\alpha p}{c\rho S}u$, $0<x<l$, $t>0$, $u|_{t=0}=f(x)$, $0\leqslant x\leqslant l$, $u|_{x=0}=u_0$, $(u_x+hu)|_{x=l}=0$, $t>0$, p is the perimeter of the cross section of the rod, $h=\alpha/k$, and $a^2=k/c\rho$.

1.38. $u_t=a^2u_{xx}+\frac{q}{c}\delta(x-v_0t)$, $-\infty<x<+\infty$, $t>0$, $u(x, 0)=\varphi(x)$, $a^2=k/c\rho$.

1.39. $u_t=a^2u_{xx}-b(u-u_0)$, $0<x<l$, $t>0$, $u(x, 0)=f(x)$, $0\leqslant x\leqslant l$, $u|_{x=0}=u|_{x=l}$, $u_x|_{x=0}=u_x|_{x=l}$, $a^2=k/c\rho$, $b=\alpha P/c\rho S$, where P is the perimeter of the cross section of the ring, $x=R\theta$, and θ is the angular coordinate.

1.40. $u_t=Du_{zz}-vu_z$, $z>z_0$, $t>0$, $(Du_z-vu)|_{z=z_0}=0$, $t>0$, where v is the precipitation rate.

1.41. $\left(1-\frac{x}{H}\right)^2 \frac{\partial u}{\partial t}=a^2 \frac{\partial}{\partial x}\left[\left(1-\frac{x}{H}\right)^2 \frac{\partial u}{\partial x}\right]-\frac{2\alpha(1-x/H)}{c\rho r_0 \cos\gamma} u,\ 0<$ $<x<l,\ t>0,\ u|_{t=0}=u_0,\ 0\leqslant x\leqslant l,\ u_x|_{x=0}=u_x|_{x=l}=0,$ $t>0$, where $a^2=k/c\rho$, H is the overall height of the cone, γ the half of the opening span of the angle of the cone, r_0 the radius of the larger base, and l the height of the frustum of the cone.

1.42. $c_t=Dc_{xx},\ 0<x<l,\ t>0,\ c(x,0)=\begin{cases} c_0, & 0<x<h, \\ 0, & h<x<l. \end{cases}$

1.43. $u_t=a^2\left(u_{rr}+\frac{2}{r}u_r\right)+\frac{Q}{c\rho},\quad 0\leqslant r<R,\ t>0,\quad u|_{t=0}=f(r),$ $0\leqslant r\leqslant R,\ |u(0,t)|<\infty$; the boundary conditions are (a) $u(R,t)=0$, (b) $(u_r+Hu)|_{r=R}=0$, $H=\alpha/k$, and $a^2=k/c\rho$.

1.44. $u_t=a^2\left(u_{rr}+\frac{2}{r}u_r\right),\ 0\leqslant r<R,\ t>0,\ u|_{t=0}=0,\ 0\leqslant r\leqslant R;$ the boundary conditions are (a) $|u(0,t)|<\infty,\ u_r(R,t)=q/k,$ $t>0$, (b) $|u(0,t)|<\infty,\ (u_r+Hu)|_{r=R}=\varphi(t),\ t>0,\ H=$ $=\alpha/k,\ a^2=k/c\rho.$

1.45. (a) $u_t=a^2u_{xx},\quad -h<x<h,\quad t>0,\quad u|_{t=0}=0,$ $(ku_x+q)|_{x=-h}=0,\ (-ku_x+q)|_{x=h}=0$, (b) $u_t=a^2u_{xx}+$ $+Q/c\rho,\ -h<x<h,\ t>0,\ u|_{t=0}=0,\ u|_{x=\pm h}=0,\ a^2=$ $=k/c\rho.$

1.49. $u_t=a(x)u_{xx},\quad x\neq 0,\ t>0,\quad u(x,0)=f(x),\quad u(-0,t)=$ $=u(+0,t),\quad k_1u_x(-0,t)=k_2u_x(+0,t),\quad a(x)=\begin{cases} a_1^2, & x<0, \\ a_2^2, & x>0, \end{cases}$ $a_i^2=k_i/c_i\rho_i,\ i=1,2.$

2 Classification of Second-order Equations

The equation

$$\sum_{i,j=1}^{n} a_{ij}(x)\,u_{x_ix_j}+\Phi(x,u,\operatorname{grad}u)=0$$

can be reduced at each point x_0 to canonical form by applying the nonsingular linear transformation $\xi=B^Tx$, where B is a matrix such that the transformation $\mathbf{y}=B\eta$ reduces the quadratic form $\sum\limits_{i,j=1}^{n} a_{ij}(x_0)\,y_iy_j$ to canonical form. (Every quadratic form can be reduced to canonical form by, say, isolating perfect squares.)

2.1. Reduce the following equations to canonical form:

1. $u_{xx}+2u_{xy}-2u_{xz}+2u_{yy}+6u_{zz}=0.$
2. $4u_{xx}-4u_{xy}-2u_{yz}+u_y+u_z=0.$
3. $u_{xy}-u_{xz}+u_x+u_y-u_z=0.$

4. $u_{xx}+2u_{xy}-2u_{xz}+2u_{yy}+2u_{zz}=0.$
5. $u_{xx}+2u_{xy}-4u_{xz}-6u_{yz}-u_{zz}=0.$
6. $u_{xx}+2u_{xy}+2u_{yy}+2u_{yz}+2u_{yt}+2u_{zz}+3u_{tt}=0.$
7. $u_{xy}-u_{xt}+u_{zz}-2u_{zt}+2u_{tt}=0.$
8. $u_{xy}+u_{xz}+u_{xt}+u_{zt}=0.$
9. $u_{xx}+2u_{xy}-2u_{xz}-4u_{yz}+2u_{yt}+u_{zz}=0.$
10. $u_{xx}+2u_{xz}-2u_{xt}+u_{yy}+2u_{yz}+2u_{yt}+2u_{zz}+2u_{tt}=0.$

11. $u_{x_1x_1}+2\sum_{k=2}^{n} u_{x_kx_k}-2\sum_{k=1}^{n-1} u_{x_kx_{k+1}}=0.$

12. $u_{x_1x_1}-2\sum_{k=2}^{n}(-1)^k u_{x_{k-1}x_k}=0.$

13. $\sum_{k=1}^{n} ku_{x_kx_k}+2\sum_{l<k} lu_{x_lx_k}=0.$

14. $\sum_{k=1}^{n} u_{x_kx_k}+\sum_{l<k} u_{x_lx_k}=0.$

15. $\sum_{l<k} u_{x_lx_k}=0.$

The equation

$$a(x,y)\,u_{xx}+2b(x,y)\,u_{xy}+c(x,y)\,u_{yy}=\Phi(x,y,u,u_x,u_y),\quad (2.1)$$

where $|a|+|b|+|c|\neq 0$, belongs (at a point or in a region) to the

hyperbolic type if $b^2-ac>0$,
parabolic type if $b^2-ac=0$,
elliptic type if $b^2-ac<0$.

The characteristic equation

$$a(x,y)(dy)^2-2b(x,y)\,dx\,dy+c(x,y)(dx)^2=0$$

of Eq. (2.1) splits into two equations,

$$a\,dy-\left(b+\sqrt{b^2-ac}\right)dx=0, \qquad (2.2)$$

$$a\,dy-\left(b-\sqrt{b^2-ac}\right)dx=0. \qquad (2.3)$$

Equations of the hyperbolic type: $b^2-ac>0$. The general solutions $\varphi(x,y)=c_1$ and $\psi(x,y)=c_2$ of Eqs. (2.2) and (2.3) are real and distinct. They determine two different families of real characteristics of Eq. (2.1). By a change of variables, $\xi=\varphi(x,y)$ and $\eta=\psi(x,y)$, Eq. (1.1) is reduced to canonical form

$$u_{\xi\eta}=\Phi_1(\xi,\eta,u,u_\xi,u_\eta).$$

Equations of the parabolic type: $b^2-ac=0$. Equations (2.2) and (2.3) coincide. The general solution $\varphi(x,y)=c$ of Eq. (2.2) determines the family of real characteristics of Eq. (2.1). A change of variables $\xi=\varphi(x,y)$ and $\eta=\psi(x,y)$, where $\psi(x,y)$ is any smooth function such that the change of variables is unique in the region considered, reduces Eq. (2.1) to canonical form

$$u_{\eta\eta}=\Phi_1(\xi,\eta,u,u_\xi,u_\eta).$$

Equations of the elliptic type: $b^2 - ac < 0$. Let $\varphi(x, y) + i\psi(x, y) = c$ be the general solution of Eq. (2.2), where $\varphi(x, y)$ and $\psi(x, y)$ are real-valued functions. (If a, b, and c are analytic functions, the existence of a general solution for Eq. (2.2) follows from Kovalevskaya's theorem.) Then the change of variables $\xi = \varphi(x, y)$ and $\eta = \psi(x, y)$ reduces Eq. (2.1) to canonical form

$$u_{\xi\xi} + u_{\eta\eta} = \Phi_1(\xi, \eta, u, u_\xi, u_\eta).$$

2.2. Reduce to canonical form the equations given below in regions that preserve the type of the equation.

1. $u_{xx} - 2u_{xy} - 3u_{yy} + u_y = 0$.
2. $u_{xx} - 6u_{xy} + 10u_{yy} + u_x - 3u_y = 0$.
3. $4u_{xx} + 4u_{xy} + u_{yy} - 2u_y = 0$. 4. $u_{xx} - xu_{yy} = 0$.
5. $u_{xx} - yu_{yy} = 0$. 6. $xu_{xx} - yu_{yy} = 0$. 7. $yu_{xx} - xu_{yy} = 0$.
8. $x^2u_{xx} + y^2u_{yy} = 0$. 9. $y^2u_{xx} + x^2u_{yy} = 0$.
10. $y^2u_{xx} - x^2u_{yy} = 0$. 11. $(1 + x^2)u_{xx} + (1 + y^2)u_{yy} + yu_y = 0$.
12. $4y^2u_{xx} - e^{2x}u_{yy} = 0$.
13. $u_{xx} - 2\sin x u_{xy} + (2 - \cos^2 x)u_{yy} = 0$.
14. $y^2u_{xx} + 2yu_{xy} + u_{yy} = 0$.
15. $x^2u_{xx} - 2xu_{xy} + u_{yy} = 0$.

Suppose the coefficients of Eq. (2.1) are continuous in a region D. The function $u(x, y)$ is said to be a *solution* of Eq. (2.1) if it belongs to the class $C^2(D)$ and satisfies Eq. (2.1) in D. The collection of all the solutions of Eq. (2.1) is said to be the *general solution* of Eq. (2.1).

2.3. Find the general solution of each of the equations with constant coefficients given below:

1. $u_{xy} = 0$. 2. $u_{xx} - a^2u_{yy} = 0$.
3. $u_{xx} - 2u_{xy} - 3u_{yy} = 0$. 4. $u_{xy} + au_x = 0$.
5. $3u_{xx} - 5u_{xy} - 2u_{yy} + 3u_x + u_y = 2$.
6. $u_{xy} + au_x + bu_y + abu = 0$.
7. $u_{xy} - 2u_x - 3u_y + 6u = 2e^{x+y}$.
8. $u_{xx} + 2au_{xy} + a^2u_{yy} + u_x + au_y = 0$.

2.4. Prove that the equation with constant coefficients

$$u_{xy} + au_x + bu_y + cu = 0$$

is reduced to the form $v_{xy} + (c - ab)v = 0$ by the change of variables $u(x, y) = v(x, y)e^{-bx-ay}$.

2.5. Prove that the general solution of the equation $u_{xy} = u$ has the form

$$u(x, y) = \int_0^x f(t) J_0\left(2i\sqrt{y(x-t)}\right) dt$$

$$+ \int_0^y g(t) J_0\left(2i\sqrt{x(y-t)}\right) dt + [f(0) + g(0)] J_0\left(2i\sqrt{xy}\right),$$

where $J_0(z)$ is the Bessel function of the zeroth order, and f and g are arbitrary functions of the C^1 class.

2.6. Prove that the general solution of the equation $u_{xy} = F(x, y)$, where $F \in C(|x - x_0| < a, |y - y_0| < b)$, has the form

$$u(x, y) = f(x) + g(y) + \int_{x_0}^{x} \int_{y_0}^{y} F(\xi, \eta)\, d\eta\, d\xi,$$

where f and g are arbitrary functions of the C^2 class.

2.7. Prove that the general solution of the equation $u_{xy} + A(x, y) u_x = 0$, where $A(x, y) \in C^1(|x - x_0| < a, |y - y_0| < b)$, has the form

$$u(x, y) = f(y) + \int_{x_0}^{x} g(\xi) \exp\left\{-\int_{y_0}^{y} A(\xi, \eta)\, d\eta\right\} d\xi,$$

where f and g are arbitrary functions of classes C^2 and C^1, respectively.

2.8. Prove that the general solution of the equation

$$u_{xy} - \frac{1}{x-y} u_x + \frac{1}{x-y} u_y = 0$$

has the form $u(x, y) = \frac{f(x) + g(y)}{x-y}$, where f and g are arbitrary functions of the C^2 class.

2.9. Prove that the general solution of the equation

$$u_{xy} - \frac{n}{x-y} u_x + \frac{m}{x-y} u_y = 0,$$

where n and m are positive integers, has the form

$$u(x, y) = \frac{\partial^{n+m-2}}{\partial x^{m-1} \partial y^{n-1}} \left[\frac{f(x) + g(y)}{x-y}\right],$$

where f and g are arbitrary functions of classes C^{m+1} and C^{n+1}, respectively.

2.10. Prove that the general solution of the equation

$$u_{xy} + \frac{n}{x-y} u_x - \frac{m}{x-y} u_y = 0,$$

where n and m are nonnegative integers, has the form

$$u(x, y) = (x-y)^{n+m+1} \frac{\partial^{n+m}}{\partial x^n \partial y^m} \left[\frac{f(x) + g(y)}{x-y}\right],$$

where f and g are arbitrary functions of classes C^{n+2} and C^{m+2}, respectively.

2.11. Find the general solution of each of the equations below in a region that preserves the type of the equation.

1. $yu_{xx}+(x-y)u_{xy}-xu_{yy}=0$. 2. $x^2u_{xx}-y^2u_{yy}=0$.
3. $x^2u_{xx}+2xyu_{xy}-3y^2u_{yy}-2xu_x=0$.
4. $x^2u_{xx}+2xyu_{xy}+y^2u_{yy}=0$.
5. $u_{xy}-xu_x+u=0$. 6. $u_{xy}+2xyu_y-2xu=0$.
7. $u_{xy}+u_x+yu_y+(y-1)u=0$.
8. $u_{xy}+xu_x+2yu_y+2xyu=0$.

Answers to Problems of Sec. 2

2.1. 1. $u_{\xi\xi}+u_{\eta\eta}+u_{\zeta\zeta}=0$; $\xi=x$, $\eta=y-x$, $\zeta=x-\frac{1}{2}y+\frac{1}{2}z$. 2. $u_{\xi\xi}-u_{\eta\eta}+u_{\zeta\zeta}+u_\eta=0$; $\xi=\frac{1}{2}x$, $\eta=\frac{1}{2}x+y$, $\zeta=-\frac{1}{2}x-y+z$. 3. $u_{\xi\xi}-u_{\eta\eta}+2u_\xi=0$; $\xi=x+y$, $\eta=y-x$, $\zeta=y+z$. 4. $u_{\xi\xi}+u_{\eta\eta}=0$; $\xi=x$, $\eta=y-x$, $\zeta=2x-y+z$. 5. $u_{\xi\xi}-u_{\eta\eta}-u_{\zeta\zeta}=0$; $\xi=x$, $\eta=y-x$, $\zeta=\frac{3}{2}x-\frac{1}{2}y+\frac{1}{2}z$. 6. $u_{\xi\xi}+u_{\eta\eta}+u_{\zeta\zeta}+u_{\tau\tau}=0$; $\xi=x$, $\eta=y-x$, $\zeta=x-y+z$, $\tau=2x-2y+z+t$. 7. $u_{\xi\xi}-u_{\eta\eta}+u_{\zeta\zeta}+u_{\tau\tau}=0$; $\xi=x+y$, $\eta=y-x$, $\zeta=z$, $\tau=y+z+t$. 8. $u_{\xi\xi}-u_{\eta\eta}+u_{\zeta\zeta}-u_{\tau\tau}=0$; $\xi=x+y$, $\eta=x-y$, $\zeta=-2y+z+t$, $\tau=z-t$. 9. $u_{\xi\xi}-u_{\eta\eta}+u_{\zeta\zeta}=0$; $\xi=x$, $\eta=y-x$, $\zeta=2x-y+z$, $\tau=x+z+t$.
10. $u_{\xi\xi}+u_{\eta\eta}=0$; $\xi=x$, $\eta=y$, $\zeta=-x-y+z$, $\tau=x-y+t$.
11. $\sum\limits_{k=1}^{n}u_{\xi_k\xi_k}=0$, $\xi_k=\sum\limits_{l=1}^{k}x_l$; $k=1, 2, \ldots, n$. 12. $\sum\limits_{k=1}^{n}(-1)^{k+1}u_{\xi_k\xi_k}=0$, $\xi_k=\sum\limits_{l=1}^{k}x_l$, $k=1, 2, \ldots, n$. 13. $\sum\limits_{k=1}^{n}u_{\xi_k\xi_k}=0$, $\xi_1=x_1$, $\xi_k=x_k-x_{k-1}$, $k=2, 3, \ldots, n$. 14. $\sum\limits_{k=1}^{n}u_{\xi_k\xi_k}=0$, $\xi_k=\sqrt{\frac{2k}{k+1}}\times\left(x_k-\frac{1}{k}\sum\limits_{l<k}x_l\right)$, $k=1, 2, \ldots, n$. 15. $u_{\xi_1\xi_1}-\sum\limits_{k=2}^{n}u_{\xi_k\xi_k}=0$, $\xi_1=\frac{3-n}{\sqrt{2(n-1)}}x_1+\sqrt{\frac{2}{n-1}}\sum\limits_{k=2}^{n}x_k$, $\xi_k=\frac{1}{\sqrt{2}}x_1-\sqrt{2}x_k$, $k=2, 3, \ldots, n$.

2.2. 1. $u_{\xi\eta}-\frac{1}{16}(u_\xi-u_\eta)=0$, $\xi=x-y$, $\eta=3x+y$. 2. $u_{\xi\xi}+u_{\eta\eta}+u_\xi=0$, $\xi=x$, $\eta=3x+y$. 3. $u_{\eta\eta}+u_\xi=0$, $\xi=x-2y$, $\eta=x$. 4. $u_{\xi\eta}+\frac{1}{6(\xi+\eta)}(u_\xi+u_\eta)=0$, $\xi=\frac{2}{3}x^{3/2}+y$, $\eta=\frac{2}{3}x^{3/2}-y$, $x>0$; $u_{\xi\xi}+u_{\eta\eta}+\frac{1}{3\xi}u_\xi=0$, $\xi=\frac{2}{3}(-x)^{3/2}$, $\eta=y$, $x<0$. 5. $u_{\xi\eta}+\frac{1}{2(\xi-\eta)}(u_\xi-u_\eta)=0$, $\xi=x+2\sqrt{y}$, $\eta=x-2\sqrt{y}$, $y>0$; $u_{\xi\xi}+u_{\eta\eta}-\frac{1}{\eta}u_\eta=0$, $\xi=x$, $\eta=2\sqrt{-y}$, $y<0$. 6. $u_{\xi\xi}-u_{\eta\eta}-$

$-\frac{1}{\xi}u_\xi+\frac{1}{\eta}u_\eta=0$, $\xi=\sqrt{|x|}$, $\eta=\sqrt{|y|}$ ($x>0$, $y>0$ or $x<0$, $y<0$); $u_{\xi\xi}+u_{\eta\eta}-\frac{1}{\xi}u_\xi-\frac{1}{\eta}u_\eta=0$, $\xi=\sqrt{|x|}$, $\eta=\sqrt{|y|}$ ($x>0$, $y<0$ or $x<0$, $y>0$). 7. $u_{\xi\xi}-u_{\eta\eta}+\frac{1}{3\xi}u_\xi-\frac{1}{3\eta}u_\eta=0$, $\xi=|x|^{3/2}$, $\eta=|y|^{3/2}$ ($x>0$, $y>0$ or $x<0$, $y<0$); $u_{\xi\xi}+u_{\eta\eta}+\frac{1}{3\xi}u_\xi+\frac{1}{3\eta}u_\eta=0$, $\xi=|x|^{3/2}$, $\eta=|y|^{3/2}$ ($x>0$, $y<0$ or $x<0$, $y>0$). 8. $u_{\xi\xi}+u_{\eta\eta}-u_\xi-u_\eta=0$, $\xi=\ln|x|$, $\eta=\ln|y|$ (in each quadrant). 9. $u_{\xi\xi}+u_{\eta\eta}+\frac{1}{2\xi}u_\xi+\frac{1}{2\eta}u_\eta=0$, $\xi=y^2$, $\eta=x^2$ (in each quadrant). 10. $u_{\xi\eta}+\frac{1}{2(\eta^2-\xi^2)}(\eta u_\xi-\xi u_\eta)=0$, $\xi=y^2-x^2$, $\eta=y^2+x^2$ (in each quadrant). 11. $u_{\xi\xi}+u_{\eta\eta}-\tanh\xi u_\xi=0$, $\xi=\ln(x+\sqrt{1+x^2})$, $\eta=\ln(y+\sqrt{1+y^2})$. 12. $u_{\xi\eta}-\frac{1}{2(\xi-\eta)}(u_\xi-u_\eta)+\frac{1}{4(\xi+\eta)}(u_\xi+u_\eta)=0$, $\xi=y^2+e^x$, $\eta=y^2-e^x$ ($y>0$ or $y<0$) 13. $u_{\xi\xi}+u_{\eta\eta}+\cos\xi u_\eta=0$, $\xi=x$, $\eta=y-\cos x$. 14. $u_{\eta\eta}-2u_\xi$, $\xi=2x-y^2$, $\eta=y$. 15. $u_{\eta\eta}-\xi u_\xi=0$; $\xi=xe^y$, $\eta=y$.

2.3. 1. $f(x)+g(y)$. 2. $f(y+ax)+g(y-ax)$. 3. $f(x-y)+g(3x+y)$. 4. $f(y)+g(x)e^{-ay}$. 5. $x-y+f(x-3y)+g(2x+y)e^{(3y-x)/7}$. 6. $[f(x)+g(y)]e^{-bx-ay}$. 7. $e^{x+y}+[f(x)+g(y)]e^{3x+2y}$. 8. $f(y-ax)+g(y-ax)e^{-x}$.

2.11. 1. $f(x+y)+(x-y)g(x^2-y^2)$ ($x>-y$ or $x<-y$). 2. $f(xy)+\sqrt{|xy|}\,g(x/y)$ (in each quadrant). 3. $f(xy)+|xy|^{3/4}g(x^3/y)$ (in each quadrant). 4. $f(\varphi)+rg(\varphi)$, $x=r\cos\varphi$, $y=r\sin\varphi$, $(x^2+y^2\neq 0)$. 5. $xf(y)-f'(y)+\int_0^x(x-\xi)g(\xi)e^{\xi y}d\xi$. *Hint.* Introducing the notation $u_x=v$, find the relationships $u=xv-v_y$, $v_{xy}-xv_x=0$. 6. $2yg(x)+\frac{1}{x}g'(x)+\int_0^y(y-\xi)f(\xi)e^{-x^2\xi}d\xi$. *Hint.* Introducing the notation $u_y=v$, find the relationships $u=\frac{1}{2x}v_x+yv$, $v_{xy}+2xyv_y=0$. 7. $e^{-y}[yf(x)+f'(x)+\int_0^y(y-\eta)g(\eta)e^{-x\eta}d\eta]$. *Hint.* Introducing the notation, $u_y+u=v$, find the relationships $u=v_x+yv$, $v_{xy}+v_x+yv_y+yv=0$. 8. $e^{-xy}[yf(x)+f'(x)+\int_0^y(y-\eta)g(\eta)e^{-x\eta}d\eta]$. *Hint.* Introducing the notation $u_y+xu=v$, find the relationships $u=v_x+2yv$, $(v_y+xv)_x+2y(v_y+xv)=0$.

Chapter II

Function Spaces and Integral Equations

3 Measurable Functions. The Lebesgue Integral

The set $E \subset R^n$ is called a *set of (n-dimensional) measure zero* if for each positive ε there exists a countable set of open (n-dimensional) cubes that cover E and whose total volume is less than ε.

Suppose $Q \subset R^n$ is a region. If a certain property is valid everywhere in Q except, perhaps, on a set of measure zero, we say that this property is valid *almost everywhere* in Q. A function $f(x)$ defined in Q is said to be *measurable* in Q if it is the limit of a sequence of functions from $C(\bar{Q})$ that converges almost everywhere in Q. If $f(x) = g(x)$ almost everywhere in Q, the two functions are said to be *equivalent* in Q.

3.1. See whether the following sets are sets of measure zero:
(1) a finite set of points,
(2) a countable set of points,
(3) the intersection of a countable set of sets of measure zero,
(4) the union of a countable set of sets of measure zero,
(5) a smooth $(n-1)$-dimensional surface,
(6) a smooth k-dimensional surface $(k \leqslant n-1)$.

In Problems 3.2-3.9 prove the propositions.

3.2. Prove that the Dirichlet function $\chi(x)$ (which takes the value 1 at rational points and 0 at irrational points) is zero almost everywhere.

3.3. The function $f(x) = \dfrac{1}{1-|x|}$ is continuous in R^n almost everywhere

3.4. The sequence of the functions $f_n(x) = |x|^n$ converges to zero in the ball $|x| \leqslant 1$ almost everywhere.

3.5. Theorem *A set E is a set of measure zero if and only if there exists a cover of E consisting of a countable system of open cubes with a finite sum of their volumes and such that each point of E proves to be covered by an infinite set of cubes.*

3.6. A function that belongs to $C(Q)$ is measurable.

3.7. If $f(x)$ and $g(x)$ are equivalent and $g(x)$ is measurable in Q, then $f(x)$ is measurable in Q, too.

3.8. The limit of a sequence of measurable functions that converges almost everywhere is a measurable function.

3.9. A function that is continuous on Q everywhere except on a subset consisting of a finite or countable number of smooth k-dimensional surfaces ($k \leqslant n-1$) is measurable in Q.

3.10. Establish whether the following functions defined on $[-1, 1]$ are measurable:

$$\text{(a)}\ y = \operatorname{sign} x; \quad \text{(b)}\ y = \begin{cases} \sin \frac{1}{x} & \text{if } x \neq 0, \\ 0 & \text{if } x = 0; \end{cases}$$

$$\text{(c)}\quad y = \begin{cases} \operatorname{sign}\left(\sin \frac{1}{x}\right) & \text{if } x \neq 0, \\ 0 & \text{if } x = 0; \end{cases}$$

$$\text{(d)}\quad y = \begin{cases} \frac{1}{n} \text{ if } x = m/n, \text{ with } m \text{ and } n \text{ relatively prime,} \\ 0 \text{ if } x \text{ is irrational.} \end{cases}$$

3.11. Let the functions $f(x)$ and $g(x)$ be measurable in Q. Establish whether the functions given below are measurable.

(a) $f(x)\,g(x)$, (b) $f(x)/g(x)$ (provided $g(x) \neq 0$, $x \in Q$), (c) $|f(x)|$; (d) $(f(x))^{g(x)}$ if $f(x) > 0$.

3.12. Suppose $f(x) \in C(Q)$ and at each point x of Q the function $f(x)$ has a derivative, f_{x_i}. Prove that f_{x_i} is measurable in Q.

3.13. (a) Suppose $f(x)$ and $g(x)$ are measurable in Q. Prove that $\max\{f(x), g(x)\}$ and $\min\{f(x), g(x)\}$ are measurable in Q.

(b) Prove that every measurable function $f(x)$ is the difference of two nonnegative measurable functions, $f^+(x) = \max\{f(x), 0\}$, and $f^-(x) = \max\{0, -f(x)\}$.

3.14. Prove that a function that is nondecreasing (nonincreasing) in $[a, b]$ is measurable.

3.15. Prove that if $f(x)$ is measurable in Q, then there is a sequence of polynomials that converges to $f(x)$ almost everywhere in Q.

We will assume that a function $f(x)$ defined in a region Q belongs to the class $L^+(Q)$ if there is a nondecreasing sequence of finite functions $f_n(x)$, $n = 1, 2, \ldots$, that are continuous in $\overline{Q}$ such that it converges to $f(x)$ almost everywhere in Q and such that the sequence of the (Riemann) integrals $\int_Q f_n(x)\,dx$ is bounded from above. The

Lebesgue integral of $f(x) \in L^+(Q)$ is then defined thus:

$$(L) \int_Q f\,dx = \sup_n \int_Q f_n\,dx = \lim_{n\to\infty} \int_Q f_n\,dx.$$

A function $f(x)$ is said to be *Lebesgue integrable over a region* Q if it can be represented in the form of the difference $f(x) = f_1(x) - f_2(x)$ of two functions $f_1(x)$ and $f_2(x)$ that belong to $L^+(Q)$. The Lebesgue integral of $f(x)$ is then defined as follows:

$$(L) \int_Q f\,dx = (L) \int_Q f_1\,dx - (L) \int_Q f_2\,dx.$$

A complex-valued function $f(x) = \operatorname{Re} f(x) + i \operatorname{Im} f(x)$ is said to be *Lebesgue integrable over a region* Q if the functions $\operatorname{Re} f(x)$ and $\operatorname{Im} f(x)$ are Lebesgue integrable. By definition we assume that

$$(L) \int_Q f\,dx = (L) \int_Q \operatorname{Re} f\,dx + i\,(L) \int_Q \operatorname{Im} f\,dx.$$

We denote the set of complex-valued functions that are Lebesgue integrable over a domain Q and are identified if they are equivalent by $L_1(Q)$.

Functions that belong to $L_1(Q)$ are finite almost everywhere in Q. If a function is Riemann integrable, it is Lebesgue integrable, too, and the Riemann and Lebesgue integrals coincide. Because of this we will drop the (L) in front of the integral sign; an integral will be always understood as a Lebesgue integral and the integrand as a Lebesgue integrable function. Moreover, if a function is an absolutely improperly Riemann integrable, it is Lebesgue integrable, too, and the Riemann and Lebesgue integrals coincide.

The following theorems play an important role in the theory of Lebesgue integrable functions:

(a) *if a function* $f(x)$ *is measurable in a region* Q *and* $|f(x)| \leqslant g(x)$, *where* $g(x) \in L_1(Q)$, *then* $f(x)$ *belongs to* $L_1(Q)$. *In particular, a measurable bounded function belongs to* $L_1(Q)$ *in a bounded region* Q.

(b) Lebesgue's theorem *If a sequence of functions that are measurable in* Q, *say* $f_1(x), f_2(x), \ldots, f_n(x), \ldots, n = 1, 2, \ldots$, *converges to a function* $f(x)$ *almost everywhere in* Q *and if* $|f_n(x)| \leqslant g(x)$, *where* g *belongs to* $L_1(Q)$, *then* f *also belongs to* $L_1(Q)$ *and* $\int_Q f_n(x)\,dx \to \int_Q f\,dx$ *as* $n \to \infty$.

(c) Fubini's theorem *If* $f(x, y) \in L_1(Q \times P)$, $x = (x_1, \ldots, x_n) \in Q$ *and* $y = (y_1, \ldots, y_m) \in P$, *where* Q *and* P *are regions, then*

$$\int_Q f(x, y)\,dx \in L_1(P), \quad \int_P f(x, y)\,dy \in {}_1(Q)$$

and

$$\int_{Q\times P} f(x, y)\, dx\, dy = \int_Q dx \int_P f(x, y)\, dy = \int_P dy \int_Q f(x, y)\, dx.$$

If $f(x, y)$ is measurable in $Q \times P$ and for almost all $x \in Q$ the function $|f(x, y)| \in L_1(P)$ and $\int_P |f(x, y)|\, dy \in L_1(Q)$, then $f(x, y) \in L_1(Q \times P)$.

In Problems 3.16-3.20 prove the validity of the propositions.

3.16. If $f(x) \geqslant 0$ and $\int_Q f(x)\, dx = 0$, then $f(x) = 0$ almost everywhere in Q.

3.17. If $f(x) = 0$ almost everywhere in Q, then $\int_Q f\, dx = 0$.

3.18. If $f, g \in L_1(Q)$, then $\alpha f + \beta g \in L_1(Q)$ for all constant α and β.

3.19. If $f \in L_1(Q)$, then $|f| \in L_1(Q)$ and

$$\left| \int_Q f\, dx \right| \leqslant \int_Q |f|\, dx.$$

3.20. If $f \in L_1(Q)$, then for every positive ε there exists a finite function $g_\varepsilon \in C(\bar{Q})$ such that $\int_Q |f - g_\varepsilon|\, dx < \varepsilon$.

3.21. Prove that the Dirichlet function

$$f(x) = \begin{cases} 1 & \text{for } x \text{ rational}, \\ 0 & \text{for } x \text{ irrational} \end{cases}$$

is Lebesgue integrable on $[0, 1]$ but is not Riemann integrable. What is its Lebesgue integral equal to?

3.22. Find the integral over the segment $[0, 1]$ of each of the functions given below (first prove the finiteness of the integral):

(a) $f(x) = \begin{cases} x^2 & \text{for } x \text{ irrational}, \\ 0 & \text{for } x \text{ rational}; \end{cases}$

(b) $f(x) = \begin{cases} x^2 & \text{for } x \text{ irrational and greater than } 1/3, \\ x^3 & \text{for } x \text{ irrational and less than } 1/3, \\ 0 & \text{for } x \text{ rational}; \end{cases}$

(c) $f(x) = \begin{cases} \sin \pi x & \text{for } x \text{ irrational and less than } 1/2, \\ x^2 & \text{for } x \text{ irrational and greater than } 1/2, \\ 0 & \text{if } x \text{ is rational}; \end{cases}$

(d) $f(x)=\begin{cases} 1/n \text{ for } x=m/n, \text{ where } m \text{ and } n \text{ are relatively prime,} \\ 0 \text{ for } x \text{ irrational;} \end{cases}$

(e) $f(x)=\begin{cases} x^{-1/3} \text{ for } x \text{ irrational,} \\ x^3 \text{ for } x \text{ rational;} \end{cases}$

(f) $f(x)=\operatorname{sign}\left(\sin\frac{\pi}{x}\right)$.

3.23. At which values of α are the following functions integrable over the ball $|x|<1$:

(a) $f(x)=\frac{1}{|x|^{\alpha}}$; (b) $f(x)=\frac{1}{(1-|x|)^{\alpha}}$; (c) $f(x)=\frac{x_1x_2\ldots x_n}{|x|^{\alpha}}$.

3.24. Let $g(x)$ be a measurable and bounded function in a region Q. Show that the function $f(x)=\int\limits_Q \frac{g(\xi)}{|x-\xi|^{\alpha}}\,d\xi$ belongs to $C^k(R^n)$ at $k<n-[\alpha]$.

3.25. Suppose f belongs to $L_1(Q)$. Show that the function

$$f_N(x)=\begin{cases} f(x) \text{ if } |f(x)|<N \text{ at point } x, \\ N \text{ if } |f(x)|\geqslant N \text{ at point } x \end{cases}$$

is integrable over Q and that

$$\lim_{N\to\infty}\int\limits_Q f_N(x)\,dx=\int\limits_Q f(x)\,dx.$$

3.26. Suppose $Q=(0<x_1<1,\ 0<x_2<1)$ and let $f(x)$ be defined in Q in the following manner:

(a) $f(x)=\begin{cases} x_1x_2/|x|^4 & \text{for } (x_1, x_2)\neq(0, 0), \\ 0 & \text{for } x_1=x_2=0; \end{cases}$

(b) $f(x)=\begin{cases} \frac{x_1^2-x_2^2}{|x|^4} & \text{for } (x_1x_2)\neq(0, 0), \\ 0 & \text{for } x_1=x_2=0; \end{cases}$

(c) $f(x)=\begin{cases} \frac{1}{x_2^2} & \text{for } 0<x_1<x_2<1, \\ -\frac{1}{x_1^2} & \text{for } 0<x_2<x_1<1, \\ 0 & \text{at other points.} \end{cases}$

(1) Do these functions belong to $L_1(Q)$?

(2) Do the functions $\int\limits_0^1 f(x_1, x_2)\,dx_1$ and $\int\limits_0^1 f(x_1, x_2)\,dx_2$ belong to $L_1(0, 1)$?

(3) Check whether

$$\int_0^1 dx_1 \int_0^1 f(x_1, x_2)\, dx_2 = \int_0^1 dx_2 \int_0^1 f(x_1, x_2)\, dx_1.$$

3.27. Suppose that a sequence of steep functions

$$f_n(x) = \begin{cases} 1 \text{ for } \frac{i}{2^k} \leqslant x \leqslant \frac{i+1}{2^k}, \\ 0 \text{ at other points } x \in [0, 1], \end{cases}$$

$n = 1, 2, \ldots$, is defined on the segment $[0, 1]$, with the integers n, k, and i related thus:

$$n = 2^k + i, \quad 0 \leqslant i \leqslant 2^k - 1.$$

Show that $\lim\limits_{n\to\infty} \int_0^1 f_n\, dx = 0$ and that $f_n(x)$ does not tend to zero as $n \to \infty$ at a single point in $[0, 1]$.

The set of all functions that are measurable on Q and whose modulus squared belongs to $L_1(Q)$ is denoted by $L_2(Q)$ (here, just as in the case with $L_1(Q)$, equivalent functions are assumed to be identified).

In Problems 3.28-3.33 prove the proposition.

3.28. If $f_1, f_2 \in L_2(Q)$, then $\alpha f_1 + \beta f_2 \in L_2(Q)$ for all constants α and β.

3.29. If $f \in L_2(Q)$, where Q is a bounded region (or a region with a limited volume), then $f \in L_1(Q)$.

3.30. Not one of the following inclusions is true: $L_1(R^n) \subset L_2(R^n)$, $L_2(R^n) \subset L_1(R^n)$.

3.31. If $f, g \in L_2(Q)$, then $f \cdot g \in L_1(Q)$.

3.32. If $f, g \in L_2(Q)$, then

$$\left| \int_Q f \cdot g\, dx \right| \leqslant \left(\int_Q |f|^2\, dx \right)^{1/2} \left(\int_Q |g|^2\, dx \right)^{1/2}$$

Bunyakovskii's inequality.

3.33. If $f, g \in L_2(Q)$, then

$$\left(\int_Q |f+g|^2\, dx \right)^{1/2} \leqslant \left(\int_Q |f|^2\, dx \right)^{1/2} + \left(\int_Q |g|^2\, dx \right)^{1/2}$$

Minkowski's inequality.

3.34. Establish whether the following functions belong to $L_2(Q)$:

(a) $y=x^{-1/3}$, $Q=[0, 1]$; (b) $y=\dfrac{\sin x}{x^{3/4}}$, $Q=(0, 1)$;

(c) $$y=\begin{cases} x^{-1/3}\cos x \text{ for } x \text{ irrational}, \\ x^{-1/3} \text{ for } x \text{ rational}, x\neq 0, \\ 0 \text{ for } x=0; \end{cases} \quad Q=[-1, 1],$$

(d) $$y=\begin{cases} \dfrac{\sin(x_1x_2)}{x_1^2+x_2^2} & |x|\neq 0, \\ 0 & |x|=0; \end{cases} \quad Q=(|x|<1),$$

(e) $$y=\begin{cases} \dfrac{1}{x^{1/3}\,\mathrm{sign}\,(\sin \pi/x)} & x\neq 0\ x\neq 1/k, \\ 0 & x=0,\ x=1/k. \end{cases} \quad Q=[0, 1],$$

3.35. For which values of α and β does the function $f(x)= =\dfrac{1}{|x_1|^{\alpha}+|x_2|^{\beta}}$ belong to $L_2(Q)$ if $Q=\{|x_1|+|x_2|>1\}$?

3.36. For which values of α does the function $r^{-\alpha}$, $r=(x_1^2+x_2^2)^{1/2}$, belong to $L_2(Q)$ if (a) $Q=(r<1)$ and (b) $Q=(r>1)$?

3.37. For which values of α does the function

$$f(x)=\begin{cases} \dfrac{\sin|x|^2}{|x|^{7-2\alpha}} & \text{at } |x|\neq 0 \\ 0 & \text{at } |x|=0 \end{cases} \quad |x|=(x_1^2+x_2^2+x_3^2)^{1/2},$$

belong to $L_2(Q)$ if $Q=(|x|<1)$?

3.38. For which values of α does the function $|x|^{-\alpha}$, where $|x|= =(x_1^2+\ldots+x_n^2)^{1/2}$, belong to $L_2(Q)$ if

(a) $Q=(|x|<1)$, (b) $Q=(|x|>1)$, (c) $Q=R^n$?

3.39. Suppose $g\in L_2(Q)$, where Q is a bounded region. Show that the function $f(x)=\displaystyle\int_Q \frac{g(y)}{|x-y|^{\alpha}}\,dy$ belongs to the space $C^k(\overline{Q})$ if $\alpha<n/2$ and $k<n/2-[\alpha]$.

3.40. Show that if $f\in L_2(Q)$, with Q a bounded region, then for every positive ε there exists a function $f_\varepsilon\in C(\overline{Q})$ such that

$$\int_Q |f-f_\varepsilon|^2\,dx<\varepsilon.$$

Answers to Problems of Sec. 3

3.21. 0. **3.22.** (a) 1/3, (b) 35/108, (c) $1/\pi+7/24$, (d) 0, (e) 3/2, (f) $1-2\ln 2$. **3.23.** (a) $\alpha<n$, (b) $\alpha<1$, (c) $\alpha<2n$.

3.26. (a) 1. No. 2. No. 3. No. (b) 1. No. 2. Yes. 3. No. (c) 1. No. 2. Yes. 3. No.

3.35. $\alpha > 0$, $\beta > 0$, $\frac{1}{2\alpha} + \frac{1}{2\beta} < 1$. **3.36.** (a) $\alpha < 1$, (b) $\alpha > 1$. **3.37.** $\alpha > 7/4$. **3.38.** (a) $\alpha < n/2$, (b) $\alpha > n/2$, (c) not for a single value of α.

4 Function Spaces

A *complex* (*real*) *linear space* is a set M for whose elements the operations of addition and multiplication by complex (real) numbers have been defined. These operations must not take us outside M and must possess the following properties: (a) $f_1 + f_2 = f_2 + f_1$, (b) $(f_1 + f_2) + f_3 = f_1 + (f_2 + f_3)$, (c) there is an element 0 of M such that $0f = 0$ for every $f \in M$, (d) $(c_1 + c_2) f = c_1 f + c_2 f$, (e) $c(f_1 + f_2) = cf_1 + cf_2$, (f) $(c_1c_2) f = c_1 (c_2 f)$, and (g) $1f = f$ for all f_1, f_2, f_3, and f from M and all complex (real) numbers c, c_1, and c_2.

A system of elements $f_1, \ldots, f_k$ from M is said to be *linearly independent* if $c_1 f_1 + \ldots + c_k f_k = 0$ only if $c_1 = \ldots = c_k = 0$. Otherwise the system is linearly dependent. An infinite system $f_1, f_2, \ldots$ is said to be linearly independent if each of its finite subsystems is linearly independent.

A linear space is said to be *normed* if each of its elements f has corresponding to it a real number $\| f \|$ called the *norm of* f. This number must satisfy the following conditions:

(a) $\| f \| \geqslant 0$, with $\| f \| = 0$ only if $f = 0$,

(b) $\| f + g \| \leqslant \| f \| + \| g \|$ (the triangle inequality),

(c) $\| cf \| = | c | \, \| f \|$ for an arbitrary constant c.

For a normed linear space one can introduce the idea of the *distance* between elements, $\rho(f, g) = \| f - g \|$, and that of *convergence in the norm*, which means that a sequence $f_1, f_2, \ldots$ converges to an element f, or $f_n \to f$ as $n \to \infty$, if $\rho(f, f_n) \to 0$ as $n \to \infty$.

A sequence of functions $f_1, f_2, \ldots$ belonging to a normed linear space is said to be *fundamental* if for each positive ε there exists a positive integer $N = N(\varepsilon)$ such that $\| f_m - f_n \| < \varepsilon$ for $m, n > N$.

A normed linear space is said to be *complete* if the limit of every fundamental sequence of elements belongs to this space. A complete normed linear space B is known as a *Banach space*.

A set $R \in B$ is said to be *dense* in B if for each element $f \in B$ there exists a sequence $f_1, f_2, \ldots$ belonging to R that converges to f, or $f_n \to f$ as $n \to \infty$.

4.1. Establish whether the following sets are linear spaces:

(a) the set $C^k(Q)$, $0 \leqslant k \leqslant \infty$,

(b) the set of points of the n-dimensional space R^n and the set of points of the complex plane $\mathbb{C}$,

(c) the set of functions that are finite in Q,

(d) the set of functions that are bounded in Q,

(e) the set of functions that are analytic in a region Q of the complex plane $\mathbb{C}$,

(f) the set of functions from $C(\bar{Q})$ that vanish on a set $E \in \bar{Q}$,
(g) the set $C(Q \setminus \{x^0\})$, with $x^0 \in Q$,
(h) the set of functions f from $C(\bar{Q})$ for which $\int_Q f\varphi \, dx = 0$, where φ is a bounded function from $C(\bar{Q})$, with Q a bounded region,
(i) the set of functions f from $C(\bar{Q})$ for which $\int_S f\varphi \, ds = 0$, where φ is a function from $C(\bar{Q})$, and S is a bounded section of a smooth surface in Q,
(j) the set of functions that are Riemann integrable (over Q),
(k) the set of functions from $C^k(\bar{Q})$ that are solutions of the linear differential equation

$$\sum_{|\alpha| \leqslant k} A_\alpha(x) D^\alpha f = 0, \quad \text{where } A_\alpha \in C(\bar{Q}), \quad |\alpha| \leqslant k,$$

(l) the set of functions that are measurable in Q,
(m) the space $L_1(Q)$,
(n) the space $L_2(Q)$.

4.2. Verify that each set of functions given below is not a linear space:
(a) the set of functions from $C(\bar{Q})$ that are equal to unity at a point $x^0 \in Q$,
(b) the set of functions $f \in C(\bar{Q})$ for which $\int_Q f \, dx = 1$, with Q a bounded region,
(c) the set of functions that are solutions of the differential equation $\nabla^2 u = 1$.

4.3. Prove that each of the systems of functions given below constitutes a linearly independent system:
(a) $1, x, x^2, \ldots$ on the segment $[a, b]$ $(a < b)$,
(b) x^α, $|\alpha| = 0, 1, 2, \ldots$, in a region Q,
(c) e^{ikx}, $k = 0, 1, 2, \ldots$, on the segment $[a, b]$,
(d) $[f(x)]^k$, $k = 0, 1, 2, \ldots$, in a domain Q, with $f(x)$ a function belonging to $C(Q)$ and not a constant.

4.4. Prove that the set $C(\bar{Q})$ is a normed linear space with

(1) $\|f\|_{C(\bar{Q})} = \max_{x \in \bar{Q}} |f(x)|$,

(2) $\|f\|'_{C(\bar{Q})} = 13 \max_{x \in \bar{Q}} |f(x)|$.

4.5. Prove that the set $C^k(\bar{Q})$ is a normed linear space with

$$\|f\|_{C^k(\bar{Q})} \sum_{\alpha \leqslant |k|} \max_{x \in \bar{Q}} |D^\alpha f(x)|. \qquad (4.1)$$

4.6. Suppose E is a set belonging to $\bar{Q}$. Show that the set of functions $f(x)$ that are continuous in $\bar{Q}$ and vanish in E constitute a normed linear space with norm (4.1) and $k = 0$.

4.7. Establish whether the following sets of functions defined in a bounded region Q constitute normed linear spaces with norm (4.1) and $k = 0$:

(a) the set of functions that belong to $C(\bar{Q})$ and are finite in Q,
(b) the set $C^\infty(\bar{Q})$,
(c) the set of functions that are analytic in Q and continuous in $\bar{Q}$.

4.8. Establish whether the following norms can be introduced in R^n:

$$\text{(a) } \| x \|_\infty = \max_{1 \leqslant i \leqslant n} |x_i|, \text{ (b) } \| x \|_2 = \Big(\sum_{i=1}^{n} x_i^2\Big)^{1/2}, \text{ (c) } \| x \|_1 = \sum_{i=1}^{n} |x_i|.$$

4.9. Establish whether for all p's that are positive integers the following norm can be introduced in R^n

$$\| x \|_p = \Big(\sum_{1}^{n} |x|^p\Big)^{1/p}.$$

Find $\lim\limits_{p \to \infty} \| x \|_p$.

4.10. Show that for all p's that are positive integers the following norm can be introduced in $C(\bar{Q})$:

$$\| f \|_p = \Big(\int_G |f|^p \, dx \Big)^{1/p} \tag{4.2}$$

(region Q is bounded). Find $\lim\limits_{p \to \infty} \| f \|_p$.

4.11. Verify that the normed linear spaces of Problems 4.4, 4.5, 4.6, 4.8 and 4.9 are Banach spaces (i.e. complete in the corresponding norms), while the normed linear spaces of Problems 4.7 and 4.10 are incomplete.

4.12. Show that in spaces $L_1(Q)$ and $L_2(Q)$ the following norms can be introduced:

$$\| f \|_{L_1(Q)} = \int_Q |f| \, dx, \tag{4.3}$$

$$\| f \|_{L_2(Q)} = \Big(\int_Q |f|^2 \, dx \Big)^{1/2}. \tag{4.4}$$

Theorem *Space $L_1(Q)$ with norm* (4.3) *and space $L_2(Q)$ with norm* (4.4) *are Banach spaces.*

A subset B' of a Banach space B is said to be a *(Banach) subspace of* B if it is a Banach space with the norm of B.

4.13. Suppose Q is a bounded region. Show that

(a) the set $\dot{C}(\bar{Q})$ of functions that belong to $C(\bar{Q})$ and vanish at the boundary of Q is a Banach subspace of $C(\bar{Q})$ (with norm (4.1) and $k = 0$),

(b) the subset of functions f from (1) $C(\bar{Q})$, (2) $L_1(Q)$, and (3) $L_2(Q)$ for which $\int_Q f(x)\,\varphi_i(x)\,dx = 0$, $i = 1, 2, \ldots, s$, where $\varphi_1, \ldots, \varphi_s$ are functions belonging to $C(\bar{Q})$, constitutes a Banach subspace of $C(\bar{Q})$ (with norm (4.1) and $k = 0$), of $L_1(Q)$ (with norm (4.3)), and of $L_2(Q)$ (with norm (4.4)), respectively.

4.14. Show that the countable set consisting of linear combinations of monomials x^α, $x = (x_1, \ldots, x_n)$, $\alpha = (\alpha_1, \ldots, \alpha_n)$, $|\alpha| = 0, 1, 2, \ldots$, with rational coefficients is dense everywhere in (a) $C(\bar{Q})$ (with norm (4.1) and $k = 0$), (b) $L_1(Q)$ (with norm (4.3)), and (c) $L_2(Q)$ (with norm (4.4)), where Q is a bounded region.

Suppose any two elements f and g of a complex (real) linear space H have corresponding to them a complex (real) number (f, g), called the *scalar product of f and g*, with the following properties:

(a) $(f, g) = \overline{(g, f)}$, (b) $(f + g, f_1) = (f, f_1) + (g, f_1)$, (c) $(cf, g) = c(f, g)$, and (d) the number (f, f) is real and nonnegative for every $f \in H$, with $(f, f) = 0$ only if $f = 0$.

Space H can be assigned a *norm*, say $\|f\| = (f, f)^{1/2}$. This particular norm is called the *norm generated by the scalar product.*

Space H is called a *Hilbert space* if it is complete in a norm generated by the scalar product.

The sequence of elements $f_1, f_2, \ldots$ belonging to H is said to be *weakly convergent to element $f \in H$* if for every $h \in H$ we have

$$(f_k, h) \to (f, h) \text{ as } k \to \infty.$$

Elements f and g are said to be *orthogonal* if $(f, g) = 0$. Element f is called *normed* if $\|f\| = 1$. The system $e_1, e_2, \ldots$ is said to be *orthonormal* if $(e_i, e_k) = \delta_{ik}$, $i, k = 1, 2, \ldots$.

Suppose $f \in H$ and $e_1, e_2, \ldots$ is an orthonormal system in H. The numbers $f_k = (f, e_k)$, $k = 1, 2, \ldots$, are called the *Fourier coefficients of element f*, and the series $\sum_1^\infty (f, e_k)\,e_k$ convergent in the norm of H is called the *Fourier series of element f* over the orthonormal system $e_1, e_2, \ldots$.

The system $e_1, e_2, \ldots$ is said to be an *orthonormal basis* or a *complete orthonormal system* if it is orthonormal and if the set of elements $c_1e_1 + c_2e_2 + \ldots + c_ke_k$ for all possible constants $c_1, \ldots, c_k$, and k is dense everywhere in H.

The Fourier series of element f over an orthonormal basis is convergent to f in the norm of H.

4.15. Show that $L_2(Q)$ is a Hilbert space with the scalar product

$$(f, g) = \int_Q f\bar{g}\,dx. \tag{4.5}$$

4.16. Show that the subset of functions $f \in L_2(Q)$ that are orthogonal to functions $\varphi_1, \ldots, \varphi_k$ belonging to $L_2(Q)$ constitutes a subspace of space $L_2(Q)$.

Suppose a continuous and positive function $\rho(x)$ (a *weight function*) is defined in a region Q. We denote the set of functions $f(x)$ that are measurable in Q and such that $\rho |f|^2 \in L_1(Q)$ by $L_{2,\rho}(Q)$.

4.17. Show that $L_{2,\rho}(Q)$ is a Hilbert space with the scalar product

$$(f, g) = \int_Q \rho f\bar{g}\,dx. \tag{4.6}$$

4.18. Prove that

(a) $L_2(Q) \subset L_{2,\rho}(Q)$ if $\rho(x)$ is a bounded function in Q.
(b) $L_{2,\rho}(Q) \subset L_2(Q)$ if $\rho(x) \geqslant \rho_0 > 0$ in Q.

4.19. Establish whether the trigonometric system $1, \sin x, \cos x, \sin 2x, \cos 2x, \ldots$ is orthogonal in $L_2(0, 2\pi)$.

4.20. Prove that the systems of functions $\sin(n + 1/2)x$, $n = 1, 2, \ldots$, and $\cos(n + 1/2)x$, $n = 1, 2, \ldots$, are orthogonal in $L_2(0, \pi)$.

4.21. Prove that the Legendre polynomials

$$P_n(x) = \frac{1}{2^n n!}\sqrt{\frac{2n+1}{2}}\,\frac{d^n}{dx^n}[(x^2-1)^n], \quad n = 0, 1, 2, \ldots,$$

constitute an orthonormal system in $L_2(-1,1)$.

4.22. Prove that the functions

$$T_n(x) = \sqrt{\frac{2}{\pi}}\cos n(\arccos x), \quad n = 0, 1, 2, \ldots$$

constitute a system of polynomials (the Chebyshev polynomials) orthonormal in $L_{2,(1-x^2)^{-1/2}}(-1, 1)$.

4.23. Prove that the functions

$$H_n(x) = (-1)^n e^{x^2}\frac{d^n}{dx^n}e^{-x^2}, \quad n = 0, 1, \ldots,$$

constitute a system of polynomials (the Hermite polynomials) orthogonal in $L_{2,e^{-x^2}}(-\infty, \infty)$.

4.24. Suppose the operator $-d^2/dx^2$ is defined on functions that belong to $C^2((0, 1)) \cap C^1([0, 1])$ with the boundary conditions $(hu - u_x)|_{x=0} = u|_{x=1} = 0$ (h a constant). Show that the eigenfunctions of this operator corresponding to different eigenvalues constitute an orthogonal system in $L_2(0, 1)$.

4.25. Suppose the operator $-\nabla^2$ is defined on functions that belong to $C^2(Q)\cap C^1(\overline{Q})$ with the boundary condition $\left(\frac{\partial u}{\partial n}+g(x)u\right)\Big|_\Gamma=0$ or $u|_\Gamma=0$, where $g\in C(\Gamma)$. Show that the eigenfunctions of this operator corresponding to different eigenvalues constitute an orthogonal system in $L_2(Q)$.

4.26. Let $\rho\in C(\overline{Q})$ and $\rho(x)\geqslant\rho_0>0$. Suppose the operator $-\frac{1}{\rho(x)}\nabla^2$ is defined on functions that belong to $C^2(Q)\cap C^1(\overline{Q})$ with the boundary conditions of Problem 4.25. Show that the eigenfunctions of this operator corresponding to different eigenvalues constitute an orthogonal system in $L_{2,\rho}(Q)$.

4.27. Let $p\in C^1[0, 1]$, $q\in C[0, 1]$, and $\rho\in C[0, 1]$, with $\rho(x)\geqslant\rho_0>0$. Suppose the operator $-\frac{1}{\rho(x)}\frac{d}{dx}\left[p(x)\frac{d}{dx}\right]+q(x)$ is defined on $C^2((0, 1))\cap C^1([0, 1])$ with the boundary conditions $u_x|_{x=0}=0$ and $(u_x+hu)|_{x=1}=0$ (h a constant). Show that the eigenfunctions of this operator corresponding to different eigenvalues constitute an orthogonal system in $L_{2,\rho}(0, 1)$.

4.28. Let $p\in C^1(\overline{Q})$, $q\in C(\overline{Q})$, and $\rho\in C(\overline{Q})$, with $\rho(x)\geqslant\rho_0>0$. Suppose the operator $-\frac{1}{\rho(x)}\operatorname{div}(p\ \operatorname{grad})+q(x)$ is defined on $C^2(Q)\cap C^1(\overline{Q})$ with the boundary conditions of Problem 4.25. Show that the eigenfunctions of this operator corresponding to different eigenvalues constitute an orthogonal system in $L_{2,\rho}(Q)$.

4.29. Show that the solutions of the equation $\nabla^2u=0$ that belong to $C^2(Q)\cap C^1(\overline{Q})$ and satisfy in Q the boundary condition $\left(\frac{\partial u}{\partial n}+\lambda u\right)\Big|_\Gamma=0$ constitute an orthogonal system in $L_2(\Gamma)$.

4.30. Show that the sequence $\sin kx$, $k=1, 2, \ldots$, is weakly convergent to zero in $L_2(0, 2\pi)$ but does not converge in the norm of $L_2(0, 2\pi)$.

In Problems 4.31-4.39 prove the propositions:

4.31. If the sequence of functions $f_n(x)$, $n=1, 2, \ldots$, that belong to $L_2(Q)$ converges to $f(x)$ in the norm of $L_2(Q)$, then it is weakly convergent to $f(x)$, too.

4.32. If the sequence of functions $f_n(x)$, $n=1, 2, \ldots$, that belong to $L_2(Q)$ converges to $f(x)$ in the norm of $L_2(Q)$, then $\int\limits_Q f_n\,dx\to\int\limits_Q f\,dx$ as $n\to\infty$ (Q is a bounded region).

4.33. If $u_k \in L_2(Q)$, $k=1, 2, \ldots$, and the series $\sum_1^\infty u_k(x)$ is convergent to $u(x)$ in the norm of $L_2(Q)$, then $\sum_{k=1}^\infty \int_Q u_k\, dx = \int_Q u\, dx$ (Q a bounded region).

4.34. If the sequence of functions $f_n(x)$, $n = 1, 2, \ldots$, that belong to $C(\bar{Q})$ is convergent to $f(x)$ uniformly in $\bar{Q}$, then it is convergent to $f(x)$ in the norm of $L_2(Q)$, too (Q a bounded region).

4.35. If the sequence of functions $f_n(x)$, $n = 1, 2, \ldots$, that belong to $L_2(Q)$ is weakly convergent to $f(x) \in L_2(Q)$, then the sequence of norms $\| f_n(x) \|_{L_2(Q)}$, $n = 1, 2, \ldots$, is bounded.

4.36. If the sequence of functions $f_n(x)$, $n = 1, 2, \ldots$, that belong to $L_2(Q)$ is weakly convergent to $f(x) \in L_2(Q)$ and $\| f_n(x) \| \to \| f(x) \|$ as $n \to \infty$, then it is convergent to $f(x)$ in the norm of $L_2(Q)$, too.

4.37. For every function $f(x) \in L_2(Q)$,

$$\sum_{k=1}^\infty |f_k|^2 \leqslant \| f \|^2$$

(Bessel's inequality), where the f_k, $k = 1, 2, \ldots$, are the Fourier coefficients of f over an orthonormal system $e_1, e_2, \ldots$.

4.38. Every orthonormal system $e_1, \ldots, e_n, \ldots$ in $L_2(Q)$ is weakly convergent to zero but is not convergent in the norm of $L_2(Q)$.

4.39. For every $f \in L_2(Q)$,

$$\min_{c_1, \ldots, c_n} \left\| f - \sum_1^n c_k e_k \right\| = \left\| f - \sum_1^n f_k e_k \right\|$$

(i.e. the nth partial sum of the Fourier series is the best approximation of $f(x)$ in $L_2(Q)$).

4.40. Find the quadratic polynomial that is the best approximation of (a) x^3, (b) $\sin \pi x$, or (c) $|x|$ in $L_2(-1, 1)$.

4.41. Find the linear trigonometric polynomial that is the best approximation in $L_2(-\pi, \pi)$ of (a) $|x|$ or (b) $\sin(x/2)$.

4.42. Find the linear polynomial that is the best approximation of $x_1^2 - x_2^2$ in $L_2(Q_i)$, where Q_i is (a) the circle $x_1^2 + x_2^2 < 1$ and (b) the square $0 < x_1, x_2 < 1$.

4.43. Establish whether the following systems of functions are complete in $L_2(Q)$:

(a) $\sin kx$, $k = 1, 2, \ldots, Q = [0, \pi]$,
(b) $\sin (2k + 1) x$, $k = 0, 1, \ldots, Q = [0, \pi/2]$.

In Problems 4.44-4.50 prove the propositions:

4.44. The Legendre polynomials (Problem 4.21) and the Chebyshev polynomials (Problem 4.22) constitute orthonormal bases in the spaces $L_2(-1, 1)$ and $L_{2,(1-x^2)^{-1/2}}(-1, 1)$, respectively.

4.45. A system $e_1, e_2, \ldots$ that is orthonormal in $L_2(Q)$ is an orthonormal basis in $L_2(Q)$ if and only if for every $f \in L_2(Q)$ the Parseval-Steklov equality

$$\| f \|^2 = \sum_1^{\infty} |f_k|^2$$

is true.

4.46. If $f \in L_2(a, b)$ and $\int_a^b x^k f(x)\, dx = 0$ for $k = 0, 1, \ldots,$ then $f(x) = 0$ almost everywhere in (a, b).

4.47. If $f \in L_2(Q)$ and $\int_Q x^{\alpha} f(x)\, dx = 0$ for all α, $|\alpha| = 0, 1, \ldots,$ then $f(x) = 0$ almost everywhere in Q.

4.48. If f_k and g_k, $k = 1, 2, \ldots,$ are the Fourier coefficients of the functions $f, g \in L_2(Q)$ over an orthonormal basis, then

$$(f, g) = \sum_1^{\infty} f_k \bar{g}_k.$$

4.49. Every orthonormal system $e_1, e_2, \ldots, e_n$ is linearly independent.

4.50. A system of functions $\varphi_1, \ldots, \varphi_n$ belonging to $L_2(Q)$ is linearly independent if and only if Gram's determinant $\det \| (\varphi_i, \varphi_j) \|$, $i, j = 1, \ldots, n$, is nonzero.

Suppose $\varphi_1, \ldots, \varphi_n$ constitutes a linearly independent system of functions belonging to $L_2(Q)$ (or $L_{2,\rho}(Q)$). We define the function $e_1(x)$ in the following manner: $e_1 = \varphi_1/\| \varphi_1 \|$. We select constants c_1 and c_2 in such a manner that the function $e_2 = c_1 e_1 + c_2 \varphi_2$ is normed and orthogonal in $L_2(Q)$ (or in $L_{2,\rho}(Q)$) to the function e_1. This process can be continued, that is, if we have constructed the functions $e_1, \ldots, e_{n-1}$, the function e_n is sought in the form $e_n = \beta_1 e_1 + \beta_2 e_2 + \beta_{n-1} e_{n-1} + \beta_n \varphi_n$ with the constants $\beta_1, \ldots, \beta_n$ selected in such a manner that e_n is normed and orthogonal to $e_1, \ldots, e_{n-1}$. This procedure is known as the *Gram-Schmidt orthogonalization process.*

4.51. Find the explicit expression for the functions e_k, $k = 1, 2, \ldots, n$, in terms of the functions $\varphi_1, \ldots, \varphi_n$.

4.52. Employ the Gram-Schmidt orthogonalization process to orthonormalize in $L_{2,\rho}(Q)$ the following sequences of functions (first verify their linear independence):

(a) $1,\ x,\ x^2,\ x^3$ $(\rho\equiv 1,\quad Q=(-1,\ +1))$,

(b) $1-x,\ 1+x^2,\ 1+x^3$ $(\rho\equiv 1,\quad Q=(-1,\ +1))$,

(c) $\sin^2 \pi x,\ 1,\ \cos x\pi$ $(\rho\equiv 1,\quad Q=(-1,\ +1))$,

(d) $1,\ x,\ x^2$ $(\rho=e^{-x},\quad Q=(0,\ \infty))$,

(e) $1,\ x,\ x^2$ $(\rho=e^{-x^2/2},\quad Q=(-\infty,\ +\infty))$,

(f) $1,\ x,\ x^2$ $(\rho=\sqrt{1-x^2},\ Q=(-1,\ 1))$,

(g) $1,\ x,\ x^2$ $\left(\rho=\dfrac{1}{\sqrt{1-x^2}},\ Q=(-1,\ 1)\right)$

4.53. Show that as a result of orthonormalization of the system of functions $1,\ x,\ x^2,\ \ldots$ by the Gram-Schmidt orthogonalization process in the scalar product $(f,\ g)=\int\limits_0^1 \frac{f\bar{g}}{\sqrt{1-x^2}}\,dx$, we arrive at an orthonormal basis of $L_{2,\,(1-x^2)^{-1/2}}(-1,\ 1)$ consisting of the Chebyshev polynomials $T_n(x),\ n=1,\ 2,\ \ldots$.

4.54. Orthonormalize the system of polynomials $1,\ x_1,\ x_2$ in the circle $|x|<1$ in the scalar product

$$(u,\ v)=\int\limits_{|x|<1} u\bar{v}\,dx.$$

4.55. Orthonormalize the system of polynomials $1,\ x_1,\ x_2,\ x_3$ in the ball $|x|<1,\ x=(x_1,\ x_2,\ x_3)$ in the scalar product

$$(u,\ v)=\int\limits_{|x|<1} u\bar{v}\,dx.$$

4.56. We denote the set of functions $f(x)\in L_{2,\,\mathrm{loc}}(-\infty,\ \infty)$ for which $\lim\limits_{k\to\infty}\frac{1}{2k}\int\limits_{-k}^{k}|f|^2\,dx$ is finite by $L_2'(-\infty,\ \infty)$. Show that $L_2'(-\infty,\ \infty)$ is a Hilbert space with the scalar product

$$(f,\ g)=\lim_{k\to\infty}\frac{1}{2k}\int\limits_{-k}^{k} f\bar{g}\,dx.$$

4.57. Prove that the system of functions $e^{i\alpha x}$, where α is an arbitrary real number, is orthonormal in $L_2'(-\infty,\ \infty)$ (see the previous problem).

Suppose Q is a bounded region of space R^n with a piecewise smooth boundary Γ. Let $\alpha = (\alpha_1, \ldots, \alpha_n)$ be a multi-index (see Symbols and Definitions). The function $f^{(\alpha)} \in L_{1,\mathrm{loc}}(Q)$ is called a *generalized derivative of order* α of a function $f \in L_{1,\mathrm{loc}}(Q)$ if

$$\int_Q f D^\alpha g \, dx = (-1)^{|\alpha|} \int_Q f^{(\alpha)} g \, dx \tag{4.7}$$

for every function $g \in C^{|\alpha|}(\bar{Q})$ finite in Q (for a more general definition see Vladimirov [2]). If $f \in C^{|\alpha|}(Q)$, the generalized derivative $f^{(\alpha)}(x)$ exists and $f^{(\alpha)}(x) = D^\alpha f(x)$ almost everywhere. For this reason we will always denote the generalized derivative of order α of a function $f(x)$ by $D^\alpha f$.

The set of the functions (we will assume that they are all real-valued) $f \in L_2(Q)$ that have all the generalized derivatives up to order k inclusive and belong to $L_2(Q)$ is known as a *Sobolev space* $H^k(Q)$, which is also a Hilbert space. The scalar product in such a space can be defined thus:

$$(f, g) = \int_Q \Big(\sum_{|\alpha| \leqslant k} D^\alpha f D^\alpha g \Big) dx, \tag{4.8}$$

while the corresponding norm can be defined thus:

$$\| f \|_{H^k(Q)} = \Big[\int_Q \Big(\sum_{|\alpha| \leqslant k} |D^\alpha f|^2 \Big) dx \Big]^{1/2}. \tag{4.8'}$$

For $k = 0$ the space $H^k(Q)$ coincides with $L_2(Q)$, or $H^0(Q) = L_2(Q)$. If Γ is sufficiently smooth, then $H^k(Q)$ is the completion of set $C^k(\bar{Q})$ in norm (2′).

Suppose f belongs to $H^1(Q)$ and the f_k, $k = 1, 2, \ldots$, constitute a sequence of functions from $C^1(\bar{Q})$ that converges to $f(x)$ in the norm of $H^1(Q)$. Every $(n-1)$-dimensional smooth surface S consisting of a finite number of pieces each of which has a unique map on a coordinate plane and lying in $\bar{Q}$ has a positive constant c that depends neither on $f(x)$ nor on $f_k(x)$, $k = 1, 2, \ldots$, and such that

$$\int_S |f_k - f_m|^2 \, ds \leqslant c \, \| f_k - f_m \|^2_{H^1(Q)}.$$

This inequality and the fact that the space $L_2(S)$ is complete imply that the sequence of traces of $f_k(x)$ on S converges to a function $g \in L_2(S)$ in the norm of $L_2(S)$. The function $g(x)$ does not depend on the choice of the sequence that approximates $f(x)$ and is called the *trace* $f|_S$ *of* $f(x)$ *on* $S \in \bar{Q}$.

We denote the set of functions that belong to $H^1(Q)$ and whose traces on Γ are zero almost everywhere on Γ by $\mathring{H}^1(Q)$. It can be obtained by completion in the norm (4.8′) at $k = 1$ of the set of functions that have continuous first-order derivatives in Q and vanish on Γ.

The convolution $f_h(x)=\int_Q \omega_h(|x-y|)\,f(y)\,dy$, where $\omega_h(|x-y|)$ is the *averaging kernel* (see Symbols and Definitions), with $f \in L_1(Q)$ is known as the *average function for f*.

Let $x_i = \varphi_i(y)$, $i = 1, 2, \ldots, n$, $y = (y_1, \ldots, y_n)$, be a one-to-one map of region Q onto region Q' such that it has a nonzero Jacobian in $\bar{Q}$ and is k times continuously differentiable in $\bar{Q}$. Then, if $f \in H^k(Q)$, we have

$$F(y) = f(\varphi_1(y), \ldots, \varphi_n(y)) \in H^k(Q').$$

Two scalar products, $(u, v)_{\mathrm{I}}$ and $(u, v)_{\mathrm{II}}$, and the corresponding norms, $\| u \|_{\mathrm{I}}$ and $\| u \|_{\mathrm{II}}$, are said to be *equivalent* in a Hilbert space H if there are positive constants c_1 and c_2 such that $c_1 \| u \|_{\mathrm{I}} \leqslant \| u \|_{\mathrm{II}} \leqslant c_2 \| u \|_{\mathrm{I}}$ for every $u \in H$.

4.58. Establish whether a mixed generalized derivative depends on the order in which the derivatives are taken.

4.59. Show that the existence of the generalized derivative $D^\alpha f$ does not necessarily imply that there is a generalized derivative $D^{\alpha'} f$ at $\alpha_i' \leqslant \alpha_i$, $i = 1, 2, \ldots, n$, $|\alpha'| < |\alpha|$.

Hint. Consider the function $f(x_1, x_2) = f_1(x_1) + f_2(x_2)$, where $f_i(x_i)$ has no first-order generalized derivative.

4.60. Show that if a function $f(x)$ has a generalized derivative $D^\alpha f$ in a region Q, then it also has $D^\alpha f$ in any subregion $Q' \subset Q$.

4.61. Suppose a function $f_1(x)$ is defined in a region Q_1 and has a generalized derivative $D^\alpha f_1$, while another function $f_2(x)$ is defined in a region Q_2 and has a generalized derivative $D^\alpha f_2$. Prove that if $Q_1 \cup Q_2$ is a region and $f_1(x) = f_2(x)$ for $x \in Q_1 \cap Q_2$, then the function

$$f(x)=\begin{cases} f_1(x), & x \in Q_1, \\ f_2(x), & x \in Q_2, \end{cases}$$

has a generalized derivative $D^\alpha f$ in $Q_1 \cup Q_2$ that is equal to $D^\alpha f_1$ in Q_1 and to $D^\alpha f_2$ in Q_2.

4.62. Let

$$f(x_1, x_2)=\begin{cases} 1 \text{ if } |x|<1, & x_2>0, \\ -1 \text{ if } |x|<1, & x_2<0. \end{cases}$$

Prove that $f(x_1, x_2)$ has first-order generalized derivatives in each of the semicircles but does not have a generalized derivative with respect to x_2 in the circle $|x| < 1$.

4.63. Prove the validity of the following properties of average functions:

(a) $f_h \in C^\infty(R^n)$,

(b) $f_h(x)$ converge to $f(x)$ in $L_2(Q)$ as $h \to 0$ if $f \in L_2(Q)$.

(c) $(D^{\alpha}f)_h = D^{\alpha}f_h$ in every strictly internal subregion $Q' \Subset Q$ for a sufficiently small h, that is, the generalized derivative of an average function is equal to the average function for the generalized derivative.

In Problems 4.64-4.72 prove the propositions:

4.64. If a function $f(x)$ has a generalized derivative $D^{\alpha}f = \omega(x)$ in a region Q and the function $\omega(x)$ has a generalized derivative $D^{\beta}\omega$, then there is a generalized derivative $D^{\alpha+\beta}f$.

4.65. (a) $y = \operatorname{sign} x \notin H^1(-1, 1)$; (b) $y = |x| \in H^1(-1, 1)$, but $y = |x| \notin H^2(-1, 1)$.

4.66. If $f \in H^1(a, b)$ and the generalized derivative $f(x)$ is zero, then $f(x) = \text{const}$ almost everywhere.

4.67. If $f \in H^1(a, b)$, then on $[a, b]$ $f(x)$ is equivalent to a continuous function.

4.68. If $f(x) \in H^1(-\infty, \infty)$, then $f(x)$ tends to zero as $|x| \to \infty$.

4.69. Let us denote the subspace of $H^1(0, 2\pi)$ consisting of all functions $f(x)$ belonging to $H^1(0, 2\pi)$ for which $f(0) = f(2\pi)$ by $\widetilde{H}^1(0, 2\pi)$. Prove that a function $f(x)$ (belonging to $H^1(0, 2\pi)$) belongs to $\widetilde{H}^1(0, 2\pi)$ if and only if the number series with the general term $n^2(a_n^2 + b_n^2)$ is convergent, with

$$a_n = \int_0^{2\pi} f(x)\cos nx\, dx, \quad b_n = \int_0^{2\pi} f(x)\sin nx\, dx, \quad n = 0, 1, 2, \ldots .$$

The relationship

$$\|f\|^2_{\widetilde{H}^1(0, 2\pi)} = \sum_{k=0}^{\infty} (a_k^2 + b_k^2)(k^2 + 1)$$

defines one of the equivalent norms of $\widetilde{H}^1(0, 2\pi)$.

4.70. A function $f \in L_2(0, \pi)$ belongs to $\mathring{H}^1(0, \pi)$ if and only if the series with the general term $k^2 b_k^2$ converges, with $b_k = \frac{2}{\pi}\int_0^{\pi} f(x)\sin kx\, dx$.

The norm is defined thus:

$$\|f\|^2_{\mathring{H}^1(0, \pi)} = \int_0^{\pi} (f^2 + f'^2)\, dx = \frac{\pi}{2}\sum_1^{\infty} (k^2 + 1)\, b_k^2.$$

4.71. For every $f \in \mathring{H}^1(a, b)$,

$$\int_a^b f^2\, dx \leqslant \left(\frac{b-a}{\pi}\right)^2 \int_a^b f'^2\, dx$$

(the one-dimensional case of the Steklov inequality).

4.72. Find the function $f_0(x) \not\equiv 0$ for which the inequality of Problem 4.71 turns into an equality. Show that if $f(x) \not\equiv cf_0(x)$, where c is a constant, then $f(x)$ obeys the strict inequality.

4.73. Prove that for every function $f \in H^1(0, 2\pi)$ for which $f(0) = f(2\pi)$ the following inequality is valid:

$$\int_0^{2\pi} f^2\,dx \leqslant \int_0^{2\pi} (f')^2\,dx + \frac{1}{2\pi}\left(\int_0^{2\pi} f(x)\,dx\right)^2.$$

4.74. Prove that

$$\int_0^{2\pi} f^2\,dx \leqslant 4\int_0^{2\pi} (f')^2\,dx + \frac{1}{2\pi}\left(\int_0^{2\pi} f\,dx\right)^2$$

(the one-dimensional case of Poincaré's inequality) for every function $f \in H^1(0, 2\pi)$. *Hint.* Employ the fact that the system of functions $\cos(kx/2)$, $k = 0, 1, 2, \ldots$, forms an orthogonal basis in $H^1(0, 2\pi)$.

4.75. Prove that in $H^1(0, 2\pi)$ there is a two-dimensional subspace such that Poincaré's inequality of Problem 4.74 becomes an equality for every element of the subspace. Find the subspace and prove that for all the elements of $H^1(0, 2\pi)$ that do not belong to this subspace the inequality of Problem 4.74 is strict.

4.76. Let $f \in \overset{\circ}{H}{}^1(|x| < 1)$, $x_1 = |x|\cos\varphi$, $x_2 = |x|\sin\varphi$. Prove that $\lim\limits_{|x|\to 1-0} \int_0^{2\pi} f^2(|x|, \varphi)\,d\varphi = 0$.

4.77. Let $f \in H^1(|x| < 1)$, $x_1 = |x|\cos\varphi$, $x_2 = |x|\sin\varphi$, $f\,|_{|x|=1} = h(\varphi)$, $0 \leqslant \varphi < 2\pi$. Prove that

$$\lim_{|x|\to 1-0} \int_0^{2\pi} |h(\varphi) - f(|x|, \varphi)|^2\,d\varphi = 0.$$

4.78. Let $f \in \overset{\circ}{H}{}^1(0 < x_1 < 1,\ 0 < x_2 < 1)$. Prove that

$$\int_0^1 f^2(x_1, x_2)\,dx_1 = o(x_2) \text{ as } x_2 \to 0.$$

4.79. Let $x = (x_1, x_2) = (\rho\cos\varphi, \rho\sin\varphi)$ and suppose the function

$$f(x) = \frac{a_0}{2} + \sum_{k=1}^{\infty} \rho^k (a_k \cos k\varphi + b_k \sin k\varphi)$$

belongs to $H^1(|x| < 1)$. Express the integral $\int\limits_{\rho<1} (|\operatorname{grad} f|^2 + |f|^2)\,dx$ in terms of a_k and b_k.

4.80. Suppose

$$\psi(\varphi)=\frac{a_0}{2}+\sum_1^{\infty}(a_k\cos k\varphi+b_k\sin k\varphi)$$

and $\sum_{k=1}^{\infty} k(a_k^2+b_k^2)<\infty$. Prove that there is a function $f(\rho,\varphi)\in \in H^1(|x|<1)$ such that $f|_{\rho=1}=\psi(\varphi)$.

4.81. For what values of α does the function $f=|x|^{-\alpha}\sin|x|$ belong to $H^2(|x|<1)$, $x=(x_1, x_2)$?

4.82. Prove that $|x_1|(|x|^2-1)\in \mathring{H}^1(|x|<1)$, $x=(x_1, x_2, x_3)$.

4.83. For what values of α does the function $f=|x|^{-\alpha}e^{x_1-x_2}$ belong to $H^1(|x|<1)$, $x=(x_1, x_2, x_3)$?

4.84. Suppose $f(x_1, x_2)=\sum_1^{\infty} a_k \sin kx_1 e^{-kx_2}$, $0\leqslant x_1\leqslant\pi$, $x_2>0$. For what values of a_k does the function f belong to $H^1(0<x_1<\pi, x_2>0)$?

4.85. Suppose $f\in H^1(|x|<1)$, $x=(x_1, x_2, \ldots, x_n)$, $n\geqslant 2$. Must the function $f(x)$ be equivalent to a function continuous in the ball $|x|<1$? (Compare the result with that of Problem 4.67.)

In Problems 4.86-4.90 prove the propositions:

4.86. If $f\in H^1(Q)$ and $f(x)\equiv$ const almost everywhere in $Q'\subset Q$, then grad $f=0$ almost everywhere in Q'.

4.87. If $f\in H^1(Q)$ and $|\text{grad} f|=0$ almost everywhere in Q, then $f(x)=$ const almost everywhere in Q.

4.88. If $f\in H^1(Q)$ and $g\in\mathring{H}^1(Q)$, then $\int_Q fg_{x_i}\,dx=-\int_Q gf_{x_i}\,dx$ for all $i=1, 2, \ldots, n$ (the integration-by-parts formula).

4.89. If $f\in H^1(Q)$ and $g\in H^1(Q)$, then

$$\int_Q fg_{x_i}\,dx=-\int_Q gf_{x_i}\,dx+\int_\Gamma fg\cos(\mathbf{n}x_i)\,ds$$

for all $i=1, 2, \ldots, n$, where in the second integral on the right-hand side the f and g are the traces of the respective functions on Γ.

4.90. $\mathring{H}^1(Q)$ is a subspace of $H^1(Q)$.

Let us assume that a function $f\in L_2(Q)$ is continued by, say, zero in the exterior of Q. The function $\delta_i^h f=\frac{f(x_1, \ldots, x_i+h, \ldots, x_n)-f(x)}{h}$, $h\neq 0$, is called the *finite-difference ratio* for $f(x)$ in the variable x_i, $i=1, 2, \ldots, n$, and also belongs to $L_2(Q)$.

In Problems 4.91-4.96 prove the propositions:

4.91. For every function $f \in L_2(a, b)$ that is finite on (a, b) and an arbitrary function g that also belongs to $L_2(a, b)$ the integration-by-parts formula

$$(\delta^h f, g) = -(f, \delta^{-h} g)$$

is valid for sufficiently small values of $|h|$.

4.92. For sufficiently small nonzero values of $|h|$ and a function $f \in L_2(Q)$ that is finite in Q and an arbitrary function g that also belongs to $L_2(Q)$, the following integration-by-parts formula is valid:

$$(\delta_i^h f, g) = -(f, \delta_i^h g), \ i = 1, 2, \ldots, n.$$

4.93. If a function f that is finite on (a, b) belongs to $H^1(a, b)$, then $\delta^h f(x) \to f'(x)$ in the norm of $L_2(a, b)$ as $h \to 0$.

4.94. If $\delta^h f \to \tilde{f}(x)$ in the norm of $L_2(a, b)$ as $h \to 0$ and $f \in L_2(a, b)$ is finite on (a, b), then $f(x)$ belongs to $H^1(a, b)$ and $\tilde{f}(x)$ is the generalized derivative of $f(x)$.

4.95. If a function $f \in L_2(Q)$ is finite in Q and has a generalized derivative $f_{x_i} \in L_2(a, b)$ for a certain value of i $(i = 1, 2, \ldots, n)$, then $\delta_i^h f \to f_{x_i}$ in the norm of $L_2(Q)$ as $h \to 0$.

4.96. If a function f that is finite in Q belongs to $L_2(Q)$ and $\delta_i^h f \to \tilde{f}_i(x)$ in the norm of $L_2(Q)$ as $h \to 0$ for a certain value of i $(i = 1, 2, \ldots, n)$, then $f(x)$ has in Q a generalized derivative in x_i and this derivative coincides with $\tilde{f}_i(x)$.

4.97. Using the result of Problem 4.71, show that the scalar products

$$(f, g)_1 = \int_0^\pi (fg + f'g')\, dx, \quad (f, g)_{II} = \int_0^\pi f'g'\, dx$$

are equivalent in space $\mathring{H}^1(0, \pi)$.

4.98. Using the result of Problem 4.74, show that the scalar products

$$(f, g)_I = \int_0^{2\pi} (fg + f'g')\, dx, \ (f, g)_{II} = \int_0^{2\pi} f'g'\, dx + \left(\int_0^{2\pi} f\, dx\right)\left(\int_0^{2\pi} g\, dx\right)$$

are equivalent in space $H^1(0, 2\pi)$.

4.99. The set $\tilde{H}^1(0, 2\pi)$ of functions $f \in H^1(0, 2\pi)$ for which $\int_0^{2\pi} f(x)\, dx = 0$ is a subspace of $H^1(0, 2\pi)$. Show that in $\tilde{H}^1(0, 2\pi)$ the scalar product can be defined thus:

$$(fg)_{\tilde{H}^1(0, 2\pi} \quad \int_0^{2\pi} f'g'\, dx.$$

4.100. Suppose $\rho(x) \in C(\bar{Q})$ and $\rho(x) \geqslant \rho_0 > 0$. Show that the formula $(f, g)_{\mathrm{I}} = \int_Q \rho fg\, dx$, with $f, g \in L_2(Q)$, defines a scalar product in $L_2(Q)$ that is equivalent to the scalar product $\int_Q fg\, dx$.

4.101. Let us assume that $\rho \in C(\bar{Q})$, $\rho(x) > 0$ in $\bar{Q} \setminus x^0$, and $\rho(x^0) = 0$, where x^0 is a point of $\bar{Q}$. Then the formula for $(f, g)_{\mathrm{I}}$ of Problem 4.100 defines a scalar product in $L_2(Q)$ that is not equivalent to the scalar product $\int_Q fg\, dx$.

4.102. Suppose $\rho \in C(\bar{Q} \setminus x^0)$, where x^0 is a point of $\bar{Q}$, $\rho(x)$ is positive for $x \in \bar{Q} \setminus x^0$ and tends to ∞ as $x \to x^0$, $x \in \bar{Q}$. Show that in $L_{2,\rho}(Q)$ the scalar product $\int_Q fg\, dx$ is not equivalent to the scalar product $\int_Q \rho fg\, dx$.

4.103. Suppose $f \in H^1(|x| < 1)$, $x = (x_1, x_2)$ and $f(x)\,|_{|x|=1} = h(\varphi)$, $x_1 = |x| \cos \varphi$, $x_2 = |x| \sin \varphi$. Prove that there is a positive constant c that does not depend on $f(x)$ and

$$\int_{|x|<1} f^2\, dx \leqslant c \left[\int_0^{2\pi} h^2(\varphi)\, d\varphi + \int_{|x|<1} |\operatorname{grad} f|^2\, dx \right].$$

4.104. Prove that there is a positive constant c such that

$$\int_Q f^2\, dx \leqslant c \int_Q |\operatorname{grad} f|^2\, dx$$

for every function $f \in \mathring{H}^1(Q)$ (the Steklov inequality).

4.105. Show that $\int_Q (\operatorname{grad} f, \operatorname{grad} g)\, dx$ defines in $\mathring{H}^1(Q)$ a scalar product that is equivalent to the scalar product $\int_Q [fg + (\operatorname{grad} f, \operatorname{grad} g)]\, dx$.

4.106. Suppose $p, q \in C(\bar{Q})$, $p(x) \geqslant p_0 > 0$, $q(x) \geqslant 0$. Prove that the scalar products

$$(f, g) = \int_Q [fg + (\operatorname{grad} f, \operatorname{grad} g)]\, dx,$$

$$(f, g)_{\mathrm{I}} = \int_Q [qfg + p(\operatorname{grad} f, \operatorname{grad} g)]\, dx$$

are equivalent in $\mathring{H}^1(Q)$.

4.107. Let the real-valued functions p_{ij}, $p_{ij}(x) = p_{ji}(x)$, $i, j = 1, 2, \ldots, n$, and q belong to $C(\overline{Q})$, with $q \geqslant 0$, and $\sum_{i,j=1}^{n} p_{ij}(x)\xi_i\xi_j \geqslant \gamma_0 \sum_{i=1}^{n} \xi_i^2$ for all real vectors $\xi = (\xi_1, \ldots, \xi_n) \in R_n$, with γ_0 a positive constant. Prove that

$$(f, g)_{\text{I}} = \int_Q \left(\sum_{i,j=1}^{n} p_{ij} f_{x_i} g_{x_j} + qfg \right) dx$$

can be defined as a scalar product in $\overset{\circ}{H}{}^1(Q)$ that is equivalent to the scalar product

$$(f, g) = \int_Q [fg + (\operatorname{grad} f, \operatorname{grad} g)]\, dx.$$

4.108. Suppose $p, q \in C(\overline{Q})$, $p(x) \geqslant p_0 > 0$, and $q(x) \geqslant q_0 > 0$. Prove that the scalar products

$$(f, g) = \int_Q [fg + (\operatorname{grad} f, \operatorname{grad} g)]\, dx,$$

$$(f, g)_{\text{I}} = \int_Q [qfg + p(\operatorname{grad} f, \operatorname{grad} g)]\, dx$$

are equivalent in $H^1(Q)$.

In solving Problems 4.109, 4.113, 4.114, and 4.118 it is expedient to employ the following

Theorem *For a set $M \subset H^1(Q)$ to be compact in $L_2(Q)$, it is sufficient that M be bounded in the norm of $H^1(Q)$, that is, there is a positive constant c such that $\| u \|_{H^1(Q)} \leqslant c$ for every $u \in M$.*

The compactness of M in L_2 means that out of any infinite sequence of elements of M we can select a sequence that is fundamental in L_2.

4.109. Let x^0 be an arbitrary point of $\overline{Q}$ and suppose $U = Q \cap \{|x - x_0| < r\}$ for a positive r. Prove that there is a positive constant c such that

$$\int_Q f^2\, dx \leqslant c \left[\int_Q |\operatorname{grad} f|^2\, dx + \int_U f^2\, dx \right]$$

for all $f \in H^1(Q)$.

4.110. Use the result of Problem 4.108 to prove that the scalar products

$$(f, g)_{\text{I}} = \int_Q [fg + (\operatorname{grad} f, \operatorname{grad} g)]\, dx,$$

$$(f, g)_{\text{II}} = \int_Q [qfg + p(\operatorname{grad} f, \operatorname{grad} g)]\, dx$$

are equivalent in $H^1(Q)$ only if the functions $p(x)$ and $q(x)$ are continuous in $\bar{Q}$ and satisfy the following conditions: $p \geqslant p_0 > 0$, $q(x) \geqslant 0$, and $q(x) \not\equiv 0$ in Q.

4.111. If $q(x) \geqslant q_0 > 0$ in the hypothesis of Problem 4.107, then $\int_Q \left(\sum_{i,j=1}^{n} p_{ij} f_{x_i} g_{x_j} + qfg \right) dx$ can be taken as a scalar product in $H^1(Q)$, and this scalar product is equivalent to the scalar product $\int_Q [(\text{grad } f, \text{grad } g) + fg]\, dx$.

4.112. If $q(x) \geqslant 0$ in $\bar{Q}$ and is not identically equal to zero in the hypothesis of Problem 4.107, then

$$\int_Q \left(\sum_{i,j=1}^{n} p_{ij} f_{x_i} g_{x_j} + qfg \right) dx$$

can be taken as a scalar product in $H^1(Q)$, and this scalar product is equivalent to

$$\int_Q [fg + (\text{grad } f, \text{ grad } g)]\, dx.$$

4.113. Show that there is a positive constant c such that

$$\int_Q f^2\, dx \leqslant c \left[\int_Q |\text{grad } f|^2\, dx + \int_{\partial Q} f^2\, ds \right]$$

for every $f \in H^1(Q)$.

4.114. Suppose x^0 is an arbitrary point on a boundary ∂Q and $U = \partial Q \cap \{|x - x^0| < r\}$ for a positive r. Prove that there is a positive constant c such that

$$\int_Q f^2\, dx \leqslant c \left[\int_Q |\text{grad } f|^2\, dx + \int_U f^2\, ds \right]$$

for every $f \in H^1(Q)$.

4.115. Prove that if $\sigma \in C(\partial Q)$ and $\sigma(x) > 0$, then $(f, g)_{\text{I}} = \int_Q (\text{grad } f, \text{ grad } g)\, dx + \int_{\partial Q} \sigma f g\, ds$ defines a scalar product in $H^1(Q)$, and this scalar product is equivalent to

$$(f, g) = \int_Q [fg + (\text{grad } f, \text{ grad } g)]\, dx.$$

4.116. Prove that if $\sigma \in C(\partial Q)$, $\sigma(x) \geqslant 0$ and $\sigma(x) \not\equiv 0$, then

$$(f, g)_{\text{I}} = \int_Q (\text{grad } f, \text{ grad } g)\, dx + \int_{\partial Q} \sigma f g\, ds$$

defines a scalar product in $H^1(Q)$ and this scalar product is equivalent to

$$(f, g) = \int_Q [fg + (\operatorname{grad} f, \operatorname{grad} g)]\, dx.$$

4.117. Let $p \in C(\bar{Q})$, $q \in C(\bar{Q})$, $\sigma \in C(\partial Q)$, $p(x) \geqslant p_0 > 0$ and $q(x) \geqslant 0$ in $\bar{Q}$, $\sigma(x) \geqslant 0$ on ∂Q, and either $q(x) \not\equiv 0$ or $\sigma(x) \not\equiv 0$. Then the scalar products

$$(f, g)_{\mathrm{I}} = \int_Q [p(\operatorname{grad} f, \operatorname{grad} g) + qfg]\, dx + \int_{\partial Q} \sigma f g\, ds,$$

$$(f, g) = \int_Q [fg + (\operatorname{grad} f \quad \operatorname{grad} g)]\, dx$$

in $H^1(Q)$ are equivalent.

4.118. Show that if there is a positive constant c such that

$$\int_Q f^2\, dx \leqslant c\left[\left(\int_Q f\, dx\right)^2 + \int_Q |\operatorname{grad} f|^2\, dx\right]$$

(Poincaré's inequality) for every $f \in H^1(Q)$ $(\partial Q \in C^1)$.

4.119. Using the result of Problem 4.118, show that in $H^1(Q)$ the scalar products

$$(f, g) = \int_Q [fg + (\operatorname{grad} f, \operatorname{grad} g)]\, dx,$$

$$(f, g)_{\mathrm{I}} = \int_Q (\operatorname{grad} f, \operatorname{grad} g)\, dx + \int_Q f\, dx \int_Q g\, dx$$

are equivalent.

4.120. Show that the set $\tilde{H}^1(Q)$ of functions $f \in H^1(Q)$ for which $\int_Q f\, dx = 0$ constitutes a subspace of $H^1(Q)$.

4.121. Show that $(f, g)_{\mathrm{I}} = \int_Q (\operatorname{grad} f, \operatorname{grad} g)\, dx$ defines a scalar product $\tilde{H}^1(Q)$that is equivalent to

$$(f, g) = \int_Q [fg + (\operatorname{grad} f, \operatorname{grad} g)]\, dx.$$

Answers to Problems of Sec. 4

4.9. $\max_i |x_i|$. **4.10.** $\max |f(x)|$. **4.40.** (a) $\frac{3}{5}x$, (b) $\frac{3}{\pi}x$, (c) $\frac{15x^2}{16} + \frac{3}{16}$. **4.41.** (a) $\frac{\pi}{2} - \frac{4}{\pi}\cos x$, (b) $\frac{8}{3\pi}\sin x$. **4.42.** (a) 0, (b) $x_1 - x_2$.

4.51. $$e_n = \frac{\varphi_n - \sum_1^{n-1} e_k(\varphi_n, e_k)}{\left\| \varphi_n - \sum_1^{n-1} e_k(\varphi_n, e_n) \right\|}.$$

4.52. (a) P_0, P_1, P_2, P_3, where the P_n are Legendre polynomials (see Problem 4.21), (b) $\sqrt{\frac{3}{8}}(1-x)$, $\sqrt{\frac{15}{16}}x(x+1)$, $\sqrt{\frac{21}{136}}\times$ $\times(2-2x-5x^2+5x^3)$, (c) $\frac{2}{\sqrt{3}}\sin^2\pi x$, $\sqrt{\frac{8}{3}}\left(\sin^2\pi x-\frac{3}{4}\right)$, $\cos\pi x$, (d) 1, $x-1$, $1-2x+\frac{x^2}{2}$, (e) $\frac{1}{\sqrt[4]{2\pi}}$, $\frac{x}{\sqrt[4]{2\pi}}$, $\frac{x^2-1}{\sqrt{2}\sqrt[4]{2\pi}}$, (f) Q_0, Q_1, Q_2, where the $Q_n(x)$ are Chebyshev polynomials of the second kind, (g) T_0, T_1, T_2, where the $T_n(x)$ are Chebyshev polynomials of the first kind.

4.54. $\frac{1}{\sqrt{\pi}}$, $\frac{2x_1}{\sqrt{\pi}}$, $\frac{2x_2}{\sqrt{\pi}}$.

4.55. $\frac{\sqrt{3}}{2\sqrt{\pi}}$, $\frac{\sqrt{15}\,x_1}{2\sqrt{\pi}}$, $\frac{\sqrt{15}\,x_2}{2\sqrt{\pi}}$, $\frac{\sqrt{15}}{2\sqrt{\pi}}x_3$.

4.72. $\sin\frac{\pi(x-a)}{b-a}$. **4.75.** A subspace with basis elements 1 and $\cos(x/2)$.

4.79. $\frac{\pi}{2}\left(\frac{a_0^2}{2}+\sum_1^\infty\frac{a_k^2+b_k^2}{k+1}+2\sum_1^\infty k(a_k^2+b_k^2)\right)$.

4.81. $\alpha<0$. **4.83.** $\alpha<1/2$. **4.84.** $\sum_1^\infty k(a_k^2+b_k^2)<\infty$. **4.85.** No.

5 Integral Equations

Equations of the type

$$\varphi(x)=\lambda\int_G \mathcal{K}(x,y)\,\varphi(y)\,dy+f(x) \tag{5.1}$$

in the unknown function $\varphi(x)$ in a region $G\subset R^n$ are called *Fredholm integral equations of the second kind.* The known functions $\mathcal{K}(x,y)$ and $f(x)$ are called the *kernel* and the *inhomogeneous term* of integral equation (5.1), and λ is a complex-valued parameter.

The integral equation

$$\varphi(x)=\lambda\int_G \mathcal{K}(x,y)\,\varphi(y)\,dy \tag{5.2}$$

is known as the *homogeneous Fredholm integral equation of the second kind corresponding to Eq.* (5.1). The Fredholm integral equation of the second kind

$$\psi(x)=\bar{\lambda}\int_G \mathcal{K}^*(x,y)\,\psi(y)\,dy, \tag{5.3}$$

where $\mathcal{K}^*(x,y)=\overline{\mathcal{K}(y,x)}$, is said to be *adjoint* to Eq. (5.2), and the kernel $\mathcal{K}^*(x,y)$ is known as the *hermitian conjugate* (*adjoint*) *kernel* for $\mathcal{K}(x,y)$.

We will write the integral equations (5.1)-(5.3) in an abbreviated manner by using the operator notation:

$$\varphi = \lambda K\varphi + f, \quad \varphi = \lambda K\varphi, \quad \psi = \bar{\lambda} K^*\psi,$$

where the integral operators K and K^* are related to the kernels $\mathscr{K}(x, y)$ and $\mathscr{K}^*(x, y)$ thus:

$$Kg = \int_G \mathscr{K}(x, y) g(y)\, dy, \quad K^*g = \int_G \mathscr{K}^*(x, y) g(y)\, dy.$$

If for a certain complex value of λ, say λ_0, the homogeneous integral equation (5.2) has nonzero solutions belonging to $L_2(G)$, then this value is said to be an *eigenvalue* (*characteristic value*) *of the kernel* $\mathscr{K}(x, y)$ (or of the integral equation (5.2)), and the corresponding solutions are called *eigenfunctions* (*characteristic functions*) of this kernel.

The maximal number of linear independent eigenfunctions corresponding to an eigenvalue λ_0 is called the *degree of degeneracy* of λ_0.

We will also assume that G in Eq. (5.1) is a bounded region (in R^n), the function f is continuous on $\bar{G}$, and the kernel $\mathscr{K}(x, y)$ is continuous on $\bar{G} \times \bar{G}$.

In Problems 5.5-5.7 we use the following notation:

$$M = \max_{x \in \bar{G},\, y \in \bar{G}} |\mathscr{K}(x, y)|, \quad v = \int_G dy.$$

5.1. Show that an integral operator K corresponding to a kernel $\mathscr{K}(x, y)$ is bounded from $L_2(G)$ into $L_2(G)$ if

$$\int_{G \times G} |\mathscr{K}(x, y)|^2\, dx\, dy = c^2 < \infty.$$

5.2. Show that an integral operator K corresponding to a continuous kernel $\mathscr{K}(x, y)$ is a null operator in $L_2(G)$ if and only if $\mathscr{K}(x, y) = 0$, $x \in G$, $y \in G$.

5.3. Suppose the kernel $\mathscr{K}(x, y)$ of integral equation (5.1) belongs to $L_2(G \times G)$. Prove the convergence of the method of successive approximations for every function $f \in L_2(G)$ with $|\lambda| < 1/c$ (constant c is taken from Problem 5.1).

5.4. Suppose K is an integral operator corresponding to a continuous kernel. Prove that the operators $K^p = K(K^{p-1})$, $p = 2, 3, \ldots$, are integral operators corresponding to continuous kernels $\mathscr{K}_p(x, y)$ satisfying the following relationships:

$$\mathscr{K}_p(x, y) = \int_G \mathscr{K}(x, \xi)\, \mathscr{K}_{p-1}(\xi, y)\, d\xi.$$

5.5. Show that the kernel $\mathscr{K}_p(x, y)$ introduced in Problem 5.4 (they are called *iterated kernels of the kernel* $\mathscr{K}(x, y)$) satisfy the following inequalities:

$$|\mathcal{K}_p(x, y)| \leqslant M^p v^{p-1}, \quad p=1, 2, \ldots .$$

5.6. Show that the series $\sum_{m=0}^{\infty} \lambda^m \mathcal{K}_{m+1}(x, y)$, $x \in \bar{G}$, $y \in \bar{G}$, converges in the circle $|\lambda| < 1/Mv$ and that its sum $\mathcal{R}(x, y; \lambda)$, known as the *resolvent of* $\mathcal{K}(x, y)$, is continuous in $\bar{G} \times \bar{G} \times U_{1/Mv}$ and analytic in λ in the circle $|\lambda| < 1/Mv$. Show that at $|\lambda| < 1/Mv$ the solution to Eq. (5.1) is unique in the class $C(\bar{G})$ and can be represented for every $f \in C(\bar{G})$ in terms of the resolvent $\mathcal{R}(x, y; \lambda)$ thus:

$$\varphi(x) = f(x) + \lambda \int_G \mathcal{R}(x, y; \lambda) f(y)\, dy.$$

5.7. Show that the resolvent $\mathcal{R}(x, y; \lambda)$ (see Problem 5.6) of a continuous kernel $\mathcal{K}(x, y)$ satisfies each of the following equations provided $|\lambda| < 1/Mv$:

(a) $\mathcal{R}(x, y; \lambda) = \lambda \int_G \mathcal{K}(x, \xi) \mathcal{R}(\xi, y; \lambda)\, d\xi + \mathcal{K}(x, y)$,

(b) $\mathcal{R}(x, y; \lambda) = \lambda \int_G \mathcal{K}(\xi, y) \mathcal{R}(x, \xi; \lambda)\, d\xi + \mathcal{K}(x, y)$,

(c) $\dfrac{\partial \mathcal{R}(x, y; \lambda)}{\partial \lambda} = \int_G \mathcal{R}(x, \xi; \lambda) \mathcal{R}(\xi, y; \lambda)\, d\xi.$

In Problems 5.8-5.13 the integral equations are the *Volterra integral equations of the first and second kinds*:

$$\int_0^x \mathcal{K}(x, y) \varphi(y)\, dy = f(x), \tag{5.4}$$

$$\varphi(x) = \lambda \int_0^x \mathcal{K}(x, y) \varphi(y)\, dy + f(x). \tag{5.5}$$

5.8. Suppose that the following conditions are met:

(a) the functions $\mathcal{K}(x, y)$ and $\mathcal{K}_x(x, y)$ are continuous in the region $0 \leqslant x \leqslant y \leqslant a$,

(b) $\mathcal{K}(x, x) \neq 0$ for all values of x,

(c) $f \in C^1([0, a])$ and $f(0) = 0$.

Prove that under the above conditions Eq. (5.4) is equivalent to the following equation:

$$\varphi(x) = \frac{f'(x)}{\mathcal{K}(x, x)} - \int_0^x \frac{\mathcal{K}_x(x, y)}{\mathcal{K}(x, x)} \varphi(y)\, dy.$$

5.9. Show that the differential equation

$$y^{(n)} + a_1(x) y^{(n-1)} + \ldots + a_n(x) y = F(x)$$

with continuous coefficients $a_i(x)$, $i = 1, 2, \ldots, n$, and initial conditions $y(0) = C_0$, $y'(0) = C_1$, $\ldots$, $y^{(n-1)}(0) = C_{n-1}$ is equivalent to the integral equation (5.5), where

$$\mathscr{K}(x, y) = \sum_{m=1}^{n} a_m(x) \frac{(x-y)^{m-1}}{(m-1)!},$$

$$f(x) = F(x) - C_{n-1}a_1(x) - (C_{n-1}x + C_{n-2})\, a_2(x) - \ldots - \left(C_{n-1}\frac{x^{n-1}}{(n-1)!} + \ldots + C_1 x + C_0\right) a_n(x).$$

5.10. Suppose $\mathscr{K}(x)$ belongs to $C(x \geqslant 0)$ and is zero for $x < 0$. Prove that the generalized function

$$\mathscr{E}(x) = \delta(x) + \mathscr{R}(x), \text{ where } \mathscr{R} = \sum_{m=1}^{\infty} \underbrace{\mathscr{K} * \mathscr{K} * \ldots * \mathscr{K}}_{m \text{ times}},$$

is the fundamental solution of Volterra's operator of the second kind with a kernel $\mathscr{K}(x, y)$ (see Eq. (5.5)), that is,

$$\mathscr{E} - \mathscr{K} * \mathscr{E} = \delta.$$

Show that the series for $\mathscr{R}(x)$ is uniformly convergent in every finite interval and satisfies the Volterra integral equation $\mathscr{R}(x) = \int_0^x \mathscr{K}(x-y)\, \mathscr{R}(y)\, dy + \mathscr{K}(x)$, $x \geqslant 0$ (the function $\mathscr{R}(x-y)$ is the resolvent of the kernel $\mathscr{K}(x-y)$ at $\lambda = 1$).

5.11. Find the resolvent of the Volterra integral equation (5.5) with kernel $\mathscr{K}(x, y)$:

1. $\mathscr{K}(x, y) = 1$, 2. $\mathscr{K}(x, y) = x - y$.

5.12. Find the following integral equations:

1. $\varphi(x) = x + \int_0^x (y - x)\, \varphi(y)\, dy.$

2. $\varphi(x) = 1 + \lambda \int_0^x (x - y)\, \varphi(y)\, dy.$

3. $\varphi(x) = \lambda \int_0^x (x - y)\, \varphi(y)\, dy + x^2.$

5.13. Show that if $g \in C^1(x \geqslant 0)$, $g(0) = 0$, $0 < \alpha < 1$, then the function $f(x) = \frac{\sin \alpha\pi}{\pi} \int_0^x \frac{g'(y)}{(x-y)^{1-\alpha}}\, dy$ satisfies Abel's integral equation

$$\int_0^x \frac{f(y)}{(x-y)^{\alpha}}\, dx = g(x).$$

In Problems 5.14-5.30 the kernel $\mathcal{K}(x, y)$ of the integral equation is *degenerate* (or *separable*), that is,

$$\mathcal{K}(x, y)=\sum_{m=1}^{N} f_m(x)\, g_m(y),$$

where the functions $f_m(x)$ and $g_m(y)$, $m = 1, 2, \ldots, N$, are continuous in the square $a \leqslant x, y \leqslant b$ and mutually linearly independent. With this kernel Eq. (5.1) can be written in the form

$$\varphi(x)=f(x)+\lambda \sum_{m=1}^{N} c_m f_m(x),$$

where the unknown coefficients c_m can be found by solving a system of algebraic equations.

5.14. Solve the integral equation

$$\varphi(x)=\lambda \int_0^1 \mathcal{K}(x, y)\, \varphi(y)\, dy+f(x)$$

for the following cases:

1. $\mathcal{K}(x, y) = x - 1, \qquad f(x) = x.$
2. $\mathcal{K}(x, y) = 2e^{x+y}, \qquad f(x) = e^x.$
3. $\mathcal{K}(x, y) = x + y - 2xy, \qquad f(x) = x + x^2.$

5.15. Solve the integral equation

$$\varphi(x)=\lambda \int_{-1}^1 \mathcal{K}(x, y)\, \varphi(y)\, dy+f(x)$$

for the following cases:

1. $\mathcal{K}(x, y) = xy + x^2y^2, \qquad f(x) = x^2 + x^4.$
2. $\mathcal{K}(x, y) = x^{1/3} + y^{1/3}, \qquad f(x) = 1 - 6x^2.$
3. $\mathcal{K}(x, y) = x^4 + 5x^3y, \qquad f(x) = x^2 - x^4.$
4. $\mathcal{K}(x, y) = 2xy^3 + 5x^2y^2, \qquad f(x) = 7x^4 + 3.$
5. $\mathcal{K}(x, y) = x^2 - xy, \qquad f(x) = x^2 + x.$
6. $\mathcal{K}(x, y) = 5 + 4xy - 3x^2 - 3y^2 + 9x^2y^2, \quad f(x) = x.$

5.16. Solve the integral equation

$$\varphi(x)=\lambda \int_0^\pi \mathcal{K}(x, y)\, \varphi(y)\, dy+f(x)$$

for the following cases:

1. $\mathcal{K}(x, y) = \sin(2x + y), \qquad f(x) = \pi - 2x.$
2. $\mathcal{K}(x, y) = \sin(x - 2y), \qquad f(x) = \cos 2x.$

3 $\mathscr{K}(x, y) = \cos(2x + y), \quad f(x) = \sin x.$

4. $\mathscr{K}(x, y) = \sin(3x + y), \quad f(x) = \cos x.$

5. $\mathscr{K}(x, y) = \sin y + y \cos x, f(x) = 1 - \frac{2x}{\pi}.$

6. $\mathscr{K}(x, y) = \cos^2(x - y), \quad f(x) = 1 + \cos 4x.$

5.17. Solve the integral equation

$$\varphi(x) = \lambda \int_0^{2\pi} \mathscr{K}(x, y)\, \varphi(y)\, dy + f(x)$$

for the following cases:

1. $\mathscr{K}(x, y) = \cos x \cos y + \cos 2x \cos 2y, f(x) = \cos 3x.$
2. $\mathscr{K}(x, y) = \cos x \cos y + 2 \sin 2x \sin 2y, f(x) = \cos x.$
3. $\mathscr{K}(x, y) = \sin x \sin y + 3 \cos 2x \cos 2y, f(x) = \sin x.$

5.18. Find all the eigenvalues and the corresponding eigenfunctions of the following integral equations:

1. $\varphi(x) = \lambda \int_0^{2\pi} \left[\sin(x+y) + \frac{1}{2}\right] \varphi(y)\, dy.$

2. $\varphi(x) = \lambda \int_0^{2\pi} \left[\cos^2(x+y) + \frac{1}{2}\right] \varphi(y)\, dy.$

3. $\varphi(x) = \lambda \int_0^{1} \left(x^2 y^2 - \frac{2}{45}\right) \varphi(y)\, dy.$

4. $\varphi(x) = \lambda \int_0^{1} \left[\left(\frac{x}{y}\right)^{2/5} + \left(\frac{y}{x}\right)^{2/5}\right] \varphi(y)\, dy.$

5. $\varphi(x) = \lambda \int_0^{\pi} (\sin x \sin 4y + \sin 2x \sin 3y$

$+ \sin 3x \sin 2y + \sin 4x \sin y)\, \varphi(y)\, dy.$

5.19. For what values of parameters a and b can the integral equation

$$\varphi(x) = 12 \int_0^1 \left(xy - \frac{x+y}{2} + \frac{1}{3}\right) \varphi(y)\, dy + ax^2 + bx - 2$$

be solved? Find the solutions for these values of a and b.

5.20. For what values of parameter a can the integral equation

$$\varphi(x) = \sqrt{15} \int_0^1 [y(4x^2 - 3x) + x(4y^2 - 3y)]\, \varphi(y)\, dy + ax + \frac{1}{x}$$

be solved? Find the solutions for these values of a.

5.21. Establish the values of λ for which the integral equation

$$\varphi(x)=\lambda\int_0^{2\pi}\cos(2x-y)\,\varphi(y)\,dy+f(x)$$

can be solved for any function $f(x) \in C([0, 2\pi])$ and solve the equation.

5.22. Find the solutions of the following integral equations for all values of λ and for all values of the parameters a, b and c in the inhomogeneous term:

1. $\varphi(x)=\lambda\int_{-\pi/2}^{\pi/2}(y\sin x+\cos y)\,\varphi(y)\,dy+ax+b.$
2. $\varphi(x)=\lambda\int_0^{\pi}\cos(x+y)\,\varphi(y)\,dy+a\sin x+b.$
3. $\varphi(x)=\lambda\int_{-1}^{1}(1+xy)\,\varphi(y)\,dy+ax^2+bx+c.$
4. $\varphi(x)=\lambda\int_{-1}^{1}(x^2y+xy^2)\,\varphi(y)\,dy+ax+bx^3.$
5. $\varphi(x)=\lambda\int_{-1}^{1}\frac{1}{2}(xy+x^2y^2)\,\varphi(y)\,dy+ax+b.$
6. $\varphi(x)=\lambda\int_{-1}^{1}[5(xy)^{1/3}+7(xy)^{2/3}]\,\varphi(y)\,dy+ax+bx^{1/3}.$
7. $\varphi(x)=\lambda\int_{-1}^{1}\frac{1+xy}{1+y^2}\,\varphi(y)\,dx+a+x+bx^2.$
8. $\varphi(x)=\lambda\int_{-1}^{1}(\sqrt[3]{x}+\sqrt[3]{y})\,\varphi(y)\,dy+ax^2+bx+c.$
9. $\varphi(x)=\lambda\int_{-1}^{1}(xy+x^2+y^2-3x^2y^2)\,\varphi(y)\,dy+ax+b.$

5.23. Find the eigenvalues and the corresponding eigenfunctions of kernel $\mathcal{K}(x, y)$ and solve the integral equation

$$\varphi(x)=\lambda\int_{-1}^{1}\mathcal{K}(x, y)\,\varphi(y)\,dy+f(x)$$

for all values of λ a, and b if

1. $\mathcal{K}(x, y) = 3x + xy - 5x^2y^2,\quad f(x) = ax.$
2. $\mathcal{K}(x, y) = 3xy + 5x^2y^2,\quad f(x) = ax^2 + bx.$

5.24. Find the eigenvalues and the corresponding eigenfunctions of kernel $\mathcal{K}(x, y)$ and solve the integral equation

$$\varphi(x)=\lambda\int_{-\pi}^{\pi}\mathcal{K}(x, y)\,\varphi(y)\,dy+f(x)$$

for all values of λ, a, and b if

1. $\mathcal{K}(x, y) = x\cos y + \sin x\cdot\sin y$, $f(x) = a + b\cos x$.
2. $\mathcal{K}(x, y) = x\sin y + \cos x$, $f(x) = ax + b$.

5.25. Find the solution and the resolvent $\mathcal{R}(x, y; \lambda)$ for each of the following equations:

1. $\varphi(x)=\lambda\int_{0}^{\pi}\sin(x+y)\,\varphi(y)\,dy+f(x)$.
2. $\varphi(x)=\lambda\int_{-1}^{1}(1-y+2xy)\,\varphi(y)\,dy+f(x)$.
3. $\varphi(x)=\lambda\int_{-\pi}^{\pi}(x\sin y+\cos x)\,\varphi(y)\,dy+ax+b$.
4. $\varphi(x)=\lambda\int_{0}^{2\pi}(\sin x\sin y+\sin 2x\sin 2y)\,\varphi(y)\,dy+f(x)$.

5.26. Find the values of the parameters a, b and c for which the following integral equations have solutions for any values of λ:

1. $\varphi(x)=\lambda\int_{-1}^{1}(xy+x^2y^2)\,\varphi(y)\,dy+ax^2+bx+c$.
2. $\varphi(x)=\lambda\int_{-1}^{1}(1+xy)\,\varphi(y)\,dy+ax^2+bx+c$,

where $a^2+b^2+c^2=1$.

3. $\varphi(x)=\lambda\int_{-1}^{1}\frac{1+xy}{\sqrt{1-y^2}}\,\varphi(y)\,dy+x^2+ax+b$.
4. $\varphi(x)=\lambda\int_{0}^{1}\left(xy-\frac{1}{3}\right)\varphi(y)\,dy+ax^2-bx+1$.
5. $\varphi(x)=\lambda\int_{0}^{1}(x+y)\,\varphi(y)\,dy+ax+b+1$.
6. $\varphi(x)=\lambda\int_{0}^{2\pi}\cos(2x+4y)\,\varphi(y)\,dy+e^{ax+b}$

7. $\varphi(x)=\lambda\int_0^{\pi}(\sin x\sin 2y+\sin 2x\sin 4y)\,\varphi(y)\,dy+ax^2+bx+c.$

8. $\varphi(x)=\lambda\int_{-1}^{1}(1+x^2+y^3)\,\varphi(y)\,dy+ax+bx^3.$

5.27. Find all the values of parameter a for which the integral equation

$$\varphi(x)=\lambda\int_0^1(ax-y)\,\varphi(y)\,dy+f(x)$$

can be solved for all real values of λ and all functions $f(x)$ that belong to $C([0, 1])$.

5.28. Find the eigenvalues and corresponding eigenfunctions of the following integral equations:

1. $\varphi(x_1, x_2)=\lambda\int_{-1}^{1}\int_{-1}^{1}\left[x_1+x_2+\frac{3}{32}(y_1+y_2)\right]\varphi(y_1, y_2)\,dy_1\,dy_2.$

2. $\varphi(x)=\lambda\int_{|y|<1}(|x|^2+|y|^2)\,\varphi(y)\,dy,\ x=(x_1, x_2).$

3. $\varphi(x)=\lambda\int_{|y|<1}\frac{1+|y|}{1+|x|}\,\varphi(y)\,dy,\ x=(x_1, x_2, x_3).$

5.29. Establish whether the integral equation

$$\varphi(x)=\lambda\int_{|y|<1}(|x|^2-|y|^2)\,\varphi(y)\,dy,\quad x=(x_1, x_2, x_3),$$

has real eigenvalues, and if it does, find the corresponding eigenvalues.

5.30. Find the eigenvalues and the corresponding eigenfunctions of the kernel $\mathcal{K}(x, y)=x_1x_2+y_1y_2$ and solve the integral equation

$$\varphi(x_1, x_2)=\lambda\int_{-1}^{1}\int_{-1}^{1}(x_1x_2+y_1y_2)\,\varphi(y_1, y_2)\,dy_1\,dy_2+f(x_1, x_2).$$

In Problems 5.31 and 5.33-5.35, the kernel $\mathcal{K}(x, y)$ of the integral equation (5.1) is *hermitian*, that is, coincides with its hermitian conjugate kernel:

$$\mathcal{K}(x, y)=\mathcal{K}^*(x, y)=\overline{\mathcal{K}(y, x)}.$$

In particular, if a hermitian kernel is real-valued, it is symmetric, that is, $\mathcal{K}(x, y)=\mathcal{K}(y, x)$.

A hermitian continuous kernel $\mathcal{K}(x, y)\not\equiv 0$ possesses the following properties:

(1) the set of eigenvalues of this kernel is not empty, lies on the real axis, is at most countable, and has no finite limit points;

(2) the system of eigenfunctions $\{\varphi_k\}$ can be chosen orthonormal:

$$(\varphi_k, \varphi_m) = \delta_{km}.$$

5.31. Prove that if $\mathcal{K}(x, y)$ is hermitian, the eigenvalues of the second iterated kernel $\mathcal{K}_2(x, y)$ (see Problems 5.4 and 5.5) are positive.

5.32. Prove that if kernel $\mathcal{K}(x, y)$ is skew-symmetric, that is, $\mathcal{K}(x, y) = -\mathcal{K}^*(x, y)$, then its eigenvalues are pure imaginary.

In Problems 5.33-5.35 it is assumed that the eigenvalues λ_k of the hermitian continuous kernel $\mathcal{K}(x, y)$ are numbered in the order in which their absolute values increase, that is,

$$|\lambda_1| \leqslant |\lambda_2| \leqslant |\lambda_3| \leqslant \ldots,$$

and each of these numbers is repeated in this sequence according to the number of linear independent eigenfunctions corresponding to it. Then we can assume that to each eigenvalue λ_k there corresponds one eigenfunction φ_k. We will also assume that the system $\{\varphi_k\}$ is orthonormal.

5.33. Let $\mathcal{K}(x, y)$ be a hermitian continuous kernel and $\mathcal{K}_p(x, y)$ an iterated kernel of $\mathcal{K}(x, y)$. Prove the validity of the following formulas:

(1) $\displaystyle\sum_{m=1}^{\infty} \frac{|\varphi_m(x)|^2}{\lambda_m^2} = \int_a^b |\mathcal{K}(x, y)|^2\, dy;$

(2) $\displaystyle\sum_{m=1}^{\infty} \frac{1}{\lambda_m^2} = \int_a^b \int_a^b |\mathcal{K}(x, y)|^2\, dx\, dy;$

(3) $\displaystyle (Kf, f) = \sum_{m=1}^{\infty} \frac{|(f, \varphi_k)|^2}{\lambda_k}$, $f \in L_2(G)$, and K an integral operator with kernel $\mathcal{K}(x, y)$;

(4) $\displaystyle\sum_{m=1}^{\infty} \frac{1}{\lambda_m^{2p}} = \int_a^b \int_a^b |\mathcal{K}_p(x, y)|^2\, dx\, dy, \quad p = 1, 2, \ldots .$

Let $\mathcal{K}_n(x, y)$ be the nth iterated kernel of a hermitian continuous kernel $\mathcal{K}(x, y)$. The integral

$$\alpha_n = \int_a^b \mathcal{K}_n(x, x)\, dx, \quad n = 1, 2, \ldots,$$

is said to be the *nth trace of kernel* $\mathcal{K}(x, y)$.

5.34. Prove that

(1) the ratio $\alpha_{2n+2}/\alpha_{2n}$ does not decrease as $n \to \infty$ and is bounded,

(2) the limit $\lim\limits_{n\to\infty} \frac{\alpha_{2n}}{\alpha_{2n+2}}$ exists and is equal to the smallest eigenvalue of the kernel $\mathcal{K}_2(x, y)$,

(3) $\sum\limits_{m=1}^{\infty} \frac{1}{\lambda_m^n} = \alpha_n \ (n \geqslant 2)$, where the λ_m, $m = 1, 2, \ldots$, are the eigenvalues of $\mathcal{K}(x, y)$, $|\lambda_1| \leqslant |\lambda_2| \leqslant \ldots$,

(4) $\frac{1}{|\lambda_1|} = \lim\limits_{n\to\infty} \sqrt{\alpha_{2n+2}/\alpha_{2n}} = \lim\limits_{n\to\infty} \sqrt[2n]{\alpha_{2n}}$.

5.35. Suppose λ is not an eigenvalue of a hermitian continuous kernel $\mathcal{K}(x, y)$. Prove that the (unique) solution of the equation

$$\varphi(x) = \lambda \int_a^b \mathcal{K}(x, y)\, \varphi(y)\, dy + f(x)$$

can be written in the form of a series,

$$\varphi(x) = \lambda \sum_{m=1}^{\infty} \frac{(f, \varphi_m)}{\lambda_m - \lambda} \varphi_m(x) + f(x),$$

which converges uniformly on $\bar{G}$, while the resolvent $\mathcal{R}(x, y; \lambda)$ can be written in the form $\mathcal{R}(x, y; \lambda) = \sum\limits_{m=1}^{\infty} \frac{\varphi_m(x)\overline{\varphi_m}(y)}{\lambda_m - \lambda}$, where the bilinear series is convergent in $L_2(G \times G)$.

5.36. Find the eigenvalues and the corresponding eigenfunctions of the integral equation

$$\varphi(x) = \lambda \int_0^1 \mathcal{K}(x, y)\, \varphi(y)\, dy$$

for the following cases

1. $\mathcal{K}(x, y) = \begin{cases} x & \text{if } 0 \leqslant x \leqslant y \leqslant 1, \\ y & \text{if } 0 \leqslant y \leqslant x \leqslant 1. \end{cases}$

2. $\mathcal{K}(x, y) = \begin{cases} x(1-y) & \text{if } 0 \leqslant x \leqslant y \leqslant 1, \\ y(1-x) & \text{if } 0 \leqslant y \leqslant x \leqslant 1. \end{cases}$

3. $\mathcal{K}(x, y) = \begin{cases} \frac{2-y}{2} x & \text{if } 0 \leqslant x \leqslant y \leqslant 1, \\ \frac{2-x}{2} y & \text{if } 0 \leqslant y \leqslant x \leqslant 1. \end{cases}$

4. $\mathcal{K}(x, y) = \begin{cases} (x+1)(y-2) & \text{if } 0 \leqslant x \leqslant y \leqslant 1, \\ (y+1)(x-2) & \text{if } 0 \leqslant y \leqslant x \leqslant 1. \end{cases}$

5. $\mathscr{K}(x, y)=\begin{cases}(x+1)\,y & \text{if } 0\leqslant x\leqslant y\leqslant 1,\\ x\,(y+1) & \text{if } 0\leqslant y\leqslant x\leqslant 1.\end{cases}$

6. $\mathscr{K}(x, y)=\begin{cases}(e^{x}-e^{-x})\,(e^{y}+e^{2-y}) & \text{if } 0\leqslant x\leqslant y\leqslant 1,\\ (e^{x}+e^{2-x})\,(e^{y}-e^{-y}) & \text{if } 0\leqslant y\leqslant x\leqslant 1.\end{cases}$

7. $\mathscr{K}(x, y)=\begin{cases}\sin x\sin(1-y) & \text{if } 0\leqslant x\leqslant y\leqslant 1,\\ \sin(1-x)\sin y & \text{if } 0\leqslant y\leqslant x\leqslant 1.\end{cases}$

5.37. Find the eigenvalues and the corresponding eigenfunctions of the integral equation of Problem 5.36 for the following cases:

1. $\mathscr{K}(x, y)=\begin{cases}(1+x)\,(1-y) & \text{if } -1\leqslant x\leqslant y\leqslant 1,\\ (1-x)\,(1+y) & \text{if } -1\leqslant y\leqslant x\leqslant 1.\end{cases}$

2. $\mathscr{K}(x, y)=\begin{cases}\cos x\sin y & \text{if } 0\leqslant x\leqslant y\leqslant \pi,\\ \cos y\sin x & \text{if } 0\leqslant y\leqslant x\leqslant \pi.\end{cases}$

3. $\mathscr{K}(x, y)=\begin{cases}\sin x\cos y & \text{if } 0\leqslant x\leqslant y\leqslant \pi,\\ \sin y\cos x & \text{if } 0\leqslant y\leqslant x\leqslant \pi.\end{cases}$

5.38. Find the eigenvalues and the corresponding eigenfunctions of the integral equation

$$\varphi(x)=\lambda\int_{-\pi}^{\pi}\omega(x+y)\,\varphi(y)\,dy$$

for the following cases:

(1) $\omega(t)$ is an even 2π-periodic function, with $\omega(t) = t$ if $t\in[0, \pi]$;

(2) $\omega(t)$ is an even 2π-periodic function, with $\omega(t) = \pi - t$ if $t\in[0, \pi]$.

5.39. Find the eigenfunctions of the integral equation of Problem 5.38 with the kernel $\mathscr{K}(x, y) = \omega(x - y)$, with $\omega(t)$ a continuous, piecewise smooth, even, and 2π-periodic function, $0\leqslant x\leqslant 2\pi$, $0\leqslant y\leqslant 2\pi$.

5.40. Solve the integral equation

$$\varphi(x)=\lambda\int_{0}^{1}\mathscr{K}(x, y)\,\varphi(y)\,dy+f(x),$$

if $f(x)\in C^{2}_{1}([0, 1])$ and

$$\mathscr{K}(x, y)=\begin{cases}x & \text{if } 0\leqslant x\leqslant y\leqslant 1,\\ y & \text{if } 0\leqslant y\leqslant x\leqslant 1.\end{cases}$$

Let $\mathscr{K}(x, y)$ be the continuous kernel of the integral equation

$$\varphi(x)=\lambda\int_{a}^{b}\mathscr{K}(x, y)\,\varphi(y)\,dy+f(x). \tag{5.6}$$

The expression

$$\mathscr{K}\begin{pmatrix} x_1 x_2 \dots x_n \\ y_1 y_2 \dots y_n \end{pmatrix} = \begin{vmatrix} \mathscr{K}(x_1, y_1) & \mathscr{K}(x_1, y_2) & \dots & \mathscr{K}(x_1, y_n) \\ \mathscr{K}(x_2, y_1) & \mathscr{K}(x_2, y_2) & \dots & \mathscr{K}(x_2, y_n) \\ \dots & \dots & \dots & \dots \\ \mathscr{K}(x_n, y_1) & \mathscr{K}(x_n, y_2) & \dots & \mathscr{K}(x_n, y_n) \end{vmatrix}$$

is known as *Fredholm's symbol* and the function

$$D(\lambda) = 1 + \sum_{n=1}^{\infty} (-1)^n \frac{A_n}{n!} \lambda^n, \tag{5.7}$$

where

$$A_n = \int_a^b \dots \int_a^b \mathscr{K}\begin{pmatrix} t_1 & t_2 & \dots & t_n \\ t_1 & t_2 & \dots & t_n \end{pmatrix} dt_1\, dt_2 \dots dt_n, \tag{5.8}$$

is *Fredholm's determinant* of kernel $\mathscr{K}(x, y)$ or integral equation (5.6).

5.41. Prove that the coefficients A_n in Fredholm's determinant satisfy the inequalities $|A_n| \leqslant n^{n/2} M^n (b-a)^n$. Use this result to prove that $D(\lambda)$ is an entire function of λ. *Hint.* Use Hadamard's inequality (see Arsenin [1]).

The function

$$D(x, y; \lambda) = \lambda \mathscr{K}(x, y) + \sum_{n=1}^{\infty} (-1)^n \frac{B_n(x, y)}{n!} \lambda^{n+1}, \tag{5.9}$$

where

$$B_n(x, y) = \int_a^b \dots \int_a^b \mathscr{K}\begin{pmatrix} x & t_1 & t_2 & \dots & t_n \\ y & t_1 & t_2 & \dots & t_n \end{pmatrix} dt_1\, dt_2 \dots dt_n, \tag{5.10}$$

is known as *Fredholm's first minor.*

5.42. Show that if $\mathscr{K}(x, y)$ is a function that is continuous in the square L: $\{a \leqslant x, y \leqslant b\}$, then $D(x, y, \lambda)$ is continuous in x, y, and λ in $L \times \mathbb{C}$ and an entire function of λ (for x and y fixed).

5.43. Prove that the coefficients A_n, the functions $B_n(x, y)$, and the kernel $\mathscr{K}(x, y)$ (see formulas (5.7)-(5.10)) are related thus:

(1) $B_n(x, y) = A_n \mathscr{K}(x, y) - n \int_a^b B_{n-1}(x, \xi) \mathscr{K}(\xi, y)\, d\xi$,

(2) $B_n(x, y) = A_n \mathscr{K}(x, y) - n \int_a^b \mathscr{K}(x, \xi) B_{n-1}(\xi, y)\, d\xi$.

Hint. Represent the determinants in both integrands in terms of the elements and cofactors of the first column.

5.44. Prove the validity of the first and second fundamental relationships of Fredholm:

$$D(x, y; \lambda)-\lambda\mathcal{K}(x, y)D(\lambda)=\lambda\int_a^b \mathcal{K}(x, \xi)D(\xi, y; \lambda)\,d\xi,$$

$$D(x, y; \lambda)-\lambda\mathcal{K}(x, y)D(\lambda)=\lambda\int_a^b \mathcal{K}(\xi, y)D(x, \xi; \lambda)\,d\xi$$

Hint. Employ expansion (5.9), collect the coefficients of the same powers of λ on the left- and right-hand sides of the relationships, and apply the result of Problem 5.43.

5.45. Prove the validity of the following formulas:

$$A_n=\int_a^b B_{n-1}(x, x)\,dx,\quad \int_a^b D(x, x; \lambda)\,dx=-\lambda D'(\lambda)$$

5.46. Prove the validity of the formula $\frac{D'(\lambda)}{D(\lambda)}=-\sum_{n=1}^{\infty}\alpha_n\lambda^{n-1}$ (the coefficients α_n were defined on p. 76).

5.47. Suppose Fredholm's determinant $D(\lambda)$ of integral equation (5.6) is not zero. Prove that in this case the integral equation has a solution for every $f(x)\in C([a, b])$ that is unique and is given by the formula

$$\varphi(x)=f(x)+\int_a^b \frac{D(x, y; \lambda)}{D(\lambda)}f(y)\,dy$$

5.48. Using the representation of the solution of an integral equation at $|\lambda|<1/M(b-a)$ in terms of the resolvent $\mathcal{R}(x, y; \lambda)$ (see Problem 5.6) and the result of Problem 5.47, prove that

$$\mathcal{R}(x, y; \lambda)=\frac{D(x, y; \lambda)}{\lambda D(\lambda)}$$

(this formula defines the analytic continuation of a resolvent given for $|\lambda|<1/M(b-a)$ in the form of a series (see Problem 5.6)).

5.49. Prove that the eigenvalues of an integral equation with a continuous kernel coincides with the zeros of Fredholm's determinant $D(\lambda)$ of the equation.

5.50. Prove that the degree of degeneracy m of an eigenvalue λ_0 of an integral equation with a continuous kernel $\mathcal{K}(x, y)$ is finite and that

$$m\leqslant|\lambda_0|^2\int_a^b\int_a^b|\mathcal{K}(x, y)|^2\,dx\,dy.$$

5.51. Prove that Fredholm's determinants of a continuous kernel $\mathcal{K}(x, y)$ and of the adjoint kernel $\mathcal{K}^*(x, y)$ coincide and, hence, the eigenvalues of the integral equation and those of the adjoint equation also coincide (see Problem 5.49).
5.52. Prove that the degree of degeneracy of an eigenvalue of a given continuous kernel and that of the adjoint equation coincide.
5.53. Prove that at $|\lambda| < 1$ the Milne integral equation

$$\varphi(x) = \frac{\lambda}{2} \int_0^\infty \left(\int_{|x-y|}^{\infty} \frac{e^{-t}}{t} dt \right) \varphi(y)\, dy$$

has a unique solution $\varphi = 0$ in the class of functions that are bounded on $[0, \infty)$.

5.54. For Peierls's integral equation

$$\varphi(x) = \frac{\lambda}{4\pi} \int_G \frac{e^{-\alpha|x-y|}}{|x-y|^2} \varphi(y)\, dy, \quad \alpha > 0,$$

prove the validity of the estimate

$$\lambda_1 (1 - e^{-\alpha D}) \geqslant \alpha,$$

where D is the diameter of the region $G \subset R^3$, and λ_1 is the smallest eigenvalue of the kernel (in absolute value).
5.55. Prove that at $\lambda < 1/2$ the solution of the integral equation

$$\varphi(x) = \lambda \int_{-\infty}^{\infty} e^{-|x-y|} \varphi(y)\, dy + f(x)$$

is unique in the class of functions bounded in R^1 and is given by the formula

$$\varphi(x) = f(x) + \frac{\lambda}{\sqrt{1-2\lambda}} \int_{-\infty}^{\infty} e^{-\sqrt{1-2\lambda}|x-y|} f(y)\, dy.$$

Answers to Problems of Sec. 5

5.11. 1. $e^{\lambda(x-y)}$. 2. $\frac{1}{\sqrt{\lambda}} \sinh \sqrt{\lambda}(x-y)$. 5.12. 1. $\sin x$. 2. $\cosh(\sqrt{\lambda} x)$. 3. $\frac{2}{\sqrt{\lambda}}(\cosh \sqrt{\lambda}\, x - 1)$. 5.14. 1. If $\lambda = -2$, there are no solutions. If $\lambda \neq -2$, then $\varphi(x) = \frac{2x(\lambda+1)-\lambda}{\lambda+2}$. 2. If $\lambda \neq \lambda_1$, where $\lambda_1 = 1/(e^2-1)$, then $\varphi(x) = \frac{e^x}{1-\lambda(e^2-1)}$. At $\lambda = \lambda_1$ the equation has no solutions. 3. If $\lambda \neq 2$ and $\lambda \neq -6$, then

$$\varphi(x) = \frac{12\lambda^2 x - 24\lambda x - \lambda^2 + 42\lambda}{6(\lambda+6)(2-\lambda)}.$$

At $\lambda = 2$ and $\lambda = -6$ the equation has no solutions.

5.15. 1. If $\lambda \neq 3/2$ and $\lambda \neq 5/2$, then $\varphi(x) = \frac{5(7+2\lambda)}{7(5-2\lambda)} x^2 + x^4$. If $\lambda = 3/2$, then $\varphi(x) = Cx + \frac{25}{7} x^2 + x^4$, with C an arbitrary constant; if $\lambda = 5/2$, the equation has no solutions. 2. $\frac{2\lambda}{12\lambda^2-5}(5\sqrt[3]{x} + 6\lambda) + 1 - 6x^2$ if $\lambda \neq \pm\sqrt{5/12}$; at $\lambda = \pm\sqrt{5/12}$ the equation has no solutions. 3. $\frac{5(2\lambda-3)}{3(5-2\lambda)} x^4 + x^2$ if $\lambda \neq 5/2$ and $\lambda \neq 1/2$; $Cx^3 + x^2 - \frac{5}{6} x^4$ at $\lambda = 1/2$, with C an arbitrary constant; at $\lambda = 5/2$ the equation has no solutions. 4. $\frac{20\lambda}{1-2\lambda} x^2 + 7x^4 + 3$ if $\lambda \neq 5/2$ and $\lambda \neq 1/2$; $7x^4 + 3 - \frac{50}{3} x^2 + Cx$, where C is an arbitrary function, if $\lambda = 5/4$; at $\lambda = 1/2$ the equation has no solutions. 5. $\frac{3(5-2\lambda)x}{5(3+2\lambda)} + x^3$ if $\lambda \neq \pm 3/2$; $\frac{1}{5} x + x^3 + Cx^2$, where C is an arbitrary constant, if $\lambda = 3/2$; at $\lambda = -3/2$ the equation has no solutions. 6. If $\lambda = \lambda_1 = 1/8$, then $\varphi(x) = C_1 + \frac{3}{2} x$; if $\lambda = \lambda_2 = 5/8$, then $\varphi(x) = C_2(3x^2 - 1) - \frac{3}{2} x$ (C_1 and C_2 are arbitrary constants); at $\lambda = \lambda_3 = 3/8$ the equation has no solutions; if $\lambda \neq \lambda_i$ $(i = 1, 2, 3)$ then $\varphi(x) = \frac{3x}{3-8\lambda}$.

5.16. 1. $\frac{12\lambda}{3-4\lambda} \sin 2x + \pi - 2x$ if $\lambda \neq 3/4$ and $\lambda \neq -3/2$; $\pi - 2x - 2\sin 2x + C\cos 2x$, where C is an arbitrary constant, if $\lambda = -3/2$; at $\lambda = 3/4$ the equation has no solutions. 2. $\frac{3\pi\lambda}{2(2\lambda+3)} \sin x + \cos 2x$ if $\lambda \neq -3/2$ and $\lambda \neq -3/4$; $\cos 2x - \frac{3\pi}{4} \sin x + C\cos x$, where C is an arbitrary constant, if $\lambda = -3/4$; at $\lambda = -3/2$ the equation has no solutions. 3. $\sin x + \frac{3\pi\lambda}{8\lambda^2-9}\left(2\lambda\cos 2x + \frac{3}{2}\sin 2x\right)$ if $\lambda \neq \pm\frac{3}{2\sqrt{2}}$; at $\lambda = \pm\frac{3}{2\sqrt{2}}$ the equation has no solutions. 4. $\frac{\lambda\pi}{2} \sin 3x + \cos x$ for all values of λ. 5. $1 - \frac{2x}{\pi} - \frac{\lambda\pi^2}{6(1+2\lambda)} \cos x$ if $\lambda \neq \pm 1/2$; $\frac{4}{3} - \frac{2x}{\pi} + (8 + \pi^2 \cos x)C$, where C is an arbitrary constant, if $\lambda = 1/2$; at $\lambda = -1/2$ the equation has no solutions. 6. $\frac{\lambda\pi}{2-\lambda\pi} + 1 + \cos 4x$ if $\lambda \neq 2/\pi$ and $\lambda \neq 4/\pi$; $\cos 4x - 1 + C_1\cos 2x + C_2 \sin 2x$, where C_1 and C_2 are arbitrary constants, $+\lambda = 4/\pi$; at $\lambda = 2/\pi$ the equation has no solutions.

5.17. 1. $\cos 3x$ if $\lambda \neq 1/\pi$; $\cos 3x + C_1 \cos x + C_2 \cos 2x$, where C_1 and C_2 are arbitrary constants, if $\lambda = 1/\pi$. 2. $\cos x/(1 - \lambda\pi)$ if $\lambda \neq$

$\neq 1/\pi$ and $\lambda \neq 1/2\pi$; $2 \cos x + C \sin 2x$, where C is an arbitrary constant, if $\lambda = 1/2\pi$; at $\lambda = 1/\pi$ the equation has no solutions. 3. $\frac{\sin x}{1-\pi\lambda}$ if $\lambda \neq 1/\pi$ and $\lambda \neq 1/3\pi$; $(3/2) \sin x + C \cos 2x$, where C is an arbitrary constant, if $\lambda = 1/3\pi$; at $\lambda = 1/\pi$ the equation has no solutions.

5.18. 1. $\lambda_1 = 1/\pi$, $\sin x + \cos x$, 1; $\lambda_2 = -1/\pi$, $\cos x - \sin x$. 2. $\lambda_1 = 1/2\pi$, 1; $\lambda_2 = 2/\pi$, $\cos 2x$; $\lambda_3 = -2/\pi$, $\sin 2x$. 3. $\lambda_1 = -45$, $3x^2 - 2$; $\lambda_2 = 45/8$, $15x^2 - 1$. 4. $\lambda_1 = 3/8$, $3x^{2/5} + x^{-2/5}$; $\lambda_2 = -3/2$, $3x^{2/5} - x^{-2/5}$. 5. $\lambda_1 = -2/\pi$, $\sin x - \sin 4x$, $\sin 2x - \sin 3x$; $\lambda_2 = 2/\pi$, $\sin 2x + \sin 3x$, $\sin x + \sin 4x$.

5.19. $a = -12$, $b = 12$, $\varphi(x) = -12x^2 + C_1x + C_2$, where C_1 and C_2 are arbitrary constants.

5.20. $a = \sqrt{15} - 3$, $\varphi(x) = C\,[4\sqrt{15}x^2 + 3(1 - \sqrt{15})\,x] + 1/x - 3x$, where C is an arbitrary constant.

5.21. The equation has a solution for all values of λ:

$$\varphi(x) = \lambda \int_0^{2\pi} \cos(2x - y)\, f(y)\, dy + f(x).$$

5.22. 1. $\frac{\lambda a\pi^3}{12(1-2\lambda)} \sin x + \frac{2\lambda b}{1-2\lambda} + ax + b$ if $\lambda \neq 1/2$ (a and b are arbitrary); at $\lambda = 1/2$ the equation has a solution if and only if $a = b = 0$, $\varphi(x) = C_1 \sin x + C_2$, where C_1 and C_2 are arbitrary constants. 2. $\frac{2(a-2\lambda b)}{2+\lambda\pi} \sin x + b$ if $\lambda \neq \pm 2/\pi$ (a and b are arbitrary); at $\lambda = 2/\pi$ the equation has solutions for all values of a and b, and $\varphi(x) = \frac{a\pi - 4b}{2\pi} \sin x + b + C_1 \cos x$ where C_1 is an arbitrary constant, if $\lambda = -2/\pi$, then the equation has a solution if and only if $a\pi + 4b = 0$, and $\varphi(x) = b + C_2 \sin x$, where C_2 is an arbitrary constant. 3. $\frac{2\lambda a + 3c}{3(1-2\lambda)} + \frac{3b}{3-2\lambda} x + ax^2$ if $\lambda \neq 1/2$ and $\lambda \neq 3/2$ (a, b, and c are arbitrary); at $\lambda = 1/2$ the equation has a solution if $a + 3c = 0$, and $\varphi(x) = (3/2)\, bx + ax^2 + C_1$, where C_1 is an arbitrary constant; at $\lambda = 3/2$ the equation has a solution if $b = 0$ and $\varphi(x) = ax^2 - \frac{1}{2}(a + c) + C_2x$, where C_2 is an arbitrary constant. 4. $\frac{2\lambda(5a+3b)}{15-4\lambda^2} x^2 + \frac{4\lambda^2(5a+3b)}{5(15-4\lambda^2)} x + ax + bx^3$ if $\lambda \neq \pm\sqrt{15}/2$ (a and b are arbitrary); at $\lambda = \sqrt{15}/2$ the equation has a solution if $5a + 3b = 0$, and $\varphi(x) = a\left(x - \frac{5}{3}x^3\right) + C_1\left(\sqrt{\frac{5}{3}}x^2 + x\right)$, where C_1 is an arbitrary constant; at $\lambda = -\sqrt{15}/2$ the equation has a solution if $5a + 3b = 0$, and $\varphi(x) = a\left(x - \frac{5}{3}x^3\right) + C_2\left(x - \sqrt{\frac{5}{3}}x^2\right)$, where C_2 is an arbitrary constant. 5. $\frac{3a}{3-\lambda}x+\frac{5\lambda b}{3(5-\lambda)}x^2+b$ if $\lambda\neq 3$ and $\lambda\neq 5$ (with a and b arbitrary); at $\lambda=3$ the equation has a solution if $a=0$, and $\varphi(x)=b\left(\frac{5}{2}x^2+1\right)+C_1$, where C_1 is an arbitrary constant; at $\lambda=5$ the equation has a solution if $b=0$, and $\varphi(x)=C_2x^2-\frac{3}{2}ax$, where C_2 is an arbitrary constant. 6. $\frac{30\lambda a+7b}{7(1-6\lambda)}x^{1/3}+ax$ if $\lambda\neq 1/6$ (with a and b arbitrary constants); at $\lambda=1/6$ the equation has a solution if $5a+7b=0$, and $\varphi(x)=-\frac{7}{5}bx+C_1x^{1/3}+C_2x^{2/3}$, where C_1 and C_2 are arbitrary constants. 7. $\frac{2a+\lambda b(4-\pi)}{2-\lambda\pi}+\frac{2}{2-\lambda(4-\pi)}x+bx^2$ if $\lambda\neq 2/\pi$ and $\lambda\neq 2/(4-\pi)$ (with a and b arbitrary); at $\lambda=2/\pi$ the equation has a solution if $a\pi+b(4-\pi)=0$, and $\varphi(x)=\frac{\pi}{2(\pi-2)}x+bx^2+C$, where C is an arbitrary constant; at $\lambda=2/(4-\pi)$ the equation has no solutions. 8. $\frac{5\lambda(14a+36\lambda b+42c)}{21(5-12\lambda^2)}x^{1/3}+\frac{28\lambda^2a+30\lambda b+35}{7(5-12\lambda^2)}+ax^2+bx$ if $\lambda\neq\pm(1/2)\sqrt{5/3}$ (a, b, and c are arbitrary); at $\lambda=(1/2)\sqrt{5/3}$ the equation has a solution if $15\times\sqrt{3}b+7\sqrt{5}(a+3c)=0$, and $\varphi(x)=ax^2+bx+c+C_1\times\left(x^{1/3}+\sqrt{\frac{3}{5}}\right)$, where C_1 is an arbitrary constant; at $\lambda=-(1/2)\sqrt{5/3}$ the equation has a solution if $15\sqrt{3}b-7\sqrt{5}\times(a+3c)=0$, and $\varphi(x)=ax^2+bx+c+C_2\left(x^{1/3}-\sqrt{\frac{3}{5}}\right)$, where C_2 is an arbitrary constant. 9. $\frac{30(b-1)\lambda}{15+8\lambda}x^2+\frac{3a\lambda^2}{3-2\lambda}x+\frac{36\lambda^2(b-1)}{(15+8\lambda)(3-2\lambda)}$ if $\lambda\neq-15/8$ and $\lambda\neq 3/2$ (a and b are arbitrary); at $\lambda=-15/8$ the equation has a solution if $b=1$, and $\varphi(x)=\frac{17}{2}ax+1-20a+C(x^2+1)$, where C is an arbitrary constant; at $\lambda=3/2$ the equation has a solution if $a=b=0$, and $\varphi(x)=C_1x+C_2$, where C_1 and C_2 are arbitrary constants.

5.23. 1. $\lambda_1=3/2$, $\varphi_1=x$; $\lambda_2=-1/2$, $\varphi_2=3x-4x^2$; $\varphi(x)=3ax/(3-2\lambda)$ if $\lambda\neq 3/2$ and $\lambda\neq-1/2$ (a is arbitrary); at $\lambda=3/2$ the equation has a solution if $a=0$, and $\varphi(x)=C_1x$, where C_1 is an arbitrary constant; at $\lambda=-1/2$ the equation has solutions for all values of a, and $\varphi(x)=\frac{3}{4}ax+C_2(3x-4x^2)$, where C_2 is an arbitrary constant. 2. $\lambda_1=1/2$, $\varphi_1^{(1)}=x$, $\varphi_1^{(2)}=x^2$; $\varphi(x)=\frac{ax^2+bx}{1-2\lambda}$ if $\lambda\neq 1/2$; at $\lambda=1/2$ the equation has a solution if $a=b=0$, and $\varphi(x)=C_1x^2+C_2x$, where C_1 and C_2 are arbitrary constants.

5.24. 1. $\lambda_1 = 1/\pi$, $\varphi_1 = \sin x$; $\varphi(x) = a + b\cos x + \lambda b\pi x + \frac{2\pi^2\lambda^2 b}{1-\lambda\pi}\sin x$ if $\lambda \neq 1/\pi$ (a and b are arbitrary); at $\lambda = 1/\pi$ the equation has a solution if $b = 0$ and $\varphi(x) = a + C\sin x$, where C is an arbitrary constant. 2. $\lambda_1 = 1/2\pi$, $\varphi_1 = x$; $\varphi(x) = \frac{ax}{1-2\pi\lambda} + b + 2\pi b\lambda\cos x$ if $\lambda \neq 1/2\pi$ (a and b are arbitrary); at $\lambda = 1/2\pi$ the equation has a solution if $a = 0$, and $\varphi(x) = b(1+\cos x) + Cx$, where C is an arbitrary constant.

5.25. 1. $\varphi(x) = \lambda \int_0^{\pi} \frac{\sin(x+y) + \lambda\frac{\pi}{2}\cos(x-y)}{\Delta(\lambda)} f(y)\,dy + f(x)$ if

$\Delta(\lambda) \neq 0$, where $\Delta(\lambda) = 1 - \lambda^2\pi^2/4$; at $\lambda = 2/\pi$ the equation has a solution if $f_1 + f_2 = 0$, where

$$f_1 = \int_0^{\pi} f(y)\cos y\,dy, \quad f_2 = \int_0^{\pi} f(y)\sin y\,dy,$$

and

$$\varphi(x) = C_1(\sin x + \cos x) + \frac{2}{\pi} f_1 \sin x + f(x)$$

(C_1 is an arbitrary constant); at $\lambda = -2/\pi$ the equation has a solution if $f_1 - f_2 = 0$, and $\varphi(x) = C_2(\sin x - \cos x) - (2/\pi) f_1 \sin x + f(x)$, where C_2 is an arbitrary constant;

$$\mathscr{R}(x, y; \lambda) = \frac{\sin(x+y) + \frac{\lambda\pi}{2}\cos(x-y)}{\Delta(\lambda)}.$$

2. $\varphi(x) = \lambda \int_{-1}^{1} \frac{1 - \frac{4}{3}\lambda + y(2x - 4\lambda x - 1)}{\Delta(\lambda)} f(y)\,dy + f(x)$ if $\Delta(\lambda) \neq 0$,

where $\Delta(\lambda) = (1-2\lambda)\left(1 - \frac{4}{3}\lambda\right)$; at $\lambda = 1/2$ the equation has a solution if $f_1 = 3f_2$, where

$$f_1 = \int_{-1}^{1} f(x)\,dx, \quad f_2 = \int_{-1}^{1} xf(x)\,dx, \text{ and } \varphi(x) = \left(x - \frac{1}{2}\right) f_1 + f(x) + C_1$$

(C_1 is an arbitrary constant); at $\lambda = 3/4$ the equation has a solution if $f_2 = 0$, and $\varphi(x) = -\frac{3}{2} f_1 + f(x) + C_2(x+1)$, where C_2 is an arbitrary constant;

$$\mathscr{R}(x, y; \lambda) = \frac{1 - \frac{4}{3}\lambda + y(2x - 4\lambda x - 1)}{\Delta(\lambda)}.$$

3. $\varphi(x) = \lambda \int_{-\pi}^{\pi} \left(\frac{x\sin y}{1-2\pi\lambda} + \cos x\right)(ay+b)\,dy + ax + b = \frac{ax}{1-2\pi\lambda} +$

$+2\pi\lambda b\cos x+b$, if $\lambda\neq 1/2\pi$ (with a and b arbitrary); at $\lambda=1/2\pi$ the equation has a solution if $a=0$, and $\varphi(x)=b(\cos x+1)+Cx$, where C is an arbitrary constant,

$$\mathcal{R}(x, y; \lambda)=\frac{x\sin y}{1-2\pi\lambda}+\cos x.$$

4. $\varphi(x)=\lambda\int\limits_0^{2\pi}\frac{\sin x\sin y+\sin 2x\sin 2y}{1-\lambda\pi}f(y)\,dy+f(x)$ if $\lambda\neq 1/\pi$; at $\lambda=1/\pi$ the equation has a solution if

$$\int\limits_0^{2\pi} f(y)\sin y\,dy=\int\limits_0^{2\pi} f(y)\sin 2y\,dy=0,$$

and

$$\varphi(x)=f(x)+C_1\sin x+C_2\sin 2x,$$

where C_1 and C_2 are arbitrary constants,

$$\mathcal{R}(x, y, \lambda)=\frac{\sin x\cdot\sin y+\sin 2x\cdot\sin 2y}{1-\lambda\pi}.$$

5.26. 1. $b=0$, $3a+5c=0$. 2. $a=3/\sqrt{10}$, $b=0$, $c=-1/\sqrt{10}$; $a=-3/\sqrt{10}$, $b=0$, $c=1/\sqrt{10}$. 3. $a=0$, $b=-1/2$. 4. $a=6$. 5. $a=0$, $b=-1$. 6. a and b are arbitrary. 7. a, b, and c are arbitrary, 8. $7a+5b=0$.

5.27. $\frac{1}{3}<a<3$.

5.28. 1. $\lambda_1=1$, $\varphi_1=4(x_1+x_2)+1$; $\lambda_2=-1$, $\varphi_2=4(x_1+x_2)-1$. 2. $\lambda_1=\frac{4\sqrt{3}-6}{\pi}$, $\varphi_1=1+\sqrt{3}(x_1^2+x_2^2)$; $\lambda_2=-\frac{4\sqrt{3}+6}{\pi}$, $\varphi_2=\sqrt{3}(x_1^2+x_2^2)-1$. 3. $\lambda_1=3/4\pi$, $\varphi_1=1/(1+r)$, where $r=\sqrt{x_1^2+x_2^2+x_3^2}$.

5.29. The equation has no real-valued eigenvalues.

5.30. The eigenvalues are $\lambda_1=3/4$ and $\lambda_2=-3/4$, and the corresponding eigenfunctions are $\varphi_1=1+3x_1x_2$ and $\varphi_2=3x_1x_2-1$. If $\lambda=\pm 3/4$, then

$$\varphi(x_1, x_2)=\frac{\lambda}{\Delta(\lambda)}\left[(f_1+4\lambda f_2)x_1x_2+\frac{4}{9}\lambda f_1+f_2\right]+f(x_1, x_2),$$

where

$$f_1=\int\limits_{-1}^{1}\int\limits_{-1}^{1} f(y_1, y_2)\,dy_1\,dy_2,$$

$$f_2=\int\limits_{-1}^{1}\int\limits_{-1}^{1} y_1y_2f(y_1, y_2)\,dy_1\,dy_2,$$

$\Delta(\lambda) = 1 - (16/9)\lambda^2$; at $\lambda = 3/4$ the equation has a solution if $f_1 + 3f_2 = 0$, and $\varphi(x_1, x_2) = \frac{3}{4} x_1 x_2 f_1 + f(x_1, x_2) + C(3x_1x_2 + 1)$, where C_1 is an arbitrary constant; at $\lambda = -3/4$ the equation has a solution if $f_1 - 3f_2 = 0$, and $\varphi(x_1, x_2) = -\frac{3}{4} x_1 x_2 f_1 + f(x_1, x_2) + C_2(3x_1x_2 - 1)$, where C_2 is an arbitrary constant.

5.36. 1. $\lambda_n = \left(\frac{\pi}{2} + \pi n\right)^2$, $\varphi_n = \sin\left(\frac{\pi}{2} + \pi n\right) x$ $(n = 0, 1, 2, \ldots)$. 2. $\lambda_n = n^2\pi^2$, $\varphi_n = \sin \pi n x$ $(n = 1, 2, \ldots)$. 3. λ_n $(n = 1, 2, \ldots)$ are the positive roots of the equation $\tan\sqrt{\lambda} = -\sqrt{\lambda}$, and $\varphi_n = \sin\sqrt{\lambda_n}\, x$. 4. $\lambda_n = -\frac{1}{3}\mu_n^2$ $(n = 1, 2, \ldots)$, where μ_n are the positive roots of the equation $\mu - 1/\mu = 2\cot\mu$, $\varphi_n = \sin \mu_n x + \mu_n \cos \mu_n x$. 5. $\lambda_0 = 1$, $\varphi_0 = e^x$; $\lambda_n = -n^2\pi^2$ $(n = 1, 2, \ldots)$, $\varphi_n = \sin \pi n x + \pi n \cos \pi n x$. 6. $\lambda_n = \frac{\pi^2(2n+1)^2 + 4}{8(1+e^2)}$ $(n = 0, 1, 2, \ldots)$, $\varphi_n = \sin\left(n + \frac{1}{2}\right)\pi x$. 7. $\lambda_n = \frac{(n\pi)^2 - 1}{\sin 1}$, $\varphi_n = \sin \pi n x$ $(n = 1, 2, \ldots)$.

5.37. 1. $\lambda_n^{(1)} = (\pi n)^2/2$, $\varphi_n^{(1)} = \sin \pi n x$ $(n = 1, 2, \ldots)$; $\lambda_n^{(2)} = \frac{1}{2}\left(\frac{\pi}{2} + \pi n\right)^2$, $\varphi_n^{(2)} = \cos\left(\frac{\pi}{2} + \pi n\right) x$ $(n = 0, 1, 2, \ldots)$. 2. $\lambda_n = 1 - \left(n + \frac{1}{2}\right)^2$, $\varphi_n = \cos\left(n + \frac{1}{2}\right) x$ $(n = 0, 1, 2, \ldots)$. 3. $\lambda_n = \left(n + \frac{1}{2}\right)^2 - 1$, $\varphi_n = \sin\left(n + \frac{1}{2}\right) x$ $(n = 0, 1, 2, \ldots)$.

5.38. 1. $\lambda_n^{(1)} = \left(\frac{2n+1}{2}\right)^2$, $\varphi_n^{(1)} = \sin(2n+1)x$ $(n = 0, 1, 2, \ldots)$; $\lambda_n^{(2)} = -\left(\frac{2n+1}{2}\right)^2$, $\varphi_n^{(2)} = \cos(2n+1)x$ $(n = 0, 1, 2, \ldots)$; $\lambda_0 = 1/\pi^2$, $\varphi_0 = 1$. 2. $\lambda_0 = 1/\pi^2$, $\varphi_0 = 1$; $\lambda_n^{(1)} = (2n+1)^2/4$, $\varphi_n^{(1)} = \cos(2n+1)x$ $(n = 0, 1, 2, \ldots)$; $\lambda_n^{(2)} = -(2n+1)^2/4$, $\varphi_n^{(2)} = \sin(2n+1)x$ $(n = 0, 1, 2, \ldots)$.

5.39. $\lambda_n = \frac{1}{\pi a_n}$, $\varphi_n^{(1)} = \sin nx$, $\varphi_n^{(2)} = \cos nx$ $(n = 1, 2, \ldots)$ if $a_n = \frac{1}{\pi}\int_0^{2n} \omega(t) \cos nt\, dt \neq 0$; $\lambda_0 = \frac{1}{\pi a_0}$ and $\varphi_0 = 1$ if $a_0 = \frac{1}{\pi}\int_0^{2\pi} \omega(t) \times dt \neq 0$.

5.40. $\varphi(x) = \lambda \int_0^1 G(x, y) f(y)\, dy + f(x)$, where

$$G(x, y) = \begin{cases} \dfrac{\sin\sqrt{\lambda}\, x \cos\sqrt{\lambda}(y-1)}{\sqrt{\lambda}\cos\sqrt{\lambda}}, & x \leqslant y, \\[2ex] \dfrac{\sin\sqrt{\lambda}\, y \cos\sqrt{\lambda}(x-1)}{\sqrt{\lambda}\cos\sqrt{\lambda}}, & x \geqslant y. \end{cases}$$

Chapter III

Generalized Functions

6 Test and Generalized Functions

We denote by $\mathscr{D} \equiv \mathscr{D}(R^n)$ the set of all finite functions that are infinitely differentiable in R^n. A sequence of functions $\{\varphi_k\}$ from $\mathscr{D}$ *converges* to a function φ (from $\mathscr{D}$) if

(a) there is a positive R such that $\operatorname{supp} \varphi_k \subset U_R$, and

(b) $$D^\alpha \varphi_k(x) \overset{x \in R^n}{\Rightarrow} D^\alpha \varphi(x), \quad k \to \infty,$$

for each α (see Symbols and Definitions). We then write $\varphi_k \to \varphi$ as $k \to \infty$ in $\mathscr{D}$. The set $\mathscr{D}$ equipped with convergence is called the *space of test functions* $\mathscr{D}$.

We relate to the set of test functions $\mathscr{S} \equiv \mathscr{S}(R^n)$ all functions of the class $C^\infty(R^n)$ that decrease together with all their derivatives, as $|x| \to \infty$, faster than any power of $|x|^{-1}$. We define convergence in $\mathscr{S}$ as follows: a sequence of functions $\{\varphi_k\}$ from $\mathscr{S}$ is said to *converge* to a function φ (from $\mathscr{S}$) if

$$x^\beta D^\alpha \varphi_k(x) \overset{x \in R^n}{\Rightarrow} x^\beta D^\alpha \varphi(x), \quad k \to \infty,$$

for all α and β. We then write $\varphi_k \to \varphi$ as $k \to \infty$ in $\mathscr{S}$. The set $\mathscr{S}$ equipped with convergence is called the *space of test functions* $\mathscr{S}$.

6.1. Suppose $\varphi \in \mathscr{D}$. Establish whether among the sequences given below there are sequences that converge in $\mathscr{D}$:

$$1.\ \frac{1}{k}\varphi(x). \quad 2.\ \frac{1}{k}\varphi(kx). \quad 3.\ \frac{1}{k}\varphi\left(\frac{x}{k}\right), \quad k = 1, 2, \ldots$$

6.2. Let $n = 1$ and

$$\chi(x) = \begin{cases} 1 & \text{for } -2\varepsilon \leqslant x \leqslant 2\varepsilon, \\ 0 & \text{for } |x| > 2\varepsilon. \end{cases}$$

Show that the function $\eta(x) = \int_{-\infty}^{\infty} \chi(y)\,\omega_\varepsilon(x-y)\,dy$, where ω_ε is the "cap", is a test function that belongs to $\mathscr{D}(R^1)$, with $0 \leqslant \eta(x) \leqslant 1$, and $\eta(x) \equiv 1$ for $-\varepsilon \leqslant x \leqslant \varepsilon$ and $\eta(x) \equiv 0$ for $|x| > 3\varepsilon$.

6.3. Let $G_{2\varepsilon} = \bigcup_{x \in G} U(x; 2\varepsilon)$ be the 2ε-neighborhood of a bounded region G and let $\chi(x)$ be the characteristic function of $G_{2\varepsilon}$, that is, $\chi(x) = 1$, $x \in G_{2\varepsilon}$, and $\chi(x) = 0$, $x \,\overline{\in}\, G_{2\varepsilon}$. Prove that

$$\eta(x) = \int \chi(y)\,\omega_\varepsilon(x-y)\,dy$$

is a test function from $\mathscr{D}(R^n)$, with $0 \leqslant \eta(x) \leqslant 1$, and $\eta(x) \equiv 1$ for $x \in G_\varepsilon$ and $\eta(x) \equiv 0$ for $x \,\overline{\in}\, G_{3\varepsilon}$.

6.4. Suppose the function $\eta(x)$ satisfies the conditions of problem 6.2 and

$$H(x) = \sum_{\nu=-\infty}^{\infty} \eta(x-\varepsilon\nu), \quad e(x) = \frac{\eta(x)}{H(x)}.$$

Prove that $H \in C^\infty(R^1)$, $H(x) \geqslant 1$; $e \in \mathscr{D}(R^1)$, $0 \leqslant e(x) \leqslant 1$; $e(x) \equiv 1$ for $|x| \leqslant \varepsilon$ and $e(x) = 0$ for $|x| \geqslant 3\varepsilon$; and $\sum_{\nu=-\infty}^{\infty} e(x-\varepsilon\nu) \equiv 1$.

6.5. Prove that there are functions $\varphi_\delta \in \mathscr{D}(R^1)$, $\delta > 1$, such that $\varphi_\delta(x) = 1$ for $|x| \leqslant \delta - 1$, $\varphi_\delta(x) = 0$ for $|x| \geqslant \delta$, and $|\varphi_\delta^{(\alpha)}(x)| \leqslant C_\alpha$, where C_α is a constant that does not depend on δ.

6.6. Suppose $f(x)$ is a finite continuous function: $f(x) = 0$, $|x| > R$. Show that

$$f_\varepsilon(x) = \int f(y)\,\omega_\varepsilon(x-y)\,dy \quad (\varepsilon > R)$$

is a test function from $\mathscr{D}(R^n)$, with $f_\varepsilon(x) = 0$ for $|x| > R + \varepsilon$. Show that

$$f_\varepsilon(x) \overset{x \in R^n}{\rightrightarrows} f(x), \quad \varepsilon \to 0.$$

6.7. 1. Prove that

$$\psi(x) = \frac{1}{x^m}\left[\varphi(x) - \eta(x) \sum_{k=0}^{m-1} \frac{\varphi^{(k)}(0)}{k!} x^k\right], \quad m = 1, 2,$$

is a test function from $\mathscr{D}(R^1)$, where $\varphi \in \mathscr{D}(R^1)$ and $\eta \in \mathscr{D}(R^1)$, with $\eta = 1$ in the neighborhood of point $x = 0$.

2. Prove that

$$\psi(x) = \frac{\varphi(x) - \eta(x)\varphi(0)}{\alpha(x)}$$

is a test function from $\mathscr{D}(R^1)$, where $\varphi \in \mathscr{D}(R^1)$, $\eta(x)$ is a function defined in Problem 6.7.1, and $\alpha \in C^\infty(R^1)$ has only one zero of order 1 at point $x = 0$.

6.8. 1. Show that a function φ_1 from $\mathscr{D}(R^1)$ can be represented in the form of the derivative of another function φ_2 from $\mathscr{D}(R^1)$ if and only if

$$\int_{-\infty}^{\infty} \varphi_1(x)\,dx = 0.$$

2. Show that every function $\varphi(x)$ from $\mathcal{D}(R^1)$ can be represented in the form $\varphi(x) = \varphi_0(x) \int_{-\infty}^{\infty} \varphi(x')\,dx' + \varphi_1'(x)$, where $\varphi_1 \in \mathcal{D}(R^1)$, and $\varphi_0(x)$ is any test function from $\mathcal{D}(R^1)$ that satisfies the condition $\int_{-\infty}^{\infty} \varphi_0(x)\,dx = 1$. *Hint.* Use results of Problem 6.8.1.

6.9. Show that $\mathcal{D} \subset \mathcal{S}$ and that convergence in $\mathcal{D}$ implies convergence in $\mathcal{S}$.

6.10. Let $\varphi \in \mathcal{S}$. Establish whether among the sequences given below there are sequences that converge in $\mathcal{S}$.

$$1.\ \frac{1}{k}\varphi(x).\quad 2.\ \frac{1}{k}\varphi(kx).\quad 3.\ \frac{1}{k}\varphi\left(\frac{x}{k}\right),\quad k = 1, 2, \ldots$$

6.11. Suppose that $\varphi \in \mathcal{S}$ and that P is a polynomial. Prove that $\varphi P \in \mathcal{S}$.

6.12. Suppose a function $\psi \in C^\infty(R^1)$ vanishes at $x < a$ and is bounded together with all of its derivatives. Prove that $\psi(x)\,e^{-\sigma x}$ is a test function from $\mathcal{S}(R^1)$ if $\sigma > 0$.

We denote by $\mathcal{D}' \equiv \mathcal{D}'(R^n)$ the set of all linear continuous functionals over the space of test functions $\mathcal{D}$. A functional f that belongs to $\mathcal{D}'$ we call a *generalized function* (from space $\mathcal{D}'$).

We denote by $\mathcal{S}' \equiv \mathcal{S}'(R^n)$ the set of all linear continuous functionals over the space of test functions $\mathcal{S}$. A functional f that belongs to $\mathcal{S}'$ we call a *generalized function of slow growth* (from space $\mathcal{S}'$).

We denote the value of a functional f over a test function φ by (f, φ). Sometimes, in order to specify the independent variable of the test function, we will write $f(x)$ and $(f(x), \varphi(x))$ instead of f and (f, φ), respectively.

A sequence $\{f_k\}$ of generalized functions from $\mathcal{D}'$ is said to *converge to a generalized function* f (from $\mathcal{D}'$) if $(f_k, \varphi) \to (f, \varphi)$ as $k \to \infty$ for every φ that belongs to $\mathcal{D}$. In particular, the series $u_1 + u_2 + \ldots + u_k + \ldots$ consisting of generalized functions is said to *converge in $\mathcal{D}'$ to a generalized function* f if the numerical series $\sum_{k=1}^{\infty} (u_k, \varphi)$ converges to (f, φ) for every function $\varphi \in \mathcal{D}$. The convergence of sequences and series in $\mathcal{S}$ can be defined along the same lines.

It is said that a generalized function f *vanishes in a region G* if $(f, \varphi) = 0$ for every φ from $\mathcal{D}$ with supp in G. The generalized functions f_1 and f_2 are said to be *equal* in G if their difference vanishes in G, and f_1 is said to *equal* f_2 if $(f_1, \varphi) = (f_2, \varphi)$ for all $\varphi \in \mathcal{D}$.

The *support* of a generalized function f, denoted supp f, is the set of all such points in whose neighborhoods the function f does not vanish. If supp f is a bounded set, then f is said to be a *finite generalized function.*

A *regular* generalized function from $\mathcal{D}'(R^n)$ is a functional of the type

$$(f, \varphi) = \int f(x)\varphi(x)\,dx, \quad \varphi \in \mathcal{D}(R^n),$$

where f is a function that is locally integrable in R^n.

If $f(x)$ is a function of slow growth in R^n, that is

$$\int |f(x)|(1+|x|)^{-m}\,dx < \infty$$

for a nonnegative integer m, it defines a regular generalized function from $\mathcal{S}'$ (of slow growth).

A generalized function that is not regular is called a *singular* generalized function. An example of a singular generalized function is Dirac's delta function, which is defined thus:

$$(\delta, \varphi) = \varphi(0), \quad \varphi \in \mathcal{D}(R^n).$$

The surface delta function is a generalization of Dirac's delta function. Suppose S is a piecewise smooth surface and $\mu(x)$ is a continuous function on S. The generalized function $\mu\delta_S$ operating according to the rule

$$(\mu\delta_S, \varphi) = \int_S \mu(x)\varphi(x)\,dS_x, \quad \varphi \in \mathcal{D}(R^n)$$

is said to be a *simple layer*. In particular, if S is the plane $t = 0$ in $R^{n+1}(x, t)$, then $\mu\delta_{(t=0)}(x, t)$ is denoted by $\mu(x)\delta(t)$, so that

$$(\mu(x)\delta(t), \varphi) = \int_{R^n} \mu(x)\varphi(x, 0)\,dx.$$

When $n = 1$, the simple layer $\delta_{S_R}(x)$ on the sphere S_R is denoted by $\delta(R - |x|)$, so that $(\delta(R - |x|), \varphi) = \varphi(R) + \varphi(-R)$.

The *product* of a generalized function f from $\mathcal{D}'(R^n)$ and a function $\alpha(x) \in C^\infty(R^n)$ is a generalized function αf operating according to the rule $(\alpha f, \varphi) = (f, \alpha\varphi)$, $\varphi \in \mathcal{D}(R^n)$.

Let $f(x)$ belong to $\mathcal{D}'(R^n)$, A be a nonsingular linear transformation, and b a vector in R^n. We define the generalized function $f(Ay + b)$ according to the formula

$$(f(Ay+b), \varphi) = \left(f\, \frac{\varphi[A^{-1}(x-b)]}{|\det A|}\right), \quad \varphi \in \mathcal{D}(R^n).$$

At $A = I$ we have a translation of the generalized function f by vector $-b$:

$$(f(y + b), \varphi) = (f, \varphi(x - b)).$$

For instance, $(\delta(x - x_0), \varphi) = (\delta, \varphi(x + x_0)) = \varphi(x_0)$ is the translation of $\delta(x)$ by vector x_0. At $A = -I$, $b = 0$ we have reflection:

$$(f(-x), \varphi) = (f, \varphi) -x)).$$

6.13. Prove that $\delta(x)$ is a singular generalized function. Give this function a physical interpretation.

6.14. Give a physical interpretation of the following generalized functions:

1. $2\delta(x-x_0)$. 2. $\sum_{k=1}^{N} m_k\delta(x-x_k)$. 3. $\mu(x)\delta_S(x)$. 4. $|x|\delta_{S_R}\times\times(x-x_0)$. 5. $2\delta(R_1-|x-1|)+3\delta(R_2-|x-2|)$. Find the support of each function.

6.15. Prove that
1. $\delta(x-\nu)\to 0,\ \nu\to\infty$ in $\mathcal{D}'(R^1)$.
2. $\delta_{S_R}(x)\to 0,\ R\to\infty$ in $\mathcal{D}'$.

6.16. Prove that $\mathcal{S}'\subset\mathcal{D}'$ and that convergence in $\mathcal{S}'$ implies convergence in $\mathcal{D}'$.

6.17. Prove that

1. $e^x\in\mathcal{D}'(R^1)$, $e^x\notin\mathcal{S}'(R^1)$.

2. $e^{\frac{1}{x}}\notin\mathcal{D}'(R^1)$. 3. $e^x\sin e^x\in\mathcal{S}'(R^1)$.

6.18. Prove that the functional $\mathcal{P}\frac{1}{x}$, which operates according to the formula

$$\left(\mathcal{P}\frac{1}{x},\varphi\right)=\mathrm{PV}\int_{-\infty}^{\infty}\frac{\varphi(x)}{x}\,dx=\lim_{\varepsilon\to+0}\left(\int_{-\infty}^{-\varepsilon}+\int_{\varepsilon}^{\infty}\right)\frac{\varphi(x)}{x}\,dx,\quad \varphi\in\mathcal{D},$$

is a singular generalized function.

6.19. Find the limits of the following functions in $\mathcal{D}'(R^1)$ as $\varepsilon\to\to+0$:

1. $f_\varepsilon(x)=\begin{cases}\frac{1}{2\varepsilon}, & |x|\leqslant\varepsilon,\\ 0, & |x|>\varepsilon.\end{cases}$ 2. $\frac{\varepsilon}{\pi(x^2+\varepsilon^2)}$. 3. $\frac{1}{2\sqrt{\pi\varepsilon}}\,e^{-x^2/(4\varepsilon)}$.

4. $\frac{1}{x}\sin\frac{x}{\varepsilon}$. 5. $\frac{1}{\pi\varepsilon x^2}\sin^2\frac{x}{\varepsilon}$.

6.20. Prove the validity of Sochozki formulas

$$\frac{1}{x\pm i0}=\mp i\pi\delta(x)+\mathcal{P}\frac{1}{x}.$$

6.21. Find the limits of the following functions in $\mathcal{D}'(R^1)$ as $t\to\to+\infty$:

1. $\frac{e^{ixt}}{x-i0}$. 2. $\frac{e^{-ixt}}{x-i0}$. 3. $\frac{e^{ixt}}{x+i0}$.

4. $\frac{e^{-ixt}}{x+i0}$. 5. $t^m e^{ixt},\ m\geqslant 0$.

6.22. Find the limit of $\mathcal{P}\frac{\cos kx}{x}$ in $\mathcal{D}'(R^1)$ as $k \to \infty$, where

$$\left(\mathcal{P}\frac{\cos kx}{x}, \varphi\right) = \mathrm{PV}\int_{-\infty}^{\infty} \frac{\cos kx}{x}\varphi(x)\,dx$$
$$= \lim_{\varepsilon\to+0}\left(\int_{-\infty}^{-\varepsilon} + \int_{\varepsilon}^{\infty}\right)\frac{\cos kx}{x}\varphi(x)\,dx, \quad \varphi \in \mathcal{D}$$

6.23. Prove that the series $\sum_{k=-\infty}^{\infty} a_k\delta(x-k)$ is

(1) convergent in $\mathcal{D}'$ for all a_k, and
(2) convergent in $\mathcal{S}'$ if $|a_k| \leqslant C(1+|k|)^m$.

6.24. Suppose $\psi \in \mathcal{D}(R^n)$, $\psi \geqslant 0$, and $\int \psi(x)\,dx = 1$. Prove that $\varepsilon^{-n}\psi\left(\frac{x}{\varepsilon}\right) \to \delta(x)$, as $\varepsilon \to +0$ in $\mathcal{D}'(R^n)$; in particular, $\omega_\varepsilon(x) \to \delta(x)$ as $\varepsilon \to 0$ in $\mathcal{D}'(R^n)$.

6.25. Show that the functional $\mathcal{P}\frac{1}{x^2}$, which operates according to the formula

$$\left(\mathcal{P}\frac{1}{x^2}, \varphi\right) = \mathrm{PV}\int_{-\infty}^{\infty} \frac{\varphi(x)-\varphi(0)}{x^2}\,dx, \qquad \varphi \in \mathcal{D},$$

is a singular generalized function.

6.26. Show that

1. $\alpha(x)\delta(x) = \alpha(0)\delta(x)$, $\alpha \in C^\infty(R^n)$; in particular, $x\delta(x) = 0$, $x \in R^1$.

2. $x\mathcal{P}\frac{1}{x} = 1$. 3. $x^m\mathcal{P}\frac{1}{x} = x^{m-1}$, $m \geqslant 1$.

6.27. 1. Suppose the generalized function f vanishes outside the segment $[-a, a]$. Prove that $f = \eta f$, where $\eta \in C^\infty(R^1)$ and $\eta(x) \equiv 1$ in $[-a-\varepsilon, a+\varepsilon]$, with ε a positive (but otherwise arbitrary) number.

2. Suppose $f \in \mathcal{D}'(R^n)$ and $\eta \in C^\infty(R^n)$, with $\eta(x) \equiv 1$ in the neighborhood supp f. Show that $f = \eta f$ and $f \in \mathcal{S}'(R^n)$.

6.28. Prove that $\delta(ax) = \frac{1}{|a|^n}\delta(x)$, $a \neq 0$.

6.29. Prove that $(\alpha f)(x+h) = \alpha(x+h)f(x+h)$, where $\alpha \in C^\infty(R^n)$, $f \in \mathcal{D}'(R^n)$, $h \in R^n$.

6.30. Prove that the generalized function

$$\left(\mathrm{Pf}\frac{1}{x^2+y^2}, \varphi(x, y)\right) = \int_{x^2+y^2<1} \frac{\varphi(x, y)-\varphi(0, 0)}{x^2+y^2}\,dx\,dy$$
$$+ \int_{x^2+y^2>1} \frac{\varphi(x, y)}{x^2+y^2}\,dx\,dy$$

satisfies the equation $(x^2+y^2)\,\mathrm{Pf}\frac{1}{x^2+y^2} = 1$ in $\mathcal{D}'(R^2)$.

6.31. Suppose that $f \in \mathcal{S}'$ and that P is a polynomial. Show that $fP \in \mathcal{S}'$.

6.32.* Suppose f belongs to $\mathcal{D}'(R^1)$ and is finite and $\eta(x)$ is an arbitrary function from $\mathcal{D}(R^1)$ equal to unity in the neighborhood supp f. Put

$$\hat{f}(z) = \frac{1}{2\pi i}\left(f(x'), \frac{\eta(x')}{x'-z}\right), \quad z = x + iy.$$

Prove that (1) $\hat{f}(z)$ does not depend on the choice of the auxiliary function η; (2) $\hat{f}(z)$ is analytic if $z \overline{\in}$ supp f; (3) $\hat{f}(z) = O(1/|z|)$, $z \to \infty$; and (4) $\hat{f}(x+i\varepsilon) - \hat{f}(x-i\varepsilon) \to f(x)$ as $\varepsilon \to +0$ in $\mathcal{D}'(R^1)$.

6.33.* Suppose $f \in \mathcal{D}'(R^1)$, supp $f \subset [-a, a]$, and $\eta \subset \mathcal{D}(R^1)$, with $\eta(\xi) \equiv 1$ in the neighborhood supp f. Prove that

$$\bar{f}(z) = (f(\xi), \ \eta(\xi)\, e^{iz\xi}), \ z = x + iy,$$

is independent of η, is an entire function, and satisfies the estimate

$$|\bar{f}(x+iy)| \leqslant C_\varepsilon e^{(a+\varepsilon)|y|}(1+|x|)^m$$

for a nonnegative integer m and an arbitrary positive ε.

6.34.* Suppose $f \in \mathcal{D}'(R^n)$ and supp $f = \{0\}$. Prove that f has the unique representation

$$f(x) = \sum_{0 \leqslant |\alpha| \leqslant N} C_\alpha D^\alpha \delta(x).$$

6.35.* Suppose the series $\sum_{\nu=0}^{\infty} a_\nu \delta^{(\nu)}(x)$ is convergent in $\mathcal{D}'(R^1)$. Prove that $a_\nu = 0$ for $\nu > \nu_0$.

Answers to Problems of Sec. 6

6.1. Sequence (1) converges to zero; sequences (2) and (3) do not converge if $\varphi(x) \not\equiv 0$.

6.6. It is clear that $f_\varepsilon(x)$ belongs to $\mathcal{D}$. Since $f(x)$ is continuous and finite, for every positive σ and a sufficiently small positive ε we have $|f(x) - f(y)| < \sigma$ when $|x - y| \leqslant \varepsilon$, where both x and y belong to R^1, so

$$|f(x) - f_\varepsilon(x)| \leqslant \int |f(x) - f(y)|\, \omega_\varepsilon(x-y)\, dy$$

$$< \sigma \int_{|x-y|<\varepsilon} \omega_\varepsilon(x-y)\, dy = \sigma, \qquad x \in R^n.$$

6.7. 1. *Solution.* Obviously, $\psi(x)$ is finite and infinitely differentiable at $x \neq 0$. We still must prove that $\psi(x)$ is infinitely differentiable at $x = 0$. Let $\eta(x) \equiv 1$ at $|x| \leqslant \varepsilon$.

We introduce the notation

$$f(x)=\varphi(x)-\sum_{k=0}^{m-1}\frac{\varphi^{(k)}(0)}{k!}x^k.$$

We obtain

$$\psi(0)=\lim_{x\to 0}\psi(x)=\lim_{x\to 0}\frac{f(x)}{x^m}=\frac{f^{(m)}(0)}{m!},$$

$$\psi'(0)=\lim_{x\to 0}\frac{\psi(x)-\psi(0)}{x}=\lim_{x\to 0}\frac{f(x)-\dfrac{x^m f^{(m)}(0)}{m!}}{x^{m+1}}=\frac{f^{(m+1)}(0)}{(m+1)!},$$

etc. Thus, $\psi(x)\in C^\infty$ and, hence, $\psi\in\mathcal{D}$.

6.8. 1. *Hint*. It is sufficient to verify that

$$\varphi_2(x)=\int_{-\infty}^{x}\varphi_1(x)\,dx\in\mathcal{D}.$$

6.10. Sequences (1) and (3) converge to 0 in $\mathcal{S}$; seauence (2) does not converge in $\mathcal{S}$ if $\varphi(x)\not\equiv 0$.

6.19. 1. $\delta(x)$. 2. $\delta(x)$. 3. $\delta(x)$. 4. $\pi\delta(x)$. 5. $\delta(x)$.

6.21. 1. $2\pi i\delta(x)$. 2. 0. 3. 0. 4. $-2\pi i\delta(x)$. 5. 0.

6.22. 0.

7 Differentiation of Generalized Functions

The *derivative* of a generalized function f from $\mathcal{D}'(R^1)$ is a functional f' defined by the formula $(f',\varphi)=-(f,\varphi')$, $\varphi\in\mathcal{D}(R^1)$. Each generalized function has derivatives of all orders, with $f^{(m)}$, $m\geqslant 1$, being a functional that operates according to the formula

$$(f^{(m)},\varphi)=(-1)^m(f,\varphi^{(m)}). \tag{7.1}$$

For $n>1$, formula (7.1), which defines the derivative $D^\alpha f$, takes the form

$$(D^\alpha f,\varphi)=(-1)^{|\alpha|}(f,D^\alpha\varphi),\quad \varphi\in\mathcal{D}(R^n).$$

Suppose S is a piecewise smooth two-sided surface, $\mathbf{n}$ is a unit vector normal to S, and $\nu(x)$ a continuous function on S. The generalized function $-\frac{\partial}{\partial\mathbf{n}}(\nu\delta_S)$, which operates via the rule

$$\left(-\frac{\partial}{\partial\mathbf{n}}(\nu\delta_S),\varphi\right)=\int_S\nu(x)\frac{\partial\varphi(x)}{\partial\mathbf{n}}\,dS,\qquad \varphi\in\mathcal{D}(R^n),$$

is called a *double layer* on S. In particular, if S is the plane $t=0$ in the space R^{n+1} of the variables $(x,t)=(x_1,x_2,\ldots,x_n,t)$, then we denote $-\frac{\partial}{\partial \mathbf{n}}(\nu\delta_{(t=0)}(x, t))$ by $-\nu(x)\delta'(t)$ and have

$$(-\nu(x)\delta'(t), \varphi) = \int_{R^n} \nu(x)\frac{\partial\varphi(x, 0)}{\partial t}\,dx.$$

Suppose the function $f(x)$ that is locally integrable in R^n is such that its classical derivative of order $\alpha = (\alpha_1, \alpha_2, \ldots, \alpha_n)$ is a piecewise continuous function in R^n. The regular generalized function defined by this derivative is denoted by $\{D^\alpha f(x)\}$ (in contrast to the generalized derivative $D^\alpha f(x)$).

7.1. Provide a physical interpretation for the following generalized functions:

in R^1: $-\delta'(x)$, $-\delta'(x-x^0)$;

in R^3: $-\frac{\partial}{\partial \mathbf{n}}(\nu\delta_S)$, $-2\frac{\partial}{\partial \mathbf{n}}\delta_{S_R}(x-x^0)$.

7.2. Show that $(\delta^{(m)}(x-x_0), \varphi(x)) = (-1)^m\varphi^{(m)}(x_0)$, $m \geqslant 1$.

7.3. Show that in $\mathscr{D}'(R^1)$,

1. $\rho(x)\delta'(x) = -\rho'(0)\delta(x) + \rho(0)\delta'(x)$, where $\rho(x) \in C^1(R^1)$.
2. $x\delta^{(m)}(x) = -m\delta^{(m-1)}(x)$, $m = 1, 2, \ldots$.
3. $x^m\delta^{(m)}(x) = (-1)^m m!\,\delta(x)$, $m = 0, 1, 2, \ldots$.
4. $x^k\delta^{(m)}(x) = 0$, $m = 0, 1, \ldots, k-1$.
5. $\alpha(x)\delta^{(m)}(x) = \sum_{j=0}^{m}(-1)^{j+m}C_m^j\alpha^{(m-j)}(0)\delta^{(j)}(x)$,
where $\alpha(x) \in C^\infty(R^1)$.
6. $x^k\delta^{(m)}(x) = (-1)^k k!\,C_m^k\delta^{(m-k)}(x)$, $m = k, k+1, \ldots$.

7.4. Show that $\theta' = \delta$, where θ is Heaviside's unit function.

7.5. 1. Show that

$$(\theta(x)\rho(x))' = \delta(x)\rho(0) + \theta(x)\rho'(x)$$

in $\mathscr{D}'(R^1)$, with $\rho(x) \in C^1(R^1)$.

2. Show that

$$\frac{\partial}{\partial t}(\theta(t)\rho(x, t)) = \delta(t)\rho(x, 0) + \theta(t)\frac{\partial\rho(x, t)}{\partial t}$$

in $\mathscr{D}'(R^2)$, with $\rho \in C^1(t \geqslant 0)$.

Hint. Use the definition of a simple layer given in Sec. 6.

7.6. Calculate
1. $\theta'(-x)$. 2. $\theta^{(m)}(x-x_0)$, with m a positive integer.
3. $\theta^{(m)}(x_0-x)$, $m \geqslant 1$. 4. $(\operatorname{sign} x)^{(m)}$, $m \geqslant 1$. 5. $(x \operatorname{sign} x)'$.
6. $(|x|)^{(m)}$, $m \geqslant 2$. 7. $(\theta(x)\sin x)'$.
8. $(\theta(x)\cos x)'$. 9. $(\theta(x)x^{m+k})^{(m)}$, $m \geqslant 1$, $k = 0, 1, 2, \ldots$.
10. $(\theta(x)x^{m-k})^{(m)}$, $m \geqslant 1$, $k = 1, \ldots, m$.
11. $(\theta(x)e^{ax})^{(m)}$, $m \geqslant 1$.

7.7. Calculate the first-, second-, and third-order derivatives of the following functions:
1. $y = |x| \sin x$. 2. $y = |x| \cos x$.

7.8. Show that

$$(D^\alpha f)(x+h) = D^\alpha f(x+h), \; f \in \mathscr{D}', \; h \in R^n.$$

7.9. Prove that the generalized functions $\delta, \delta', \delta'', \ldots, \delta^{(m)}$ constitute a linearly independent system.

7.10. Prove that

1. $\frac{d}{dx} \ln |x| = \mathscr{P}\frac{1}{x}$, where $\mathscr{P}\frac{1}{x}$ is defined in Problem 6.18.
2. $\frac{d}{dx} \mathscr{P}\frac{1}{x} = -\mathscr{P}\frac{1}{x^2}$, where $\mathscr{P}\frac{1}{x^2}$ is defined in Problem 6.25
3. $\frac{d}{dx} \frac{1}{x \pm i0} \mp \pi\delta'(x) - \mathscr{P}\frac{1}{x^2}$. 4. $\frac{d}{dx} \mathscr{P}\frac{1}{x^2} = -2\mathscr{P}\frac{1}{x^3}$ where

$$\left(\mathscr{P}\frac{1}{x^3}, \varphi\right) = \mathrm{PV} \int_{-\infty}^{\infty} \frac{\varphi(x) - x\varphi'(0)}{x^3}\, dx, \qquad \varphi \in \mathscr{D}(R^1).$$

7.11. Show that the series $\sum_{-\infty}^{\infty} a_k \delta^{(k)}(x-k)$ is convergent in $\mathscr{D}'(R^1)$ for all values of a_k.

7.12. Show that if $|a_k| \leqslant A|k|^m + B$, then the series $\sum_{-\infty}^{\infty} a_k e^{ikx}$ is convergent in $\mathscr{S}'(R^1)$.

7.13. Let $f(x)$ be a piecewise smooth function such that

$$f \in C^1(x \leqslant x_0) \cap C^1(x \geqslant x_0).$$

Prove that

$$f' = \{f'(x)\} + [f]_{x_0}\delta(x - x_0) \text{ in } \mathscr{D}'(R^1), \tag{7.2}$$

where $[f]_{x_0} = f(x_0+0) - f(x_0-0)$ is the jump of f at point x_0. Also prove that if the classical derivative of $f(x)$ has jump discontinuities at points $\{x_k\}$, then formula (7.2) takes the form

$$f' = \{f'(x)\} + \sum_k [f]_{x_k}\delta(x - x_k).$$

7.14. Calculate $f^{(m)}$ for the following functions:
1. $\theta(a - |x|)$, $a > 0$. 2. $[x]$. 3. sign sin x. 4. sign cos x.
Here $[x]$ stands for the integral part of x, that is, the greatest integer that does not exceed x.

7.15. Let $f(x)$ be a 2π-periodic function, with $f(x) = \frac{1}{2} - \frac{x}{2\pi}$, $0 < x \leqslant 2\pi$. Find f'.

7.16. Let $f(x) = x$, $-1 < x \leqslant 1$, be a 2-periodic function. Find $f^{(m)}$, $m \geqslant 1$.

7.17. Prove that

$$\frac{1}{2\pi}\sum_{k=-\infty}^{\infty} e^{ikx} = \sum_{k=-\infty}^{\infty} \delta(x-2k\pi).$$

7.18. Prove that

$$\frac{2}{\pi}\sum_{k=0}^{\infty} \cos(2k+1)x = \sum_{k=-\infty}^{\infty} (-1)^k \delta(x-k\pi).$$

7.19. Suppose $f(x) \in C^\infty(x \leqslant x_0) \cap C^\infty(x \geqslant x_0)$. Prove that

$$f^{(m)}(x) = \{f^{(m)}(x)\} + [f]_{x_0}\delta^{(m-1)}(x-x_0)$$
$$+ [f']_{x_0}\delta^{(m-2)}(x-x_0) + \ldots + [f^{(m-1)}]_{x_0}\delta(x-x_0),$$

in $\mathscr{D}'(R^1)$, with $[f^{(k)}]_{x_0} = f^{(k)}(x_0+0) - f^{(k)}(x_0-0)$, $k = 0, 1, \ldots, m-1$, the jump of the kth derivative at point x_0.

7.20. Find all the derivatives of the following functions:

1. $y = \begin{cases} \sin x, & x \geqslant 0, \\ 0, & x \leqslant 0. \end{cases}$ 2. $y = \begin{cases} \cos x, & x \geqslant 0, \\ 0, & x < 0. \end{cases}$

3. $y = \begin{cases} x^2, & -1 \leqslant x \leqslant 1, \\ 0, & |x| > 1. \end{cases}$ 4. $y = \begin{cases} 1 & x \leqslant 0, \\ x \quad 1, & 0 \leqslant x \leqslant 1, \\ x^2+1, & x \geqslant 1. \end{cases}$

5. $y = \begin{cases} 0 & x \leqslant -1, \\ (x+1)^2, & -1 \leqslant x \leqslant 0, \\ x^2+1 & x \geqslant 0. \end{cases}$ 6. $y = \begin{cases} 0, & x \leqslant 0, \\ x^2 & 0 \leqslant x \leqslant 1, \\ (x-2)^2 & 1 \leqslant x \leqslant 2, \\ 0, & x \geqslant 2. \end{cases}$

7. $y = \begin{cases} \sin x, & -\pi \leqslant x \leqslant \pi, \\ 0 & |x| \geqslant \pi. \end{cases}$

8. $y = \begin{cases} |\sin x|, & -\pi \leqslant x \leqslant \pi, \\ 0, & |x| \geqslant \pi. \end{cases}$

7.21. Prove that

1. $|\sin x|'' + |\sin x| = 2\sum_{k=-\infty}^{\infty} \delta(x-k\pi).$

2. $|\cos x|'' + |\cos x| = 2\sum_{k=-\infty}^{\infty} \delta\left(x - \frac{2k+1}{2}\pi\right).$

Hint. Use the results of Problem 7.14 (functions (3) and (4)).

Let

$$\sum_{k=0}^{m} a_k(x)\, y^{(k)} = f \tag{7.3}$$

be a linear differential equation of order m with coefficients $a_k(x) \in C^\infty(R^1)$ and with $f \in \mathscr{D}'(R^1)$. The *generalized solution* of this equation is any generalized function $y \in \mathscr{D}'(R^1)$ that satisfies Eq. (7.3) in the generalized sense, or

$$\left(\sum_{k=0}^{m} a_k y^{(k)},\ \varphi\right) = \left(y,\ \sum_{k=0}^{m} (-1)^k (a_k\varphi)^{(k)}\right) = (f,\ \varphi)$$

for every $\varphi \in \mathscr{D}(R^1)$. (Sometimes for the sake of brevity we say that the solution satisfies the equation in $\mathscr{D}'$ instead of saying that the solution satisfies the equation in the generalized sense.) Every solution of Eq. (7.3) can be written in the form of the sum of a partial solution of this equation and the general solution of the corresponding homogeneous equation.

7.22. Find the general solutions in $\mathscr{D}'(R^1)$ to the following equations: 1. $xy = 0$. 2. $\alpha(x)\, y = 0$, where $\alpha \in C^\infty(R')$ and has a first-order (unique) zero at **point** $x = 0$. 3. $\alpha(x)\, y = 0$, where $\alpha \in C$ and $\alpha > 0$. 4. $(x-1)\, y = 0$. 5. $x(x-1)\, y = 0$. 6. $(x^2-1)\, y = 0$. 7. $xy = 1$. 8. $xy = \mathscr{P}\frac{1}{x}$. 9. $x^n y = 0$, $n = 2, 3, \ldots$ 10. $x^2 y = 2$. 11. $(x+1)^2 y = 0$. 12. $(\cos x)\, y = 0$.

7.23. Find the general solutions in $\mathscr{D}'(R^1)$ to the following equations:

1. $y' = 0$. 2. $y^{(m)} = 0$, $m = 2, 3, \ldots$

7.24. Prove that the generalized function

$$y = \sum_{k=0}^{m-1} a_k \theta(x)\, x^{m-k-1} + \sum_{k=m}^{n-1} b_k \delta^{(k-m)}(x) + \sum_{k=0}^{m-1} c_k x^k,$$

with a_k, b_k, and c_k arbitrary functions, is the general solution in $\mathscr{D}'(R^1)$ to the equation $x^n y^{(m)} = 0$, $n > m$.

7.25. Find the general solutions in $\mathscr{D}'(R^1)$ to the following equations:

1. $xy' = 1$. 2. $xy' = \mathscr{P}\frac{1}{x}$. 3. $x^2 y' = 0$.
4. $x^2 y' = 1$. 5. $y'' = \delta(x)$, 6. $(x+1)\, y'' = 0$.
7. $(x+1)^2\, y'' = 0$. 8. $(x+1)\, y''' = 0$.

7.26. Prove that the generalized function $C\delta(x) + \mathscr{P}\frac{1}{|x|}$, with

$$\left(\mathscr{P}\frac{1}{|x|},\ \varphi\right) = \int\limits_{|x|<1} \frac{\varphi(x)-\varphi(0)}{|x|}\, dx + \int\limits_{|x|>1} \frac{\varphi(x)}{|x|}\, dx,$$

is the general solution in $\mathscr{D}'(R^1)$ to the equation $xy = \operatorname{sign} x$.

7.27. Prove that if $f \in \mathscr{D}'(R^1)$ is invariant under a translation, that is, $(f, \varphi) = (f(x), \varphi(x+h))$, where h is an arbitrary real number, then $f = \text{const}$. *Hint.* Prove that $f' = 0$ and employ the result of Problem 7.23.1.

7.28. Find the solution in $\mathscr{D}'(R^1)$ to the equation

$$af'' + bf' + cf = m\delta + n\delta'.$$

where a, b, c, m, and n are given numbers. Consider the following cases:

1. $a = c = n = 1,\ b = m = 2.$
2. $b = n = 0,\ a = m = 1,\ c = 4.$
3. $b = 0,\ a = n = 1,\ m = 2,\ c = -4.$

7.29. Prove that the system of equations $dy/dx = A(x)\,y$, where matrix $A(x) \in C^\infty(R^1)$, has in $\mathscr{D}'$ only a classical solution.

7.30. Prove that the equation $u' = f$ has a solution in $\mathscr{D}'(R^1)$ for every $f \in \mathscr{D}'(R^1)$. *Hint.* Employ the result of Problem 6.8.2.

7.31. Prove that the equation $xu = f$ has a solution in $\mathscr{D}'(R^1)$ for every $f \in \mathscr{D}'(R^1)$. *Hint.* Employ the result of Problem 6.7.1.

7.32. Prove that the equation $x^3u' + 2u = 0$ has no solutions in $\mathscr{D}'(R^1)$ (except zero).

7.33. Suppose $\theta(x_1, x_2, \ldots, x_n) = \theta(x_1)\ldots\theta(x_n)$. Show that

$$\frac{\partial^n\theta}{\partial x_1\,\partial x_2\ldots\partial x_n} = \delta(x) = \delta(x_1, \ldots, x_n)$$

in $\mathscr{D}'(R^n)$.

7.34. In the (x, y) plane consider the square with its vertices at points

$$A(1, 1),\ B(2, 0),\ C(3, 1),\ D(2, 2).$$

Suppose a function f is equal to unity inside $ABCD$ and to zero outside. Calculate $f''_{yy} - f''_{xx}$.

7.35. Suppose region $G \subset R^3$ is bounded by a piecewise smooth surface S and a function is specified thus: $f \in C^1(\overline{G}) \cap C^1(\overline{G}_1)$, where $G_1 = R^n \setminus \overline{G}$. Prove the validity of the formula

$$\frac{\partial f}{\partial x_i} = \left\{\frac{\partial f}{\partial x_i}\right\} + [f]_S \cos(\mathbf{n}, x_i)\,\delta_S, \quad i = 1, 2, 3,$$

in $\mathscr{D}'(R^3)$, where $\mathbf{n} = \mathbf{n}_x$ is the outward unit vector normal to S at point $x \in S$, and $[f]_S$ is the jump of $f(x)$ in the passage through surface S inward, that is,

$$\lim_{x' \to x,\ x' \in G_1} f(x') - \lim_{x' \to x,\ x' \in G} f(x') = [f]_S(x), \quad x \in S.$$

7.36. Prove that if $f \in C^2(\overline{G}) \cap C^2(\overline{G}_1)$, with $G_1 = R^n \setminus \overline{G}$, Green's formula is valid, that is,

$$\nabla^2 f = \{\nabla^2 f\} + \left[\frac{\partial f}{\partial \mathbf{n}}\right]_S \delta_S + \frac{\partial}{\partial \mathbf{n}}([f]_S\,\delta_S).$$

7.37. Prove that if $f(x, t) \in C^2$ $(t \geqslant 0)$ and $f = 0$ for $t < 0$, then the following formulas are valid in R^{n+1}:

1. $\Box_a f = \{\Box_a f\} + \delta(t) f_t(x, 0) + \delta'(t) f(x, 0)$.
2. $\frac{\partial f}{\partial t} - a^2 \nabla^2 f = \left\{\frac{\partial f}{\partial t} - a^2 \nabla^2 f\right\} + \delta(t) f(x, 0)$.

Answers to Problems of Sec. 7

7.6. 1. $-\delta(x)$. 2. $\delta^{(m-1)}(x - x_0)$. 3. $-\delta^{(m-1)}(x - x_0)$. 4. $2\delta^{(m-1)}(x)$. 5. $\operatorname{sign} x$. 6. $2\delta^{(m-2)}(x)$. 7. $\theta(x)\cos x$. 8. $\delta(x) - \theta(x)\sin x$. 9. $\frac{(m+k)!}{k!}\theta(x)x^k$. 10. $(m-k)!\,\delta^{(k-1)}(x)$. 11. $\delta^{(m-1)}(x) + a\delta^{(m-2)}(x) + \ldots + a^{m-1}\delta(x) + a^m\theta(x)e^{ax}$.

7.7. 1. $y' = \operatorname{sign} x \sin x + |x| \cos x$, $y'' = 2 \operatorname{sign} x \cos x - |x| \sin x$, $y''' = 4\delta(x) - 3 \operatorname{sign} x \sin x - |x| \cos x$. 2. $y' = \operatorname{sign} x \cos x - |x| \sin x$, $y'' = 2\delta(x) - 2 \operatorname{sign} x \sin x - |x| \cos x$, $y''' = 2\delta'(x) - 3 \operatorname{sign} x \cos x + |x| \sin x$.

7.10. 2. *Solution.*

$$\left(\left(\mathscr{P}\frac{1}{x}\right)', \varphi\right) = -\left(\mathscr{P}\frac{1}{x}, \varphi'\right) = -\mathrm{PV}\int \frac{\varphi'(x)}{x}\,dx = -\lim_{\varepsilon\to 0}\left(\int_{-\infty}^{-\varepsilon} + \int_{\varepsilon}^{\infty}\right)\frac{\varphi'(x)}{x}\,dx = \lim_{\varepsilon\to 0}\frac{\varphi(\varepsilon)+\varphi(-\varepsilon)}{\varepsilon} - \lim_{\varepsilon\to 0}\varphi(0)\left(\int_{-\infty}^{-\varepsilon} + \int_{\varepsilon}^{\infty}\right)\frac{dx}{x^2} - \lim_{\varepsilon\to 0}\left(\int_{-\infty}^{-\varepsilon} + \int_{\varepsilon}^{\infty}\right)\frac{\varphi(x)-\varphi(0)}{x^2}\,dx = \lim_{\varepsilon\to 0}\left(\frac{\varphi(\varepsilon)-\varphi(0)}{\varepsilon} - \frac{\varphi(-\varepsilon)-\varphi(0)}{-\varepsilon}\right) - \left(\mathscr{P}\frac{1}{x^2}, \varphi\right) = \left(-\mathscr{P}\frac{1}{x^2}, \varphi\right), \quad \varphi \in \mathscr{D}.$$

7.14. 1. $\delta^{(m-1)}(x+a) - \delta^{(m-1)}(x-a)$. 2. $\sum_{k=-\infty}^{\infty} \delta^{(m-1)}(x-k)$.

3. $2\sum_{k=-\infty}^{\infty}(-1)^k\delta^{(m-1)}(x - k\pi)$. 4. $2\sum_{k=-\infty}^{\infty}(-1)^{k+1}\delta^{(m-1)}\left(x - (2k+1)\frac{\pi}{2}\right)$.

7.15. $f' = -\frac{1}{2\pi} + \sum_{k=-\infty}^{\infty}\delta(x - 2k\pi)$.

7.16. $f' = 1 - 2\sum_{k=-\infty}^{\infty}\delta(x - 2k - 1)$,

$$f^{(m)} = -2\sum_{k=-\infty}^{\infty}\delta^{(m-1)}(x - 2k - 1), \qquad m = 2, 3, \ldots.$$

7.17. *Hint.* Employ the result of Problem 7.15.

7.18. *Hint.* Employ the result of Problem 7.17.

7.20. *Hint.* Employ the results of Problems 7.13 and 7.19.

1. $y' = \theta(x)\cos x$, $y^{(m)} = \sum_{k=1}^{[m/2]} (-1)^{k-1}\delta^{(m-2k)}(x) + \theta(x)(\sin x)^{(m)}$, $m = 2, 3, \ldots$, where $[m/2]$ is the integral part of $m/2$. 2. $y' = \delta(x) - \theta(x)\sin x$, $y^{(m)} = \sum_{k=1}^{\left[\frac{m+1}{2}\right]} (-1)^{k-1}\delta^{(m-2k+1)}(x) + \theta(x)(\cos x)^{(m)}$, $m = 2, 3, \ldots$. 3. $y' = 2\theta(1-|x|)x + \delta(x-1) - \delta(x+1)$, $y'' = 2\theta(1-|x|) - 2\delta(x+1) - 2\delta(x-1) + \delta'(x+1) - \delta'(x-1)$,

$$y^{(m)} = \sum_{k=1}^{3} \frac{2}{(3-k)!}\left[(-1)^{k-}\delta^{(m-k)}(x+1) - \delta^{(m-k)}(x-1)\right], \quad m = 3, 4, \ldots.$$

4. $y' = \theta(x) - \theta(x-1) + 2\theta(x-1)x$, $y'' = \delta(x) + \delta(x-1) + 2\theta(x-1)$, $y^{(m)} = 2\delta^{(m-3)}(x-1) + \delta^{(m-2)}(x-1) + \delta^{(m-2)}(x)$, $m = 3, 4, \ldots$. 5. $y' = 2\theta(x+1)(x+1) - 2\theta(x)$, $y'' = -2\delta(x) + 2\theta(x+1)$, $y^{(m)} = -2\delta^{(m-2)}(x) + 2\delta^{(m-3)}(x+1)$, $m = 3, 4, \ldots$. 6. $y' = 2\theta(x)x - 4\theta(x-1) - 2\theta(x-2)(x-2)$, $y'' = 2\theta(x) - 2\theta(x-2) - 4\delta(x-1)$, $y^{(m)} = 2\delta^{(m-3)}(x) - 2\delta^{(m-3)}(x-2) - 4\delta^{(m-2)}(x-1)$, $m = 3, 4, \ldots$.

7. $y' = \theta(\pi - |x|)\cos x$, $y^{(m)} = \theta(\pi - |x|)(\sin x)^{(m)} + \sum_{k=1}^{\left[\frac{m}{2}\right]} (-1)^k \times \{\delta^{(m-2k)}(x+\pi) - \delta^{(m-2k)}(x-\pi)\}$, $m = 2, 3, \ldots$. 8. $y' = \theta(\pi - |x|)\operatorname{sign} x \cos x$, $y^{(m)} = \theta(\pi - |x|)' \operatorname{sign} x \cdot \sin^{(m)} x - \sum_{k=1}^{[m/2]} (-1)^k \times \{2\delta^{(m-2k)}(x) + \delta^{(m-2k)}(x+\pi) + \delta^{(m-2k)}(x-\pi)\}$, $m = 2, 3, \ldots$.

7.22. 1. *Solution.* Suppose solution $y \in \mathcal{D}'$ exists. Then

$$(y, x\varphi) = 0 \text{ for every } \varphi \in \mathcal{D}. \tag{7.4}$$

Let us find this solution. We have $(y, \varphi) = (y, \varphi(0)\eta(x) + \varphi(x) - \varphi(0)\eta(x))$, where $\eta \in \mathcal{D}$, with $\eta(x) \equiv 1$ in $[-\varepsilon, \varepsilon]$ and $\eta(x) \equiv 0$ outside $[-3\varepsilon, 3\varepsilon]$,

$$(y, \varphi) = \varphi(0), \ (y, \eta(x)) + \left(y, \ x\,\frac{\varphi(x) - \varphi(0)\eta(x)}{x}\right) = \varphi(0)\,C + (y, x\psi(x)), \tag{7.5}$$

where $C = (y, \eta)$ and $\psi(x) = \frac{\varphi(x) - \varphi(0)\eta(x)}{x} \in \mathcal{D}$ (see the solution to Problem 6.7). In view of (7.4) we can write $(y, x\psi) = 0$. Then from (7.5) we obtain $(y, \varphi) = (C\delta, \varphi)$ for every $\varphi \in \mathcal{D}$, that is, $y = C\delta(x)$. Now one only has to note that $C\delta(x)$ satisfies the equation $xy = 0$. 2. $C\delta(x)$. *Hint.* Employ the result of Problem 6.7.2. 3. 0. 4. $C\delta(x-1)$. 5. $C_1\delta(x) + C_2\delta(x-1)$. 6. $C_1\delta(x-1) + C_2\delta(x+1)$. 7. $C\delta(x) + \mathcal{P}\frac{1}{x}$. 8. $C\delta(x) + \mathcal{P}\frac{1}{x^2}$. 9. $\sum_{k=0}^{m-1} C_k\delta^{(k)}(x)$.

Hint. Reduce the problem to solving the equation of the form $xz(x) = f(x)$ by introducing the notation $x^{m-1}y(x) = z(x)$, $x^{m-2}y(x) = z(x)$, etc., and use the result of Problem 7.22.1. 10. $C_0\delta(x) + C_1\delta'(x) + 2\mathcal{P}\frac{1}{x^2}$, where $\mathcal{P}\frac{1}{x^2}$ is defined in Problem 6.25.

11. $C_0\delta(x+1) + C_1\delta'(x+1)$. 12. $\sum_{k=-\infty}^{\infty} C_k\delta\left(x - \frac{\pi}{2} - k\pi\right)$

7.23. 1. *Solution*. Suppose a solution $y \in \mathcal{D}'$ exists, that is,

$$(y, \varphi') = 0 \text{ for every } \varphi \in \mathcal{D}. \qquad (7.6)$$

In view of the result of Problem 6.8.2, we can represent any function $\varphi \in \mathcal{D}$ in the form

$$\varphi(x) = \varphi_0(x) \int_{-\infty}^{\infty} \varphi(x)\,dx + \varphi_1'(x),$$

where $\varphi_1 \in \mathcal{D}$, and $\varphi_0(x)$ is any test function that belongs to $\mathcal{D}$ and satisfies the condition $\int_{-\infty}^{\infty} \varphi_0(x)\,dx = 1$. Hence,

$$(y, \varphi) = \left(y, \varphi_0 \int_{-\infty}^{\infty} \varphi\,dx + \varphi_1'\right) = (y, \varphi_0) \int_{-\infty}^{\infty} \varphi\,dx + (y, \varphi_1').$$

Since, in view of (7.6), $(y, \varphi_1') = 0$ and $(y, \varphi_0) = C$, we can write

$$(y, \varphi) = C \int_{-\infty}^{\infty} \varphi\,dx = (C, \varphi) \text{ for every } \varphi \in \mathcal{D},$$

that is, $y = C$. 2. $C_0 + C_1x + \ldots + C_{m-1}x^{m-1}$. *Hint*. Reduce the problem to solving the equation $z' = f(x)$ by introducing the notation $y^{(m-1)} = z$, $y^{(m-2)} = z$, etc., and use the result of Problem 7.23.1.

7.25. 1. $C_1 + C_2\theta(x) + \ln|x|$. 2. $C_1 + C_2\theta(x) - \mathcal{P}\frac{1}{x}$. 3. $C_1 + C_2\theta(x) + C_3\delta(x)$. 4. $C_1 + C_2\theta(x) + C_3\delta(x) - \mathcal{P}\frac{1}{x}$. 5. $C_0 + C_1x + \theta(x)x$. 6. $C_0 + C_1x + C_2\theta(x+1)(x+1)$. 7. $C_0 + C_1x + C_2\theta(x+1) + C_3\theta(x+1)(x+1)$. 8. $C_0 + C_1x + C_2x^2 + C_3\theta(x+1)(x+1)^2$.

7.28. 1. $\theta(x)e^{-x}(1+x)$. 2. $\frac{1}{2}\theta(x)\sin 2x$. 3. $\theta(x)e^{2x}$. *Hint*. Look for the solution in the form $\theta(x)z(x)$, where $z \in C^2(R^1)$ is the sought function.

7.34. $-2\delta(x-1, y-1) + 2\delta(x-2, y) + 2\delta(x-3, y-1) - 2\delta(x-2, y-2)$.

8 The Direct Product and Convolution of Generalized Functions

The *direct product* of generalized functions $f(x) \in \mathcal{D}'(R^n)$ and $g(y) \in \mathcal{D}'(R^m)$ is a generalized function $f(x)\cdot g(y)$ from $\mathcal{D}'(R^{n+m})$ defined by the formula

$$(f(x)\cdot g(y),\ \varphi(x, y)) = (f(x),\ (g(y),\ \varphi(x, y))),\quad \varphi \in \mathcal{D}(R^{n+m}). \tag{8.1}$$

The direct product operation is commutative, that is,

$$f(x)\cdot g(y) = g(y)\cdot f(x),$$

and associative, that is,

$$[f(x)\cdot g(y)]\cdot h(z) = f(x)\cdot[g(y)\cdot h(z)].$$

If $f \in \mathcal{S}'(R^n)$ and $g \in \mathcal{S}'(R^m)$, then $f(x)\cdot g(y)$ is given by (8.1), with $\varphi \in \mathcal{S}(R^{m+n})$, and belongs to $\mathcal{S}'(R^{n+m})$.

The derivative of a direct product possesses the following property:

$$D_x^\alpha (f(x)\cdot g(y)) = D^\alpha f(x)\cdot g(y);$$
$$D_y^\alpha (f(x)\cdot g(y)) = f(x)\cdot D^\alpha g(y). \tag{8.2}$$

If $\mu(x) \in \mathcal{D}'(R^n)$ and $\nu(x) \in \mathcal{D}'(R^n)$, then the generalized functions $\mu(x)\cdot\delta(t)$ and $-\nu(x)\cdot\delta'(t)$ are called the *simple and double layers* on surface $t = 0$ with *densities* $\mu(x)$ and $\nu(x)$, respectively. When the densities are continuous functions, these definitions coincide with those given in Secs. 6 and 7: $\mu(x)\cdot\delta(t) = \mu(x)\,\delta(t)$ and $-\nu(x)\cdot\delta'(t) = -\nu(x)\,\delta'(t)$.

The generalized function $\delta(at - |x|)$, $a > 0$, from $\mathcal{D}'(R^2)$ is defined thus:

$$\delta(at - |x|) = \theta(t)\,\delta(at + x) + \theta(t)\,\delta(at - x), \tag{8.3}$$

where the generalized functions $\theta(t)\,\delta(at + x)$ and $\theta(t)\,\delta(at - x)$ are obtained via linear substitutions $t' = t$ and $\xi = at \pm x$ in $\theta(t')\,\delta(\xi)$, that is,

$$(\theta(t)\,\delta(at+x),\ \varphi) = \int_0^\infty \varphi(-at',\ t')\,dt' \tag{8.3$_1$}$$

and

$$(\theta(t)\,\delta(at-x),\ \varphi) = \int_0^\infty \varphi(at',\ t')\,dt'.$$

8.1. Prove that $\operatorname{supp}(f(x)\cdot g(y)) = \operatorname{supp} f \times \operatorname{supp} g$.

8.2. Prove that in $\mathcal{D}'(R^{n+1}(x, t))$:

1. $(u_1(x)\,\delta(t),\ \varphi) = (u_1(x),\ \varphi(x, 0))$.
2. $(u_0(x)\cdot\delta'(t),\ \varphi) = -\left(u_0(x),\ \frac{\partial\varphi(x, 0)}{\partial t}\right)$

Hint. Use formula (8.1).

8.3. Prove that

1. $\theta_t(x, t)$ is a simple layer on the $t = 0$ axis of the (x, t) plane with a density $\theta(x)$.

2. $-\theta_{tt}(x, t)$ is a double layer on the $t = 0$ axis with a density $\theta(x)$.

Hint. Employ the result of Problem 8.2.

8.4. Show that the following formulas are valid:

1. $\theta(x_1)\cdot\theta(x_2)\cdot\ldots\cdot\theta(x_n) = \theta(x_1, x_2, \ldots, x_n)$.
2. $\delta(x_1)\cdot\delta(x_2)\cdot\ldots\cdot\delta(x_n) = \delta(x_1, x_2, \ldots, x_n)$.

8.5. Show that

$$\frac{\partial^n\theta(x_1, \ldots, x_n)}{\partial x_1\,\partial x_2\ldots\partial x_n} = \delta(x_1)\cdot\delta(x_2)\cdot\ldots\cdot\delta(x_n).$$

8.6. Show that $(f\cdot g)(x + x_0, y) = f(x + x_0)\cdot g(y)$.

8.7. Show that $a(x)(f(x)\cdot g(y)) = a(x)f(x)\cdot g(y)$, with $a \in C^\infty(R^n)$.

8.8. Prove that in $\mathscr{D}'(R^2)$,

1. $\frac{\partial}{\partial t}\theta(at - |x|) = a\delta(at - |x|)$.

2. $\frac{\partial}{\partial x}\theta(at - |x|) = \theta(t)\,\delta(at + x) - \theta(t)\,\delta(at - x)$.

3. $\left(\frac{\partial^2}{\partial t^2}\theta(at - |x|), \varphi\right) = -a\left(\delta(at - |x|), \frac{\partial\varphi}{\partial t}\right)$.

4. $\left(\frac{\partial^2}{\partial x^2}\theta(at - |x|), \varphi\right) = -\left(\theta(t)\,\delta(at + x), \frac{\partial\varphi}{\partial x}\right) + $
$+\left(\theta(t)\,\delta(at - x), \frac{\partial\varphi}{\partial x}\right)$.

It is said that a generalized function of the form $f(x)\cdot 1(y)$ *does not depend on* y. It operates according to the following rule:

$$(f(x)\cdot 1(y), \varphi) = \int (f(x), \varphi(x, y))\,dy. \qquad (8.4)$$

8.9. Show that

1. $\int (f(x), \varphi(x, y))\,dy = \left(f(x), \int \varphi(x, y)\,dy\right)$.
2. $D_y^\alpha(f(x)\cdot 1(y)) = 0$, where $f \in \mathscr{D}'$ and $|\alpha| \neq 0$.

8.10. Suppose $g(y) \in \mathscr{S}'(R^m)$ and $\varphi \in \mathscr{S}(R^{n+m})$. Prove that

(1) $\psi(x) = (g(y), \varphi(x + y)) \in \mathscr{S}(R^n)$,

(2) $D^\alpha\psi(x) = (g(y), D_x^\alpha\varphi(x, y))$,

(3) if $\varphi_k \to \varphi$ as $k \to \infty$ in $\mathscr{S}(R^{n+m})$, then $\psi_k \to \psi$ as $k \to \infty$ in $\mathscr{S}(R^n)$,

(4) if $f \in \mathscr{S}'(R^n)$ and $g \in \mathscr{S}'(R^m)$, then $f(x)\cdot g(y) \in \mathscr{S}'(R^{n+m})$.

Let $f(x)$ and $g(x)$ be functions that are locally integrable in R^n, with the function

$$h(x) = \int |f(y)\,g(x - y)|\,dy$$

being also locally integrable in R^n. The function

$$(f*g)(x)=\int f(y)\,g(x-y)\,dy=\int g(y)\,f(x-y)\,dy=(g*f)(x)$$

is called the *convolution* $f*g$ of the two functions f and g.

We will say that a sequence $\{\eta_k(x)\}$ of functions belonging to $\mathcal{D}(R^n)$ *converges* to 1 in R^n if (a) for every ball U_R there is a positive integer N such that $\eta_k(x)=1$ for all $x\in U_R$ and $k\geqslant N$, and (b) the η_k are uniformly bounded in R^n together with all their derivatives, that is,

$$|D^\alpha\eta_k(x)|\leqslant C_\alpha,\ x\in R^n,\ k=1,2,\ldots,$$

with α arbitrary.

Let $\{\eta_k(x;y)\}$ be any sequence of functions from $\mathcal{D}(R^{2n})$ that converges to 1 in R^{2n}. Suppose the generalized functions $f(x)$ and $g(x)$ from $\mathcal{D}'(R^n)$ are such that for every $\varphi\in\mathcal{D}(R^n)$ the numerical sequence

$$(f(x)\cdot g(y),\quad \eta_k(x;y)\,\varphi(x+y))$$

tends to a limit as $k\to\infty$ and this limit does not depend on the choice of the sequence $\{\eta_k\}$. We denote this limit by

$$(f(x)\cdot g(y),\quad \varphi(x+y)).$$

The functional

$$\begin{aligned}(f*g,\varphi)&=(f(x)\cdot g(y),\ \varphi(x+y))\\&=\lim_{k\to\infty}(f(x)\cdot g(y),\ \eta_k(x;y)\,\varphi(x+y)),\quad \varphi\in\mathcal{D}(R^n),\end{aligned}\tag{8.5}$$

is the *convolution* $f*g$. The convolution operation possesses the property of *commutativity*, that is, $f*g=g*f$.

If the convolution $f*g$ exists, so do the convolutions $D^\alpha f*g$ and $f*D^\alpha g$, with

$$D^\alpha f*g=D^\alpha(f*g)=f*D^\alpha g.\tag{8.6}$$

The convolution is invariant under translations, that is,

$$f(x+h)*g(x)=(f*g)(x+h),\ h\in R^n.$$

Let us establish certain sufficient conditions for the existence of the convolution.

1. If f is an arbitrary generalized function from $\mathcal{D}'$ and g is a finite generalized function from $\mathcal{D}'$, then $f*g$ belongs to $\mathcal{D}'$ and can be represented in the form

$$(f*g,\varphi)=(f(x)\cdot g(y),\ \eta(y)\,\varphi(x+y)),\quad \varphi\in\mathcal{D},\tag{8.7}$$

where η is a test function equal to unity in the neighborhood supp g.

2. Let us denote the set of generalized functions from $\mathcal{D}'(R^1)$ that vanish when x is negative by $\mathcal{D}'$. If f and g belong to $\mathcal{D}'_+$, their convolution belongs to $\mathcal{D}'_+$, too, and is given by the formula

$$(f*g,\varphi)=(f(x)\cdot g(y),\quad \eta_1(x)\,\eta_2(y)\,\varphi(x+y)),\tag{8.8}$$

where

$$\eta_k(t) = \begin{cases} 1, & t \geqslant -\varepsilon_k \\ 0, & t < -2\varepsilon_k, \end{cases} \quad \eta_k \in C^\infty(R^1), \qquad k = 1, 2.$$

Thus, the set $\mathscr{D}'_+$ constitutes a convolution algebra.

8.11. Suppose $f(x)$ and $g(x)$ are locally integrable in R^n. Show that the convolution $f * g$ is a locally integrable function if

(1) f and g belong to $L_1(R^n)$,
(2) f or g is finite, or
(3) $f = 0$ and $g = 0$ at $x < 0$; $n = 1$.

In the case (1) show that $f * g$ belongs to $L_1(R^n)$ and that

$$\| f * g \|_{L_1} \leqslant \| f \|_{L_1} * \| g \|_{L_1}.$$

8.12. Show that under the hypothesis of Problem 8.10.3,

$$(f * g)(x) = \int_0^x f(y)\, g(x - y)\, dy. \tag{8.9}$$

8.13. Show that

1. $\delta * f = f * \delta = f$. 2. $\delta(x - a) * f(x) = f(x - a)$.
3. $\delta(x - a) * \delta(x - b) = \delta(x - a - b)$.
4. $\delta^{(m)} * f = f^{(m)}$. 5. $\delta^{(m)}(x - a) * f(x) = f^{(m)}(x - a)$.

8.14. Evaluate in $\mathscr{D}'(R^1)$ the following convolutions:

1. $\theta(x) * \theta(x)$. 2. $\theta(x) * \theta(x) x^2$. 3. $e^{-|x|} * e^{-|x|}$.
4. $e^{-ax^2} * xe^{-ax^2}$, $a > 0$. 5. $\theta(x)\, x^2 * \theta(x) \sin x$.
6. $\theta(x) \cos x * \theta(x)\, x^3$. 7. $\theta(x) \sin x * \theta(x) \sinh x$.
8. $\theta(a - |x|) * \theta(a - |x|)$.

In Problems 8.15-8.29 prove the validity of the propositions.

8.15. If $f_\alpha(x) = \theta(x) \dfrac{x^{\alpha - 1}}{\Gamma(\alpha)} e^{-\alpha x}$, where α is a positive integer, then $f_\alpha * f_\beta = f_{\alpha+\beta}$.

8.16. If $f_\alpha(x) = \dfrac{1}{\alpha \sqrt{2\pi}} e^{-\frac{x^2}{2\alpha^2}}$, $\alpha > 0$, then $f_\alpha * f_\beta = f_{\sqrt{\alpha^2+\beta^2}}$.

8.17. If $f_\alpha(x) = \dfrac{\alpha}{\pi(x^2 + \alpha^2)}$, $\alpha > 0$, then $f_\alpha * f_\beta = f_{\alpha+\beta}$.

8.18. $\operatorname{supp}(f * g) \subset [\operatorname{supp} f + \operatorname{supp} g]$. *Hint.* Use the result of Problem 8.1.

8.19. If f and g belong to $\mathscr{D}'_+$, then $e^{ax} f * e^{ax} g = e^{ax}(f * g)$.

8.20. If $f \in \mathscr{D}'$ and $\varphi \in \mathscr{D}$, then $f * \varphi = (f(y), \varphi(x - y)) \in C^\infty(R^1)$. *Hint.* Use formula (8.7) and the result of Problem 8.9.1.

8.21. If $f \in \mathscr{D}'$, $f * \varphi = 0$ for all $\varphi \in \mathscr{D}$, and $\operatorname{supp} \varphi \in [x < 0]$, then $f = 0$ for $x < 0$.

8.22. If the convolution $f * 1$ exists, it is a constant.

8.23. A generalized function is independent of x_i if and only if it is invariant under all translations in x_i.

8.24. A generalized function $f(x) \in \mathcal{D}'(R^n)$ is independent of x_i if and only if $\partial f/\partial x_i = 0$.

8.25. If a generalized function $f(x) \in \mathcal{D}'$ is independent of x_i, then $f * g$ is independent of x_i, too.

8.26. The solution in $\mathcal{D}'(R^1)$ of the equation $Lu = \delta$, where

$$L = \frac{d^m}{dx^m} + a_1(x)\frac{d^{m-1}}{dx^{m-1}} + \ldots + a_{m-1}(x)\frac{d}{dx} + a_m(x),$$

$a_k \in C^\infty(R^1)$, is the function $u(x) = \theta(x) Z(x)$, where $Z(x) \in \in C^m(R^1)$ is the solution of the boundary value problem

$$LZ = 0,\ Z(0) = Z'(0) = \ldots = Z^{(m-2)}(0) = 0,$$
$$Z^{(m-1)}(0) = 1.$$

8.27. The solution in $\mathcal{D}'_+$ of the equation $Lu = f$, $f \in \mathcal{D}'_+$, is the function $u = \theta Z * f$, where $Z(x)$ is defined in Problem 8.26.

8.28. The solution of Abel's equation

$$\int_0^x \frac{u(\xi)}{(x-\xi)^\alpha}\, d\xi = g(x),$$

where $g(0) = 0$, $g \in C^1(x \geqslant 0)$, $0 < \alpha < 1$, is the function

$$u(x) = \frac{\sin \pi\alpha}{\pi} \int_0^x \frac{g'(\xi)\, d\xi}{(x-\xi)^{1-\alpha}}.$$

Hint. Write the equation in the form of the convolution $u * \theta(x - \alpha) = g(x)$ (assume that $u = 0$ and $g = 0$ at $x < 0$) and employ the result of Problem 8.15 with $\beta = 1 - \alpha$.

8.29. The solution in $\mathcal{D}'(R^1)$ of the equation $\theta(x) \cos x * f = g$, with $g \in C^1(x \geqslant 0)$ and $g = 0$ at $x < 0$, is the function

$$f(x) = g'(x) + \int_0^x g(\xi)\, d\xi.$$

8.30. Let us consider an electric circuit consisting of resistance R, self-inductance L, capacitance C, and a source $E(t)$ of time dependent e.m.f. switched on at $t = 0$. Show that the current $i(t)$ in the circuit satisfies the equation $Z * i = E(t)$, where

$$Z = L\delta'(t) + R\delta(t) + \frac{\theta(t)}{C}$$

is the impedance (generalized resistance) of the circuit.

8.31. Let $f \in \mathscr{D}'(R^{n+1})$. Prove that

1. $[\delta(x-x_0)\cdot\delta(t)] * f(x, t) = f(x-x_0, t)$.

2. $[\delta(x-x_0)\cdot\delta^{(m)}(t)] * f(x, t) = \dfrac{\partial^m f(x-x_0, t)}{\partial t^m}$.

8.32. Calculate the following convolution in $\mathscr{D}'(R^n)$:

1. $f * \delta_{S_R}$, where $f(x) \in C$, and $\delta_{S_R}(x)$ is a simple layer on the sphere $|x| = R$ with the density equal to unity (see Sec. 6).

2. $f * \dfrac{\partial}{\partial \mathbf{n}} \delta_{S_R}$, where $f \in C^1$. 3. $\delta_{S_R} * |x|^2$, $n=3$.

4. $\delta_{S_R} * e^{-|x|^2}$, $n=3$. 5. $\delta_{S_R} * \sin|x|^2$, $n=3$.

6. $\delta_{S_R} * \dfrac{1}{1+|x|^2}$, $n=3$.

7. $\dfrac{1}{|x|} * \mu\delta_S$, $n=3$; $\ln\dfrac{1}{|x|} * \mu\delta_S$, $n=2$.

8. $-\dfrac{1}{|x|} * \dfrac{\partial}{\partial \mathbf{n}}(\nu\delta_S)$, $n=3$, $\ln|x| * \dfrac{\partial}{\partial \mathbf{n}}(\nu\delta_S)$, $n=2$, with S a bounded surface. For the definitions of $\mu\delta_S$ and $-\dfrac{\partial}{\partial \mathbf{n}}(\nu\delta_S)$ see Secs. 6 and 7.

8.33. Calculate in $\mathscr{D}'(R^2)$ the following convolutions:

1. $\theta(t)\,x * \theta(x)\,t$. 2. $\theta(t-|x|) * \theta(t-|x|)$.

3. $\theta(t)\,\theta(x) * \theta(t-|x|)$.

8.34. Suppose f and g belong to $\mathscr{D}'(R^{n+1})$, with $f(x, t) = 0$ at $t<0$ and $g=0$ outside $\overline{\Gamma}^+$. Prove that $g * f$ exists in $\mathscr{D}'(R^{n+1})$ and is given by the formula

$$(g * f, \varphi) = (g(\xi, t)\cdot f(y, \tau),$$
$$\eta(t)\,\eta(\tau)\,\eta(a^2t^2-|\xi|^2)\,\varphi(\xi+y, t+\tau)),$$
$$\varphi \in \mathscr{D}(R^{n+1}),$$

where $\eta(t) \in C^\infty(R^1)$, with $\eta(t) = 0$ at $t < -\delta$ and $\eta(t) \equiv 1$ at $t > -\varepsilon$ $(0<\varepsilon<\delta)$.

8.35. Suppose $g(x, t) \in \mathscr{D}'(R^{n+1})$, $g=0$ outside $\overline{\Gamma}^+$, and $u(x) \in \mathscr{D}'(R^n)$. Prove that

1. $g * u(x)\cdot\delta(t) = g(x, t) * u(x)$, where the generalized function $g(x, t) * u(x)$ operates according to the formula

$$(g(x, t) * u(x), \varphi) = (g(\xi, t)\cdot u(y),$$
$$\eta(a^2t^2-|\xi|^2)\,\varphi(\xi+y, t)), \quad \varphi \in \mathscr{D}(R^{n+1}).$$

2. $g * u(x)\cdot\delta^{(k)}(t) = \dfrac{\partial^k}{\partial t^k}(g(x, t) * u(x)) = \dfrac{\partial^k g(x, t)}{\partial t^k} * u(x)$.

8.36. Calculate in $\mathscr{D}'(R^2)$ the following convolutions:

1. $\theta(at-|x|) * [\omega(t)\cdot\delta(x)]$, $a>0$, where $\omega(t) \in C\ (t \geqslant 0)$ and $\omega(t) = 0$ at $t<0$.

2. $\theta(at-|x|)*[\theta(t)\cdot\delta(x)]$. 3. $\theta(at-|x|)*\frac{\partial}{\partial t}[\theta(t)\cdot\delta(x)]$.

4. $\theta(at-|x|)*[\theta(t)\cdot\delta'(x)]$. 5. $\theta(at-|x|)*[0(x)\,\delta(t)]$.

6. $\theta(at-|x|)*\frac{\partial}{\partial t}[\omega(x)\,\delta(t)]$, where $\omega(x)\in C(R')$.

Hint. Use the result of Problem 7.5.2.

7. $\theta(at-|x|)*\frac{\partial}{\partial x}[\theta(x)\,\delta(t)]$.

8.37. Calculate in $\mathscr{D}'(R^2)$ the following convolutions:

1. $e^x\delta(t)*\frac{\theta(t)}{2a\sqrt{\pi t}}e^{-x^2/(4a^2t)}$, $a>0$.

2. $\theta(t)\,e^tx*\frac{\theta(t)}{2\sqrt{\pi t}}e^{-x^2/(4t)}$.

3. $\theta(x)\,\delta(t)*\frac{\theta(t)}{2\sqrt{\pi t}}e^{-x^2/(4t)}$.

8.38. Suppose f belongs to $C^\infty(R^n\setminus\{0\})$ and $g\in\mathscr{D}'(R^n)$ is finite. Show that $f*g\in C^\infty(R^n\setminus\text{supp }g)$. *Hint*. Use formula (8.7).

8.39. Suppose f belongs to $\mathscr{S}'$ and $g\in\mathscr{D}'$ is finite. Prove that $f*g\in\mathscr{S}'$.

8.40. Prove that if $f\in\mathscr{D}'$, then $f*\omega_\varepsilon\to f$ as $\varepsilon\to 0$ in $\mathscr{D}'$. *Hint*. Use the result of Problem 6.24.

We introduce a generalized function $f_\alpha(x)$ that depends on parameter α, $-\infty<\alpha<\infty$, as follows:

$$f_\alpha(x)=\begin{cases}\dfrac{\theta(x)\,x^{\alpha-1}}{\Gamma(\alpha)}, & \alpha>0,\\ f^{(N)}_{\alpha+N}(x), & \alpha\leqslant 0,\ \alpha+N>0,\end{cases}$$

with N an integer (cf. Problem 8.15).

8.41. Prove that $f_\alpha*f_\beta=f_{\alpha+\beta}$.

8.42. Prove that

$$f_0*=\delta*,\quad f_{-n}*=\frac{d^n}{dx^n}*,\quad f_n*=\underbrace{\theta*\theta*\ldots*\theta}_{n\text{ times}}*.$$

The convolution operation $f_{-\alpha}*$ for α positive and not an integer, is said to be a (*fractional*) *derivative* of order α (we denote this derivative by $u^{(\alpha)}$, i.e. $u^{(\alpha)}\equiv f_{-\alpha}*u$); $f_\alpha*$ for α positive is said to be the *antiderivative* of order α (we denote it by $u_{(\alpha)}$, i.e. $u_{(\alpha)}\equiv f_\alpha*u$).

8.43. Calculate the derivative of order 3/2 of $\theta(x)$.

8.44. Calculate the antiderivative of order 3/2 of $\theta(x)$.

8.45. Calculate the derivative of order 1/2 of $f(x)$, with $f=0$ at $x<0$.

8.46. Calculate the antiderivative of order 1/2 of $f(x)$, with $f = 0$ at $x < 0$.

8.47.* We denote by $\mathcal{E}'$ the space of finite generalized functions with the convergence $f_k \to 0$ as $k \to \infty$ in $\mathcal{E}'$ if (a) $f_k \to 0$ as $k \to \infty$ in $\mathcal{D}'$ and (b) there is a number R such that supp $f_k \subset U_R$ for all integers k. Prove the following

Theorem *If a linear continuous operator L from $\mathcal{E}'$ to $\mathcal{D}'$ is commutative with the translation operation, then L is a convolution operator, or $L = f_0 *$, where $f_0 = L\delta$.*

Answers to Problems of Sec. 8

8.8. 1. *Solution.* In view of (8.3) and (8.3$_1$), $\left(\frac{\partial}{\partial t}\theta(at - |x|), \varphi\right) =$

$$= -\left(\theta(at - |x|), \frac{\partial\varphi}{\partial t}\right) = -\int_{-\infty}^{\infty}\int_{\frac{|x|}{a}}^{\infty}\frac{\partial\varphi(x, t)}{\partial t}\,dt\,dx = \int_{-\infty}^{\infty}\varphi\left(x, \frac{|x|}{a}\right)dx =$$

$$= a\int_0^{\infty}\varphi(-at', t')\,dt' + a\int_0^{\infty}\varphi(at', t')\,dt' = (a\theta(t)\,\delta(at + x) +$$

$$+ a\theta(t)\,\delta(at - x), \varphi) = (a\delta(at - |x|), \varphi).$$

8.14. 1. *Solution.* In view of (8.9), $\theta * \theta = \int_0^x \theta(y)\,\theta(x - y)\,dy =$

$= \theta(x)\int_0^x dy = \theta(x)\,x$. 2. $\theta(x)\frac{x^3}{3}$. 3 $e^{-|x|}(1 + |x|)$. 4. $\sqrt{\frac{\pi}{8a}} \times$ $\times\, x e^{-ax^2/2}$. 5. $\theta(x)\left(x^2 - 4\sin^2\frac{x}{2}\right)$. 6. $\theta(x)(3x^2 + 6\cos x - 6)$. 7. $\frac{\theta(x)}{2}(\sinh x - \sin x)$. 8. $\theta(2a - |x|)(2a - |x|)$.

8.21. *Hint.* Use the result of Problem 8.20 by applying it to $\varphi(-x)$ and putting $x = 0$.

8.30. *Hint.* Use the result of problem 1.31.

8.31. 2. *Solution.* In view of (8.2) and (8.6) and the results of Problems 8.4.2 and 8.13.2, $[\delta(x - x_0)\cdot\delta^{(m)}(t)] * f(x, t) = \frac{\partial^m}{\partial t^m}[\delta(x -$ $- x_0)\cdot\delta(t)] * f(x, t) = \frac{\partial^m}{\partial t^m}(\delta(x - x_0, t) * f(x, t)) = \frac{\partial^m f(x - x_0, t)}{\partial t^m}$.

8.32. 1. $\int_{|x-y|=R} f(y)\,dS_y$. 2. *Solution.* In view of (8.7) and the definition of a double layer (see Sec. 7), $\left(f * \frac{\partial}{\partial n}\delta_{S_R}, \varphi\right) = \Big(f(y) \times$ $\times \frac{\partial}{\partial n}\delta_{S_R}(\xi), \eta(\xi)\,\varphi(y + \xi)\Big) = \left(f(y), \left(\frac{\partial}{\partial n}\delta_{S_R}(\xi), \eta(\xi)\,\varphi(y + \xi)\right)\right) =$

$= -\int\limits_{R^n} f(y)\Big(\int\limits_{|\xi|=R} \frac{\partial\varphi(y+\xi)}{\partial\mathbf{n}_\xi}\,dS_\xi\Big)\,dy = -\int\limits_{|\xi|=R}\Big(\frac{\partial}{\partial\mathbf{n}_\xi}\int\limits_{R^n} f(x-\xi)\,\varphi\times$ $\times(x)\,dx\Big)\,dS_\xi = \Big(-\int\limits_{|\xi|=R}\frac{\partial f(x-\xi)}{\partial\mathbf{n}_\xi}\,dS_\xi,\ \varphi\Big)$. 3. $\int\limits_{|x-y|=R} |y|^2\,dS_y=$ $=\int\limits_{|y|=R}|x-y|^2\,dS_y = \int\limits_0^{2\pi}\int\limits_0^{\pi}(|x|^2+R^2-2R|x|\cos\theta)\,R^2\sin\theta\,d\theta\,d\varphi=$ $=4\pi R^2(|x|^2+R^2)$. 4. $\frac{\pi R}{|x|}(e^{-(R-|x|)^2}-e^{-(R+|x|)^2})$. 5. $\frac{2\pi R}{|x|}\sin(R^2+$ $+|x|^2)\sin 2R|x|$. 6. $\frac{\pi R}{|x|}\ln\frac{1+(|x|+R^2)^2}{1+(|x|-R)^2}$. 7. $\int\limits_S \frac{\mu(y)}{|x-y|}\,dS_y$; $\int\limits_S \mu(y)\ln\frac{1}{|x-y|}\,dl_y$. 8. $\int\limits_S \nu(y)\frac{\partial}{\partial\mathbf{n}_y}\frac{1}{|x-y|}\,dS_y$; $\int\limits_S \nu(y)\frac{\partial}{\partial\mathbf{n}_y}\times$ $\times\ln\frac{1}{|x-y|}\,dl_y$.

8.33. 1. No convolution exists. 2. $\theta(t-|x|)\frac{t^2-x^2}{2}$. 3. $\frac{1}{2}\theta(t)\times$ $\times[\theta(x+t)(x+t)^2+\theta(x-t)(x-t)^2-2\theta(x)\,x^2]$.

8.34. *Solution.* In view of the result of Problem 6.27, $f(y,\tau) = \eta(\tau)f(y,\tau)$ and $g(\xi,t)=\eta(t)\,\eta(a^2t^2-|\xi|^2)\,g(\xi,t)$, since $\eta(\tau)=1$ in the neighborhood supp $f(y,\tau)\subset[\tau\geqslant 0]$ and $\eta(t)\,\eta(a^2t^2-|\xi|^2)=1$ in the neighborhood supp $g(\xi,t)\subset\overline{\Gamma}^+$, where $\overline{\Gamma}^+$ is the region $a^2t^2-|\xi|^2\geqslant 0$, $t\geqslant 0$. In view of (8.5), $(g*f,\varphi) = \lim\limits_{k\to\infty}(g(\xi,t)\cdot f(y,\tau),\ \eta_k(\xi,t;y,\tau)\,\varphi(\xi+y,t+\tau)) = \lim\limits_{k\to\infty}(\eta(t)\,\eta(a^2t^2-|\xi|^2)\,g(\xi,t)\cdot\eta(\tau)f(y,\tau),\ \eta_k(\xi,t;y,\tau)\,\varphi(\xi+y;t+\tau)) = \lim\limits_{k\to\infty}(g(\xi,t)\cdot f(y,\tau),\ \eta(t)\,\eta(\tau)\,\eta(a^2t^2-|\xi|^2)\,\eta_k(\xi,t;y,\tau)\,\varphi(\xi+y,t+\tau)) = (g(\xi,t)\cdot f(y,\tau),\ \eta(t)\,\eta(\tau)\,\eta(a^2t^2-|\xi|^2)\,\varphi(\xi+y,t+\tau))$, since $\eta(t)\,\eta(\tau)\,\eta(a^2t^2-|\xi|^2)\,\varphi(\xi+y,t+\tau)\in\mathcal{D}(R^{2n+2})$.

8.35. *Solution.* 1. In view of the formula in Problem 8.34, the associativity of the direct product, and (8.1), $(g*[u(x)\cdot\delta(t)],\varphi) = ([g(\xi,t)\cdot u(y)]\cdot\delta(\tau),\ \eta(t)\,\eta(\tau)\,\eta(a^2t^2-|\xi|^2)\,\varphi(\xi+y),t+\tau)) = (g(\xi,t)\cdot u(y),\ \eta(t)\,\eta(a^2t^2-|\xi|^2)\,\varphi(\xi+y,t))$. Now, in view of the result of Problem 6.27, $g=\eta(t)\,g$, since supp $g(\xi,t)\subset[t\geqslant 0]$. Hence, $(g*[u(x)\cdot\delta(t)],\varphi) = (g(\xi,t)\cdot u(y),\ \eta(a^2t^2-|\xi|^2)\,\varphi(x+\xi,t)) = (g(x,t)*u(x),\varphi)$, since $\eta(a^2t^2-|\xi|^2)\,\varphi(x+\xi,t)\in\mathcal{D}(R^{2n+1})$. 2. In view of (8.2) and (8.6) and the formula in Problem 8.35. 1, $g*u(x)\cdot\delta^{(k)}(t)=g*\frac{\partial^k}{\partial t^k}(u(x)\cdot\delta(t))=$ $=\frac{\partial^k}{\partial t^k}(g(x,t)*u(x))=\frac{\partial^k g(x,t)}{\partial t^k}*u(x)$.

8.36. 1. *Solution.* In view of the formula in Problem 8.35. 1. $(I, \varphi) = (\theta(at - |x|)\cdot\omega(\tau), \quad \eta(a^2t^2 - |x|^2)\varphi(x, \quad t+\tau)) =$
$= \int \omega(\tau)\left(\int\int \theta(at-|x|)\right)\varphi(x, t+\tau)\,dx\,dt)\,d\tau = \int\int \varphi(x, t')\times$
$\times\left(\theta(at'-|x|)\int_0^{t'-|x|/a}\omega(\tau)\,d\tau\right)dx\,dt'$. Hence, $I = \theta(at-|x|)\times$
$\times\int_0^{t-|x|/a}\omega(\tau)\,d\tau$. 2. $\theta(at-|x|)\left(t-\frac{|x|}{a}\right)$. 3. $\theta(at-|x|)$.
Hint. Use the result of Problem 8.35.2. 4.$-\theta(at-|x|)\frac{\operatorname{sign} x}{a}$.
5. $a\theta(t)[\theta(x+at)(x+at)+\theta(x-at)(x-at)]$. 6. $a\theta(t)[\omega(x+at)+$ $+\omega(x-at)]$. 7. $\theta(at-|x|)$.

8.37. 1. $\theta(t)e^{x+a^2t}$. 2. $\theta(t)x(e^t-1)$.

3. $\theta(t)\frac{1}{\sqrt{2\pi}}\int_{-\infty}^{x/(2\sqrt{t})} e^{-z^2/2}\,dz = \theta(t)\Phi\left(\frac{x}{2\sqrt{t}}\right)$.

8.43. *Solution.* $\theta^{(3/2)}(x) = f_{-3/2}*\theta = f'_{-1/2}*\theta = f''_{-1/2}*\theta = (f_{1/2}*\theta)'' =$

$$= \frac{d^2}{dx^2}\left(\frac{\theta x}{\Gamma(1/2)}\int_0^x \frac{d\xi}{\sqrt{x-\xi}}\right) = \frac{d^2}{dx^2}\left(2\theta(x)\sqrt{\frac{x}{\pi}}\right) =$$

$$= \frac{d}{dx}\left(\theta(x)\frac{1}{\sqrt{\pi x}}\right).$$

8.44. *Solution.* $\theta_{(3/2)}(x) = f_{3/2}*\theta = \frac{\theta(x)}{\Gamma\left(\frac{3}{2}\right)}\int_0^x \sqrt{x-\xi}\,d\xi =$

$$= \theta(x)\frac{4x}{3}\sqrt{\frac{x}{\pi}}.$$

8.45. *Solution.* $f^{(1/2)}(x) = f_{-1/2}*f = f'_{1/2}*f = (f_{1/2}*f)' =$

$$= \frac{d}{dx}\left(\frac{\theta(x)}{\sqrt{\pi}}\int_0^x \frac{f(\xi)}{\sqrt{x-\xi}}\,d\xi\right).$$

8.46. $\frac{\theta(x)}{\sqrt{\pi}}\int_0^x \frac{f(\xi)}{\sqrt[3]{x-\xi}}\,d\xi$.

9 The Fourier Transform of Generalized Functions of Slow Growth

The operation of the Fourier transform $F[\varphi]$ on functions φ belonging to $\mathscr{S}$ is defined by the formula

$$F[\varphi](\xi) = \int e^{i(\xi, x)}\varphi(x)\,dx. \tag{9.1}$$

The *Fourier transform* $F[f]$ of an arbitrary *generalized function* f belonging to $\mathscr{S}'(R^n)$ is defined by the formula

$$(F[f], \varphi) = (f, F[\varphi]). \tag{9.2}$$

The operator

$$F^{-1}[f] = \frac{1}{(2\pi)^n} F[f(-x)], \quad f \in \mathscr{S}', \tag{9.3}$$

(the *inverse Fourier transform*) is the inverse of operator F, that is,

$$F^{-1}[F[f]] = f, \quad F[F^{-1}[f]] = f \quad f \in \mathscr{S}'.$$

The following formulas are valid (f and g belong to $\mathscr{S}'$):

$$\left.\begin{aligned} D^\alpha F[f] &= F[(ix)^\alpha f], \\ F[D^\alpha f] &= (-i\xi)^\alpha F[f], \\ F[f(x-x_0)] &= e^{i(x_0, \xi)} F[f], \\ F[f](\xi+\xi_0) &= F[f(x)\,e^{i(x, \xi_0)}](\xi), \\ F[f(cx)] &= \frac{1}{|c|^n} F[f]\left(\frac{\xi}{c}\right), \quad c \neq 0, \\ F[f(x)\cdot g(y)] &= F[f](\xi)\cdot F[g](\eta), \\ F[f * g] &= F[f]\,F[g] \text{ (either } f \text{ or } g \text{ is finite).} \end{aligned}\right\} \tag{9.4}$$

The Fourier transform F_x with respect to the variable x of the generalized function $f(x, y) \in \mathscr{S}'(R^{n+m})$, with $x \in R^n$ and $y \in R^m$, is defined by the formula

$$(F_x[f(x, y)](\xi, y)\,\varphi(\xi, y)) = \\ = (f(x, y), F[\varphi(\xi, y)](x, y)), \quad \varphi \in \mathscr{S}(R^{n+m}). \tag{8.5}$$

9.1. 1. Let $f(x) \in C^k(R^1)$, $k \geqslant 0$, and $\int |f^{(\alpha)}(x)|\,dx < \infty$, $\alpha \leqslant k$. Prove that $F[f] \in C[R^1]$ and $|\xi|^k |F[f](\xi)| \leqslant a$.

2. Let $f(x) \in C^k(R^n)$, $k \geqslant 0$, and $|x|^{n+l}\,|D^\alpha f(x)| \leqslant b$, $|\alpha| \leqslant k$, with l a positive integer. Prove that
$F[f] \in C^{l-1}(R^n)$ and $|\xi|^k\,|D^\beta F[f](\xi)| \leqslant b$, $|\beta| \leqslant l-1$.

9.2. Prove that $f = F^{-1}[F[f]]$, where F^{-1} is defined in (8.3), for the following f's:

1. $f(x) \in C(R^n)$, $|x|^{n+\varepsilon}\,|f(x)| \leqslant a$, $|\xi|^{n+\varepsilon}\,|F[f](\xi)| \leqslant a$, $\varepsilon > 0$.

2. $f(x) \in C^2(R^1)$, $\int |f^{(\alpha)}(x)|\, dx < \infty$, $\alpha \leqslant 2$.

3. $f(x) \in C^{n+1}(R^n)$, $|D^\alpha f(x)|\,|x|^{n+1} \leqslant a$, $|\alpha| \leqslant n+1$.
Verify that case 3 follows from case 1.

9.3. Prove that

$$\xi^\beta D^\alpha F[\varphi](\xi) = i^{|\alpha|+|\beta|} F[D^\beta(x^\alpha \varphi)](\xi), \qquad \varphi \in \mathcal{S}.$$

9.4. 1. Prove that if $\varphi \in \mathcal{S}$, then $F[\varphi] \in \mathcal{S}$, too.

2. Prove that the Fourier transform operation is continuous from $\mathcal{S}$ to $\mathcal{S}$, that is, $\varphi_k \to \varphi$ as $k \to \infty$ in $\mathcal{S}$ implies $F[\varphi_k] \to F[\varphi]$ in $\mathcal{S}$.
Hint. Use the result of Problem 9.3.

9.5. 1. Prove that if $f \in \mathcal{S}'$, then $F[f] \in \mathcal{S}'$, too.

2. Prove that the Fourier transform operation is continuous from $\mathcal{S}'$ to $\mathcal{S}'$, that is $f_k \to f$ as $k \to \infty$ in $\mathcal{S}'$ implies $F[f_k] \to F[f]$ in $\mathcal{S}'$.

3. Prove that if f is a function of slow growth, then

$$F[f](\xi) = \lim_{R\to\infty} \int_{|x|<R} f(x)\, e^{i(\xi, x)}\, dx \text{ in } \mathcal{S}'.$$

4. Prove that if $f \in L_2(R^n)$, then $F]f] \in L_2(R^n)$ and

$$F[f](\xi) = \lim_{R\to\infty} \int_{|x|<R} f(x)\, e^{i(\xi, x)}\, dx \text{ in } L_2(R^n)$$

(Plancherel's theorem).

5. Prove that if f and g belong to $L_2(R^n)$, Parseval's equality holds:

$$(2\pi)^n (f, g) = (F[f], F[g]).$$

6. Prove that if $f \in L_1(R^n)$, then $F[f] \in L_\infty(R^n) \cap C(R^n)$ and

$$F[f](\xi) = \int f(x)\, e^{i(\xi, x)}\, dx,$$

$$\|F[f]\|_{L_\infty(R^n)} \leqslant \|f\|_{L_1(R^n)},$$

$$F[f(\xi)] \to 0, \quad |\xi| \to \infty$$

(the Riemann-Lebesgue theorem); also $F[f * g] = F[f]\, F[g]$, $f, g \in L_1(R^n)$.

7. Prove that if $f \in \mathcal{S}'$ and $\varphi \in \mathcal{S}$, then

$$F[f * \varphi] = F[f]\, F[\varphi].$$

8. Suppose $f \in L_1(R^1)$ is a piecewise continuous function for which $\{f'(x)\}$ is piecewise continuous, too. Prove the validity of the inversion formula

$$\frac{f(x+0)+f(x-0)}{2} = \frac{1}{2\pi} \mathrm{PV} \int_{-\infty}^{\infty} F[f](\xi)\, e^{-i\xi x}\, d\xi, \quad x \in R^1,$$

9.6. Prove that in $\mathscr{S}'(R^n)$,

1. $F[\delta(x-x_0)] = e^{i(\xi, x_0)}$.

2. $F[\delta] = 1$. 3. $F[1] = (2\pi)^n \delta(\xi)$.

4. $F\left[\frac{\delta(x-x_0)+\delta(x+x_0)}{2}\right] = \cos x_0\xi, \ n=1$.

5. $F\left[\frac{\delta(x-x_0)-\delta(x+x_0)}{2i}\right] = \sin x_0\xi, \ n=1$.

9.7. Prove that in $\mathscr{S}'(R^n)$,

1. $F[D^\alpha\delta] = (-i\xi)^\alpha$. 2. $F[x^\alpha] = (2\pi)^n(-i)^{|\alpha|} D^\alpha\delta(\xi)$.

9.8. Calculate the Fourier transform of the following functions ($n = 1$):

1. $\theta(R-|x|)$. 2. $e^{-a^2x^2}$. 3. e^{ix^2}. 4. e^{-ix^2}.

5. $f(x) = 0$ at $x < 0$, $f(x) = k$, $k < x < k+1$, $k = 0, 1, \ldots$.

9.9. Prove that ($n = 1$)

1. $F[\theta(x)e^{-ax}] = \frac{1}{a-i\xi}, \quad a > 0$.

2. $F[\theta(-x)e^{ax}] = \frac{1}{a+i\xi}, \quad a > 0$.

3. $F[e^{-a|x|}] = \frac{2a}{a^2+\xi^2}, \quad a > 0$.

4. $F\left[\frac{2a}{a^2+x^2}\right] = 2\pi e^{-a|\xi|}, \quad a > 0$.

5. $F\left[\theta(x)e^{-ax}\frac{x^{\alpha-1}}{\Gamma(\alpha)}\right] = \frac{1}{(a+i\xi)^\alpha}, \quad a > 0, \ \alpha > 0$.

9.10. Using Sochozki formulas (see Problem 6.20) and the results of Problems 9.5, 9.9.1, and 9.9.2, prove that

1. $F[\theta(x)] = \pi\delta(\xi) + i\mathscr{P}\frac{1}{\xi}$.

2. $F[\theta(-x)] = \pi\delta(\xi) - i\mathscr{P}\frac{1}{\xi}$.

9.11. Calculate the Fourier transform of each of the following genaralized functions ($n = 1$):

1. $\delta^{(k)}$, $k = 1, 2, \ldots$. 2. $\theta(x-a)$. 3. $\operatorname{sign} x$. 4. $\mathscr{P}\frac{1}{x}$.

5. $\frac{1}{x \pm i0}$. 6. $|x|$. 7. $\theta(x)x^k$, $k = 1, 2, \ldots$. 8. $|x|^k$, $k = 2, 3, \ldots$

9. $x^k\mathscr{P}\frac{1}{x}$, $k = 1, 2, \ldots$. 10. $x^k\delta$, $k = 1, 2, \ldots$.

11. $x^k\delta^{(m)}(x)$, $m \geqslant k$. 12. $\mathscr{P}\frac{1}{x^2}$, where $\mathscr{P}\frac{1}{x^2}$ defined in Problem 6.25.

13. $\mathscr{P}\frac{1}{x^3}$, where $\mathscr{P}\frac{1}{x^3}$ is defined in Problem 7.10.4.

14. $\sum_{k=-\infty}^{\infty} a_k \delta(x-k)$, $|a_k| \leqslant C(1+|k|)^m$.

15. $\theta^{1/2}(x)$ (for the definition of a fractional derivative see Sec. 8).

9.12.* Prove that

$$F\left[\mathcal{P}\frac{1}{|x|}\right] = -2c - 2\ln|\xi|,$$

with $x = c = \int_0^1 \frac{1-\cos u}{u}\,du - \int_1^\infty \frac{\cos u}{u}\,du$ the Euler constant, and $\mathcal{P}\frac{1}{|x|}$, $x \in R^1$, defined in Problem 7.26.

9.13.* Prove that

$$F\left[\mathcal{P}\frac{1}{|x|^2}\right] = -2\pi\ln|\xi| - 2\pi c_0,$$

where the generalized function $\mathcal{P}\frac{1}{|x|^2}$, $x \in R^2$, is defined as follows:

$$\left(\mathcal{P}\frac{1}{|x|^2}, \varphi\right) = \int_{|x|<1} \frac{\varphi(x)-\varphi(0)}{|x|^2}\,dx + \int_{|x|>1} \frac{\varphi(x)}{|x|^2}\,dx,$$

$$c_0 = \int_0^1 \frac{1-J_0(u)}{u}\,du - \int_1^\infty \frac{J_0(u)}{u}\,du$$

and J_0 is the Bessel function of order zero.

9.14. Solve in $\mathcal{S}'$ the integral equation

$$\int_0^\infty u(\xi)\cos\xi x\,dx = \theta(1-x).$$

9.15. Evaluate the integral

$$\int_0^\infty \frac{\sin ax \sin bx}{x^2}\,dx.$$

Hint. Use Parseval's equality and the result of Problem 9.8.1.

9.16. Prove that $F\left[\frac{\theta(R-|x|)}{\sqrt{R^2-|x|^2}}\right] = 2\pi\frac{\sin R|\xi|}{|\xi|}$, $\xi \in R^2$.

9.17. Prove that

1. $F\left[\frac{1}{|x|^2}\right] = \frac{2\pi^2}{|\xi|}$, $\xi \in R^3$.

2. $F[|x|^{-k}]=2^{n-k}\pi^{\frac{n}{2}}\dfrac{\Gamma\left(\frac{n-k}{2}\right)}{\Gamma\left(\frac{k}{2}\right)}|\xi|^{k-n}, \quad \xi\in R^n, \quad 0<k<n.$

Hint. Use formula (9.2) with $f=|x|^{-k}$ in $\mathscr{S}'(R^1)$ and $\varphi=e^{-|x|^2/2}$.

9.18. Prove that $F\left[\frac{1}{z}\right]=\frac{2\pi i}{\zeta}, \quad \zeta=\xi+i\eta.$

9.19. Calculate the Fourier transform of the generalized function $\frac{1}{4\pi R}\delta_{S_R}$, $n=3$, defined in Sec. 6.

9.20. Employ the Fourier transform method to prove that in $\mathscr{S}'(R^1)$,

1. $y=c_0\delta(x)+c_1\delta^1(x)+\ldots+c_{n-1}\delta^{(n-1)}(x)$ is the general solution of the equation $x^ny=0$, $n=1, 2, \ldots$

2. $y=\sum\limits_{k=0}^{m-1}a_kx^k+\sum\limits_{k=0}^{m-1}b_k\theta(x)x^{m-k-1}+\sum\limits_{k=m}^{n-1}c_k\delta^{(k-m)}(x)$ is the general solution of the equation $x^ny^{(m)}=0$, $n>m$.

Hint. Use the results of Problems 7.23.2 and 7.24.

9.21. Prove that in $\mathscr{S}'(R^{n+1}(x, t))$, with $(x, t)=(x_1, \ldots, x_n, t)$,

1. $F_x[\delta(x, t)]=1(\xi)\cdot\delta(t)$.
2. $F_x\left[\dfrac{\partial^m f(x, t)}{\partial t^m}\right]=\dfrac{\partial^m}{\partial t^m}F_x[f(x, t)]$.
3. $F_x[\theta(at-|x|)]=2\theta(t)\sin\dfrac{a\xi t}{\xi}, \quad a>0, \quad n=1.$
4. $F_x[f(x)\delta(t)]=F[f](\xi)\delta(t), \quad f\in\mathscr{S}'(R^n).$

9.22. Prove that in $\mathscr{S}'(R^{n+m})$,

1. $D_\xi^\alpha D_y^\beta F_x[f(x, y)]=F_x[(ix)^\alpha D_y^\beta f]$.
2. $F_x[D_x^\alpha D_y^\beta f]=(-i\xi)^\alpha F_x[D_y^\beta f]$.

9.23. Prove that in $\mathscr{S}'(R^2)$,

$$F_\xi^{-1}[\theta(t)e^{-a^2\xi^2t}]=\frac{\theta(t)}{2a\sqrt{\pi t}}e^{-\frac{x^2}{4a^2t}}.$$

Hint. Employ formula (9.3) and the result of Problem 9.8.2.

9.24. Prove that in $\mathscr{S}'(R^{n+1})$,

$$F_\xi^{-1}[\theta(t)e^{-a^2|\xi|^2t}]=\theta(t)\left(\frac{1}{2a\sqrt{\pi t}}\right)^n e^{-\frac{|x|^2}{4a^2t}}.$$

Hint. Use the result of Problem 9.23.

9.25. Prove that in $\mathscr{S}'(R^2)$,

$$F_\xi^{-1}\left[\theta(t)\frac{\sin a\xi t}{a\xi}\right]=\frac{1}{2a}\theta(at-|x|).$$

Hint. Use the result of Problem 9.8.1.

9.26. Prove that in $\mathcal{S}'(R^3)$,

$$F_\xi^{-1}\left[\theta(t)\frac{\sin a|\xi|t}{a|\xi|}\right]=\frac{\theta(at-|x|)}{2\pi a\sqrt{a^2t^2-|x|^2}}$$

Hint. Use the result of Problem 9.16

9.27. Prove that in $\mathcal{S}'(R^4)$,

$$F_\xi^{-1}\left[\theta(t)\frac{\sin a|\xi|t}{a|\xi|}\right]=\frac{\theta(t)}{4\pi a^2 t}\delta_{S_{at}}(x),$$

with $S_{at}=\{x\colon |x|=at\}$. *Hint.* Use the result of Problem 9.19.

9.28. Let f be a finite generalized function and η any test function from $\mathcal{D}$ equal to unity in the neighborhood supp f. Prove that the function $\tilde{f}(z)=(f(\xi), \eta(\xi)e^{i(z,\xi)})$, $z=x+iy$, (a) is independent of η, (b) is an entire function, and (c) $\tilde{f}(x)=F[f]$.

9.29. Prove that if f and g are finite and $f*g=0$, then either $f=0$ or $g=0$. *Hint.* Use the result of Problem 9.28.

9.30. 1. Prove that $F[\delta(x)\cdot 1(y)]=(2\pi)^m 1(\xi)\delta(\eta)$.

2. Let us denote the delta function in the hyperplane $(a, x)=0$ of the space R^n by $\delta((a, x))$, so that

$$(\delta((a,x)),\varphi)=\int\limits_{(a,\,x)=0}\varphi\,ds,\quad \varphi\in\mathcal{D}(R^n).$$

Prove that $F[\delta(a_1x_1+a_2x_2)]=2\pi\delta(a_2\xi_1-a_1\xi_2)$.

Answers to Problems of Sec. 9

9.8. 1. $2\frac{\sin R\xi}{\xi}$. 2. $\frac{\sqrt{\pi}}{a}e^{-\frac{\xi^2}{4a^2}}$. 3. $\sqrt{\pi}\,e^{-\frac{i}{4}(\xi^2-\pi)}$. 4. $\sqrt{\pi}\,e^{\frac{i}{4}(\xi^2-\pi)}$. 5. *Solution.* $F[f]=\sum\limits_{k=1}^{\infty}k\int\limits_k^{k+1}e^{ix\xi}\,dx=\sum\limits_{k=1}^{\infty}\frac{k}{i\xi}e^{ik\xi}\times$ $\times(e^{i\xi}-1)=\frac{e^{i\xi}-1}{i\xi}\sum\limits_{k=1}^{\infty}ke^{ik\xi}=-\frac{e^{i\xi}-1}{\xi}\frac{d^3}{d\xi^3}\sum\limits_{k=1}^{\infty}\frac{e^{ik\xi}}{k^2}$, where the series converges in $\mathcal{D}'$, since $\sum\limits_{k=1}^{\infty}\frac{e^{ik\xi}}{k^2}$ converges uniformly in R^1.

9.11. 1. $(-i\xi)^k$. 2. $\pi\delta(\xi)+ie^{ia\xi}\mathcal{P}\frac{1}{\xi}$. 3. $2i\mathcal{P}\frac{1}{\xi}$. 4. $i\pi\,\mathrm{sign}\,\xi$. 5. $\mp i\pi+i\pi\,\mathrm{sign}\,\xi$. 6. $2\left(\mathcal{P}\frac{1}{\xi}\right)'=-2\mathcal{P}\frac{1}{\xi^2}$. 7. $(-i)^k\left[\pi\delta(\xi)+\right.$ $\left.+i\mathcal{P}\frac{1}{\xi}\right]^{(k)}$. 8. $(-i)^k 2\pi\delta^{(k)}(\xi)$ when k is even, $(-i)^{k-1}2\left(\mathcal{P}\frac{1}{\xi}\right)^{(k)}$ when k is odd. 9. $2(-i)^{k-1}\pi\delta^{(k-1)}(\xi)$. 10. 0. 11. $(-i)^{k+m}\times$

$\times \frac{m}{(m-k)!} \xi^{m-k}$. 12. $-\pi |\xi|$. 13. *Solution.* In view of the result of Problem 7.10.4, the second formula in (9.4), and the result of Problem 9.11.12, $F\left[\mathcal{P}\frac{1}{x^3}\right] = F\left[-\frac{1}{2}\frac{d}{dx}\mathcal{P}\frac{1}{x^2}\right] = \frac{i\xi}{2} F\left[\mathcal{P}\frac{1}{x^2}\right] = -\frac{i\pi\xi|\xi|}{2}$. 14. *Solution.* In view of the results of Problems 6.25.2, 7.12, and 9.6.1, $\left(F\left[\sum_{k=-\infty}^{\infty} a_k\delta(x-k)\right], \varphi(\xi)\right) = \left(\sum_{k=-\infty}^{\infty} a_k\delta \times \times (x-k), F[\varphi(\xi)]\right) = \sum_{k=-\infty}^{\infty} (a_k e^{ik\xi}, \varphi(\xi)) = \left(\sum_{k=-\infty}^{\infty} a_k e^{ik\xi}, \varphi(\xi)\right)$, $\varphi \in \mathcal{S}(R^1)$. 15. *Solution.* $F[\theta^{(1/2)}] = F[f_{-1/2} * \theta] = F[f'_{1/2} * \theta] =$

$$= F[(f_{1/2} * \theta)'] = F\left[\left(\frac{\theta x^{-\frac{1}{2}}}{\Gamma\left(\frac{1}{2}\right)} * \theta\right)'\right] = \frac{2}{\sqrt{\pi}} F[(\theta \cdot \sqrt{x})'] =$$

$$= \frac{1}{\sqrt{\pi}} F\left[\frac{\theta}{\sqrt{x}}\right] = \frac{1}{\sqrt{\pi}}\left(1 + \frac{d}{i\,d\xi}\right) \int_0^\infty \frac{1}{\sqrt{x}(1+x)} e^{ix\xi}\, dx.$$

9.14. $\frac{2}{\pi}\frac{\sin\xi}{\xi}$. **9.15.** $\frac{\pi}{2}\min(a, b)$. **9.19.** $\frac{\sin R|\xi|}{|\xi|}$.

9.20. *Solution.* In view of the first formula in (9.4), $F[x^n y^{(m)}] = 0$ implies $F^{(n)}[y^{(m)}] = 0$. This fact combined with the results of Problems 7.23.2 and 9.7.2 and with formula (9.3) yields $F[y^{(m)}(x)] = \alpha_0 + \alpha_1\xi + \ldots + \alpha_{n-1}\xi^{n-1}$, $y^{(m)} = \beta_0\delta(x) + \beta_1\delta'(x) + \ldots + \beta_{n-1}\delta^{(n-1)}(x)$. Whence, in view of the results of Problems 7.23.2 and 7.6.10,

$$y = \sum_{k=0}^{m-1} a_k x^k + \sum_{k=0}^{m-1} b_k \theta(x) x^{m-k-1} + \sum_{k=m}^{n-1} c_k \delta^{(k-m)}(x).$$

9.21. 1. *Solution.* In view of (9.5) and the definition of the direct product (see Sec. 8), $(F_x[\delta(x, t)](\xi, t), (\varphi, (\xi, t)) = (\delta(x, t), F_\xi[\varphi(\xi, t)](x, t)) = (\delta(x, t), \int e^{i(\xi, x)}\varphi(\xi, t)\, d\xi) = \int \varphi(\xi, 0)\, d\xi = (1(\xi)\cdot\delta(t), \varphi(\xi, t))$. 2. *Solution.* In view of (9.5) and the definition of the derivative of a generalized function (Sec. 7), $\left(F_x\left[\frac{\partial^m f(x, t)}{\partial t^m}\right](\xi, t)\, \varphi\right) = (-1)^m \left(f(x, t), \frac{\partial^m}{\partial t^m} F_\xi[\varphi(\xi, t)]\right) = (-1)^m \left(f(x, t), F_\xi\left[\frac{\partial^m \varphi(\xi, t)}{\partial t^m}\right]\right) = \left(\frac{\partial^m}{\partial t^m} F_x[f(x, t)], \varphi\right)$.

9.27. $F\left[\frac{\sin t|\xi|}{|\xi|}\right]$ for $t > 0$ and $n = 3$ is calculated thus:

$$F\left[\frac{\sin t|\xi|}{|\xi|}\right] = \lim_{R\to\infty} 2\pi \int_0^r \rho \sin t\rho \int_0^\pi e^{ir\rho\cos\theta} \sin\theta\, d\theta\, d\rho =$$

$$-\frac{2\pi}{r}\lim_{R\to\infty}\frac{\partial}{\partial t}\int_0^R \cos t\rho \int_{-r}^{r} e^{i\rho u}\,du\,d\rho = -\frac{2\pi}{r}\lim_{R\to\infty}\frac{\partial}{\partial t}\int_{-r}^{r}\int_0^R [\cos t\rho e^{i\rho u} d\rho\, du =$$

$$= -\frac{\pi}{r}\lim_{R\to\infty}\frac{\partial}{\partial t}\int_{-r}^{r}\int_0^R [\cos\rho(u-t)+\cos\rho(u+t)]\,d\rho\,du =$$

$$-\frac{\pi}{r}\lim_{R\to\infty}\frac{\partial}{\partial t}\int_{-r}^{r}\left[\frac{\sin R(u-t)}{u-t}+\frac{\sin R(u+t)}{u+t}\right]du = -\frac{2\pi^2}{r}\frac{\partial}{\partial t}\theta(r-t) =$$

$= \frac{2\pi^2}{r}\delta(r-t) = \frac{2\pi^2}{t}\delta_t(x)$, $r = |x|$. *Hint.* In passing to the limit, use the result of Problem 6.19.4.

10 The Laplace Transform of Generalized Functions

We denote by $\mathscr{D}'_+(a)$ the set of generalized functions $f(t)$ from $\mathscr{D}'(R^1)$ that vanish at $t < 0$ and such that $f(t)e^{-\sigma t} \in \mathscr{S}'$ for all $\sigma > a$.

The *Laplace transform* $\mathscr{F}(p)$ of a generalized function $f(t)$ from $\mathscr{D}'_+(a)$ is defined through the following formula:

$$\mathscr{F}(p) = F[f(t)e^{-\sigma t}](-\omega), \quad \sigma > a.$$

The (generalized) function $f(t)$ is known at the *original function* and $\mathscr{F}(p)$ as the *image function.* This fact is denoted thus:

$$f(t) \leftrightarrow \mathscr{F}(p), \quad \sigma > a;$$

here $p = \sigma + i\omega$. The function $\mathscr{F}(p)$ is analytic in the half-plane $\sigma > a$ and satisfies the following restriction on growth: for every positive ε and $\sigma_0 > a$ there are numbers $c_\varepsilon(\sigma_0) \geqslant 0$ and $m = m(\sigma_0) \geqslant 0$ such that

$$|\mathscr{F}(p)| \leqslant c_\varepsilon(\sigma_0)\, e^{\varepsilon\sigma}(1+|p|)^m, \quad \sigma > \sigma_0.$$

The following formulas are valid

$$(-t)^m f(t) \leftrightarrow \mathscr{F}^{(m)}(p), \quad \sigma > a, \quad m = 0, 1, \ldots;$$

$$f^{(m)}(t) \leftrightarrow p^m \mathscr{F}(p), \quad \sigma > a, \quad m = 0, 1, \ldots;$$

$$f(t)e^{\lambda t} \leftrightarrow \mathscr{F}(p-\lambda), \quad \sigma > a + \operatorname{Re}(\lambda);$$

$$f(kt) \leftrightarrow \frac{1}{k}\mathscr{F}\left(\frac{p}{k}\right), \quad \sigma > ka, \quad k > 0;$$

$$f(t-\tau) \leftrightarrow e^{-\tau p}\mathscr{F}(p), \quad \sigma > a;$$

$$f_{(m)}(t) \leftrightarrow \frac{\mathscr{F}(p)}{p^m}, \quad \sigma > a, \quad m = 0, 1, \ldots$$

(where $f_{(m)}$ is the mth derivative of f from $\mathscr{D}'_+(a)$);

$$(f * g)(t) \leftrightarrow \mathscr{F}(p)\mathscr{G}(p), \quad \sigma > a,$$

if

$$g(t) \leftrightarrow \mathscr{G}(p), \quad \sigma > a;$$

and

$$f(t)=\frac{1}{2\pi i}\left(\frac{d}{dt}-a\right)^{m+2}\int_{\sigma-i\infty}^{\sigma+i\infty}\frac{\mathcal{F}(p)\,e^{pt}}{(p-a)^{m+2}}\,dp$$

is the formula for the *inverse Laplace transform*, with the integral being independent of $\sigma>\sigma_0>a$ and of $m=m(\sigma_0)$.

In Problems 10.1-10.9 and 10.11-10.14 prove the validity of the propositions.

10.1. If $f(t)$ is locally integrable in R^1, $f(t)=0$, $t<0$, and $f(t)=O(e^{at})$, $t\to\infty$, then $f\in\mathcal{D}'_+(a)$ and

$$\mathcal{F}(p)=\int_0^\infty f(t)\,e^{-pt}\,dt,\quad \sigma>a.$$

10.2. If $f\in\mathcal{D}'_+(a)$, $f(t)\leftrightarrow\mathcal{F}(p)$, $\sigma>a$, and the function $\mathcal{F}(\sigma+i\omega)$ is absolutely integrable with respect to ω over R^1 for a certain $\sigma>a$, then the inversion formula takes the form

$$f(t)=\frac{1}{2\pi i}\int_{\sigma-i\infty}^{\sigma+i\infty}\mathcal{F}(p)\,e^{pt}\,dp.$$

10.3. 1. $\mathcal{D}'_+(a_1)\subset\mathcal{D}'_+(a_2)$ if $a_1\leqslant a_2$.
2. If $f\in\mathcal{S}'\cap\mathcal{D}'_+$, then $f\in\mathcal{D}'_+(0)$.

10.4. If $f\in\mathcal{D}'_+(a)$, then
1. $pf\in\mathcal{D}'_+(a)$, where p is a polynomial. 2. $f(kt)\in\mathcal{D}'_+(ka)$, $k>0$. 3. $f(t)\,e^{\lambda t}\in\mathcal{D}'_+(a+\operatorname{Re}\lambda)$.

10.5. If f and g belong to $\mathcal{D}'_+(a)$, then $f*g$ belongs to $\mathcal{D}'_+(a)$ and

$$(f*g)\,e^{-\sigma t}=(fe^{-\sigma t})*(ge^{-\sigma t}),\quad \sigma>a.$$

Hint. Use the result of Problem 8.20.

10.6. If $f\in\mathcal{D}'_+(a)$, then
1. $f(t-\tau)\in\mathcal{D}'_+(a)$, $\tau\geqslant 0$. 2. $f^{(m)}\in\mathcal{D}'_+(a)$, $m=1, 2, \ldots$. 3. $f_{(m)}\in\mathcal{D}'_+(a)$, $m=1, 2, \ldots$.

10.7. 1. $\delta(t)\leftrightarrow 1$.
2. $\delta^{(m)}(t-\tau)\leftrightarrow p^m e^{-\tau p}$, $\tau\geqslant 0$, p is arbitrary, $m=0, 1, \ldots$.
3. $\theta(t)\leftrightarrow\frac{1}{p}$, $\sigma>0$. 4. $\theta(t)\,e^{i\omega t}\leftrightarrow\frac{1}{p-i\omega}$, $\sigma>0$. 5. $\theta(t)\,e^{-i\omega t}\leftrightarrow\frac{1}{p+i\omega}$, $\sigma>0$. 6. $\theta(t)\cos t\leftrightarrow\frac{p}{p^2+\omega^2}$, $\sigma>0$.
7. $\theta(t)\sin t\leftrightarrow\frac{\omega}{p^2+\omega^2}$, $\sigma>0$.
8. $\frac{\theta(t)\,t^{m-1}}{\Gamma(m)}e^{\lambda t}\leftrightarrow\frac{1}{(p-\lambda)^m}$, $\sigma>\operatorname{Re}\lambda$, $m=0, 1, \ldots$.
9. $\theta(t)\,J_0(t)\leftrightarrow\frac{1}{\sqrt{1+p^2}}$, $\sigma>0$.

10.8. If f belongs to $\mathscr{D}'_+(a)$, $f \in C^n\,(t \geqslant 0)$, and $f \leftrightarrow \mathscr{F}$, then

$$\{f^{(n)}(t)\} \leftrightarrow p^n \mathscr{F}(p) - \sum_{k=0}^{n-1} f^{(k)}(+0)\, p^{n-k-1}, \quad \sigma > a.$$

10.9. If f and g belong to $\mathscr{D}'_+(a)$, $g \in C^1\,(t \geqslant 0)$, $f \leftrightarrow \mathscr{F}$, and $g \leftrightarrow \mathscr{G}$, then

$$\int_0^t f(\tau)\{g'(t-\tau)\}\, d\tau \leftrightarrow p\mathscr{F}(p)\,\mathscr{G}(p) - g(+0)\,\mathscr{F}(p), \quad \sigma > a.$$

10.10. Solve the equation $L\dfrac{di}{dt} + Ri + \dfrac{1}{C}\displaystyle\int_0^1 i(\tau)\, d\tau = e(t)$, where $e(t)$ is a locally integrable function, $e(t) = 0$, $t < 0$.

10.11. The fundamental solution $\mathscr{E}(t)$ of the equation $\mathscr{E}^{(m)} + a_1\mathscr{E}^{(m-1)} + \ldots + a_m\mathscr{E} = \delta$ exists and is unique in the class $\mathscr{D}'_+(a)$ and satisfies the following relationship:

$$\mathscr{E}(t) \leftrightarrow \frac{1}{Q(p)}, \quad \sigma > a,$$

where $\theta(p) = p^m + a_1 p^{m-1} + \ldots + a_m$, $a = \max \operatorname{Re} \lambda_j$, with λ_j the roots of the polynomial Q.

10.12. If $f_\alpha(t)$, $-\infty < \alpha < \infty$, is the generalized function introduced in Sec. 8 for Problems 8.41 and 8.42, then

1. $f_\alpha(t) \leftrightarrow 1/p^\alpha$, $\sigma > 0$, where p^α is the branch on which $p^\alpha > 0$ at $p > 0$;

2. $f_\alpha(t)\, e^{\lambda t} \leftrightarrow \dfrac{1}{(p-\lambda)^\alpha}$; $\sigma > \operatorname{Re} \lambda$.

10.13. If $|a_k| \leqslant c(1+k)^m$, $k = 0, 1, \ldots$, then

$$\sum_{k=0}^{\infty} a_k \delta(t-k) \leftrightarrow \sum_{k=0}^{\infty} a_k e^{-kp}, \quad \sigma > 0.$$

10.14. If $f(t)$ is a T-periodic function that is absolutely integrable over a period, then

$$\theta(t) f(t) \leftrightarrow \frac{1}{1 - e^{-pT}} \int_0^T f(t)\, e^{-pt}\, dt, \quad \sigma > 0.$$

10.15. Find the solutions of the following equations in the $\mathscr{D}'_+(a)$ class (for an appropriate choice of a):

1. $(\theta \cos t) * \mathscr{E} = \delta(t)$. 2. $(\theta t \cos t) * \mathscr{E} = \delta(t)$.
3. $\mathscr{E} + 2(\theta \cos t) * \mathscr{E} = \delta(t)$.
4. $$\left.\begin{aligned} \theta * u_1 + \delta' * u_2 &= \delta(t), \\ \delta * u_1 + \delta' * u_2 &= 0. \end{aligned}\right\}$$

10.16. Suppose $\mathscr{E}_1$ is the solution to the equation $g * \mathscr{E}_1 = \theta$ in $\mathscr{D}'_+(a)$, with $\mathscr{E}_1$ a locally integrable function, $\mathscr{E}_1 \in C^1\,(t \geqslant 0)$.

Prove that the solution in $\mathscr{D}'_+(a)$ of the equation $g * u = f$, where f is a locally integrable function from $\mathscr{D}'_+(a)$, is expressed by the formula

$$u(t) = \mathscr{E}_1(+0) f(t) + \int_0^t f(\tau)\{\mathscr{E}'_1(t-\tau)\}\, d\tau.$$

10.17. Calculate the Laplace transform of the function

$$a(t) = \begin{cases} 0, & t < 0, \\ 2^k, & k < t < k+1, \quad k = 0, 1, \ldots. \end{cases}$$

10.18. Solve the equation $\chi * a = \sum_{k=0}^{\infty} 2^k \delta(t-k)$ in $\mathscr{D}'_+(\ln 2)$; the function $a(t)$ is defined in Problem 10.17.

10.19. Prove the validity of the formula $\sin t = \int_0^t J_0(t-\tau) J_0(\tau)\, d\tau$.

10.20. Solve the following Cauchy problems:

1. $u' + 3u = e^{-2t}$, $u(0) = 0$.
2. $u'' + 5u' + 6u = 12$, $u(0) = 2$, $u'(0) = 0$.
3. $$\left.\begin{aligned} &u' + 5u + 2v = e^{-t}, \\ &v' + 2v + 2u = 0, \end{aligned}\right\} \quad u(0) = 1, \quad v(0) = 0.$$

Answers to Problems of Sec. 10

10.3. 2. *Solution.* Suppose η is a function of the $C^\infty(R^1)$ class such that $\eta(t) = 0$, $t < -\delta$, and $\eta(t) = 1$, $t > -\delta/2$, with δ an arbitrary positive number. Then for all $\sigma > 0$ we have $\eta(t) e^{-\sigma t} \in \in \mathscr{S}$, $f = \eta f$, and whence $f(t) e^{-\sigma t} = f(t)\eta(t) e^{-\sigma t} \in \mathscr{S}'$.

10.6. *Hint.* Use the result of Problem 10.5 and the following formulas, respectively: (1) $f(t-\tau) = f * \delta(t-\tau)$, (2) $f^{(m)} = f * \delta^{(m)}$, and (3) $f_{(m)} = \underbrace{\theta * \theta * \ldots * \theta}_{m \text{ times}} * f$.

10.7. 9. *Hint.* Use Bessel's equation.

10.10. $\frac{1}{\sqrt{d}} \int_0^t [p_+ e^{p_+(t-\tau)} - p_- e^{p_-(t-\tau)}]\, e(\tau)\, d\tau,$

$$p_\pm = -\frac{R}{2L} \pm \frac{\sqrt{d}}{2L}, \quad d = R^2 - \frac{4L}{c}.$$

10.15. 1. $\delta'(t) + \theta(t)$. $a = 0$. 2. $\delta''(t) + 3\delta(t) + 4\theta(t)\sinh t$, $a = 1$. 3. $\delta(t) - 2\theta(t) e^t (1-t)$, $a = 1$. 4. $u_1(t) = -\delta(t) - \theta(t) e^t$, $u_2(t) = \theta(t) e^t$, $a = 1$.

10.16. *Hint.* Employ the formula in Problem 10.8.

10.17. $\frac{1-e^{-p}}{p(1-2e^{-p})}$, $\sigma > \ln 2$. **10.18.** $\sum\limits_{k=0}^{\infty} \delta'(t-k)$.

10.19. *Hint*. Use the result of Problem 10.7.9.

10.20. 1. $e^{-2t}-e^{-3t}$. 2. 2. 3. $\frac{9}{25}e^{-t}+\frac{1}{5}te^{-t}+\frac{16}{25}e^{-6t}$, $\frac{8}{25}e^{-t}-$ $-\frac{2}{5}te^{-t}+\frac{8}{25}e^{-6t}$.

11 Fundamental Solutions of Linear Differential Operators

A *generalized solution* in a region $G \subset R^n$ of a linear differential equation

$$L(x, D)u \equiv \sum_{|\alpha|=0}^{m} a_\alpha(x) D^\alpha u = f(x), \tag{11.1}$$

with $a_\alpha(x) \in C^\infty(R^n)$ and $f \in \mathscr{D}'$, is any generalized function u that satisfies this equation in G in the generalized sense, that is, for every $\varphi \in \mathscr{D}$ whose support lies in G the following equality holds:

$$(u, L^*(x, D)\varphi) = (f, \varphi),$$

where

$$L^*(x, D)\varphi = \sum_{|\alpha|=0}^{m} (-1)^{|\alpha|} D^\alpha(a_\alpha\varphi).$$

A generalized function u belongs to class $C^p(G)$ if in G it coincides with a function $u_0(x)$ of class $C^p(G)$, that is, if

$$(u, \varphi) = \int u_0(x)\varphi(x)\,dx,$$

for every $\varphi \in \mathscr{D}$, supp $\varphi \in G$.

Suppose $f \in C(G) \cap \mathscr{D}'$. A generalized function u satisfies Eq. (11.1) in region G in the classical sense if and only if it belongs to the class $C^m(G)$ and satisfies this equation in G in the generalized sense.

Let $L(D) = \sum\limits_{|\alpha|=0}^{m} a_\alpha D^\alpha$ be a linear differential operator with constant coefficients, $a_\alpha(x) = a_\alpha$. A generalized function $\mathscr{E}$ that satisfies the equation

$$L(D)\mathscr{E} = \delta(x)$$

in R^n is said to be a *fundamental solution* (*influence function*) of $L(D)$.

Every linear differential operator $L(D)$ has a fundamental solution of slow growth, and this solution is given by the formula

$$L(-i\xi)F[\mathscr{E}] = 1.$$

Suppose f belongs to $\mathscr{D}'$ and is such that the convolution $\mathscr{E} * f$ exists in $\mathscr{D}'$. Then

$$u = \mathscr{E} * f$$

is a solution of the equation $L(D)\,u = f$. This solution is unique in the class of generalized functions for which the convolution with $\mathscr{E}$ exists.

11.1. Prove that the fundamental solution of the operator $\frac{d^m}{dx^m} + a_1 \frac{d^{m-1}}{dx^{m-1}} + \ldots + a_m$ that is unique in $\mathscr{D}'_+$ is given by the formula in Problem 8.26 (for the definition of $\mathscr{D}'_+$ see Sec. 8).

11.2. Prove that each function $\mathscr{E}(x)$ in the left column is a fundamental solution of the corresponding operator in the right column:

1. $\mathscr{E}(x) = \theta(x)\, e^{\pm ax}$; $\frac{d}{dx} \mp a$.

2. $\mathscr{E}(x) = \theta(x)\, \frac{\sin ax}{a}$; $\frac{d^2}{dx^2} + a^2$.

3. $\mathscr{E}(x) = \theta(x)\, \frac{\sinh ax}{a}$; $\frac{d^2}{dx^2} - a^2$.

4. $\mathscr{E}(x) = \theta(x)\, e^{\pm ax} \frac{x^{m-1}}{(m-1)!}$; $\left(\frac{d}{dx} \mp a\right)^m$, $m = 2, 3, \ldots$.

11.3. Find the fundamental solutions for the following operators that are unique in $\mathscr{D}'_+$:

1. $\frac{d^2}{dx^2} + 4\frac{d}{dx}$. 2. $\frac{d^2}{dx^2} - 2\frac{d}{dx} + 1$. 3. $\frac{d^2}{dx^2} + 3\frac{d}{dx} + 2$. 4. $\frac{d^2}{dx^2} - 4\frac{d}{dx} + 5$. 5. $\frac{d^3}{dx^3} - a^3$. 6. $\frac{d^3}{dx^3} - 3\frac{d^2}{dx^2} + 2\frac{d}{dx}$. 7. $\frac{d^4}{dx^4} - a^4$. 8. $\frac{d^4}{dx^4} - 2\frac{d^2}{dx^2} + 1$.

11.4. Prove that

1. $\mathscr{E}(x, y) = \frac{1}{\pi z} = \frac{1}{\pi(x+iy)}$ is a fundamental solution of the Cauchy-Riemann operator $\frac{\partial}{\partial \bar{z}} = \frac{1}{2}\left(\frac{\partial}{\partial x} + i\frac{\partial}{\partial y}\right)$.

2. $\mathscr{E}(x, y) = \frac{\bar{z}^{k-1} e^{\lambda \bar{z}}}{\pi \Gamma(k)\, z}$, $k = 1, 2, \ldots$, is a fundamental solution of the operator $\left(\frac{\partial}{\partial \bar{z}} - \lambda\right)^k$.

3. $\mathscr{E}(x, y) = \frac{2\bar{z}^{k-1} z^{m-1}}{\pi \Gamma(k)\, \Gamma(m)} \ln|z|$, $k, m = 1, 2, \ldots$, is a fundamental solution of the operator $\frac{\partial^{k+m}}{\partial \bar{z}^k \partial z^m}$.

4. $\mathscr{E}(x, y)=\frac{1}{2\pi i}\frac{\operatorname{sign}\operatorname{Im}\lambda}{y-\lambda x}e^{-\mu\alpha}$ is a fundamental solution of the generalized Cauchy-Riemann operator $\frac{\partial}{\partial x}+\lambda\frac{\partial}{\partial y}+\mu$, $\operatorname{Im}\lambda\neq 0$.

11.5. Prove that $\mathscr{E}(x)=\frac{1}{2\pi}\ln|x|$ is a fundamental solution of Laplace's operator in R^2. Establish the physical meaning of this solution.

11.6. Prove that

1. $\mathscr{E}(x)=-\frac{1}{4\pi|x|}$ is a fundamental solution of Laplace's operator in R^3. Establish the physical meaning of this solution.

2. $\mathscr{E}(x)=-\frac{1}{(n-2)\sigma_n|x|^{n-2}}$, $n=3, 4, \ldots$, is a fundamental solution of Laplace's operator in R^n, where $\sigma_n=\int_{S_1}dS=\frac{2\pi^{n/2}}{\Gamma(n/2)}$ is the surface area of a unit sphere in R^n.

3. $\mathscr{E}_{n,k}(x)=\frac{(-1)^k\Gamma\left(\frac{n}{2}-k\right)}{2^{2k}\pi^{\frac{n}{2}}\Gamma(k)}|x|^{2k-n}$ is a fundamental solution of Laplace's iterated operator $(\nabla^2)^k$ at $2k<n$, $k=1, 2, \ldots$, and

$$\mathscr{E}_{n,k}(x)=\frac{1}{\pi\cdot 2^{2k-1}\Gamma(k)}|x|^{2k-2}\ln|x|,\ n=2.$$

Hint. Employ the result of Problem 9.17.2.

11.7. Prore that $\mathscr{E}(x)=-\frac{e^{ik|x|}}{4\pi|x|}$ and $\overline{\mathscr{E}}(x)=-\frac{e^{-ik|x|}}{4\pi|x|}$ are fundamental solutions of Helmholtz's operator ∇^2+k^2 in R^3.

11.8. Prove that if a function $u(x)$ satisfies in R^3 the equation $\nabla^2u+k^2u=0$ and Sommerfeld's radiation conditions

$$u(x)=O(|x|^{-1}),\quad \frac{\partial u(x)}{\partial|x|}-iku(x)=o(|x|^{-1})$$

as $|x|\to\infty$, then it is identically zero.

11.9. Prove that the following functions are fundamental solutions of Helmholtz's operator ∇^2+k^2:

1. $\mathscr{E}(x)=-\frac{i}{4}H_0^{(1)}(k|x|)$ and $\overline{\mathscr{E}}(x, y)=\frac{i}{4}H_0^{(2)}(k|x|)$ in R^2, with $H_0^{(k)}$, $k=1, 2$, Hankel functions.

2. $\mathscr{E}(x)=\frac{1}{i2k}e^{ik|x|}$ and $\overline{\mathscr{E}}(x)=-\frac{1}{i2k}e^{-ik|x|}$ in R^1.

11.10. Prove that the following functions are fundamental solutions of the operator $\nabla^2 - k^2$:

1. $\mathscr{E}(x) = -\frac{e^{-k|x|}}{4\pi|x|}$ in R^3.

2. $\mathscr{E}(x) = -\frac{1}{2\pi} K_0(k|x|)$ in R^2, where $K_0(\xi) = i\frac{\pi}{2} H_0^{(1)}(i\xi)$ is a modified Hankel function.

3. $\mathscr{E}(x) = \frac{e^{-k|x|}}{2k}$ in R^1.

4. $\mathscr{E}(x) = -\left(\frac{1}{2\pi}\right)^{\frac{n}{2}} \left(\frac{k}{|x|}\right)^{\frac{n}{2}-1} K_{\frac{n}{2}-1}(k|x|)$ in R^n.

11.11. Prove that if $\mathscr{E}_1(x, t)$ is a fundamental solution of the operator $\frac{\partial}{\partial t} + L(D_x)$, then $\frac{t^{k-1}}{\Gamma(k)} \mathscr{E}_1(x, t)$ is a fundamental solution of the operator $\left(\frac{\partial}{\partial t} + L(D_x)\right)^k$.

11.12. Prove that

1. $\mathscr{E}(x, t) = \frac{\theta(t)}{(2a\sqrt{\pi t})^n} e^{-|x|^2/(4a^2t)}$ is a fundamental solution of the heat conduction operator $\frac{\partial}{\partial t} - a^2\nabla^2$ in R^n. Establish the physical meaning of this solution.

2. $\frac{\theta(t)\, t^{k-1}}{(2a\sqrt{\pi t})^n \Gamma(k)} e^{-|x|^2/(4a^2t)}$ is a fundamental solution of the operator $\left(\frac{\partial}{\partial t} - a^2\Delta\right)^k$ in R^n, $k = 1, 2, \ldots$. *Hint.* Employ the result of Problem 11.11.

11.13. Prove that $\mathscr{E}(x, t) = \frac{\theta(t)}{2a\sqrt{\pi t}} e^{ct - \frac{(x+bt)^2}{4a^2t}}$ is a fundamental solution of the operator $\frac{\partial}{\partial t} - a^2\frac{\partial}{\partial x^2} - b\frac{\partial}{\partial x} - c$.

11.14. Prove that

1. $\mathscr{E}_1(x, t) = -\frac{i\theta(t)}{2\sqrt{\pi t}} e^{i\left(\frac{x^2}{4t} - \frac{\pi}{4}\right)}$ is a fundamental solution of Schrödinger's operator $i\frac{\partial}{\partial t} + \frac{\partial}{\partial x^2}$. *Hint.* Employ the formula

$$\int_0^\infty e^{iu^2} du = \frac{\sqrt{\pi}}{2} e^{i\pi/4}.$$

2. $\mathscr{E}_n(x, t) = -\frac{i\theta(t)}{\hbar}\left(\frac{m_0}{2\pi\hbar t}\right)^{n/2} e^{i\left[\frac{|x|^2}{2\hbar t}(m+i0)-\frac{\pi n}{4}\right]}$ is a fundamental solution of Schrödinger's operator $i\hbar\frac{\partial}{\partial t}+\frac{\hbar^2}{2m_0}\nabla^2$, with n an arbitrary positive integer.

3. $\frac{\theta(t)\,t^{k-1}}{(2a\sqrt{\pi t})^n\Gamma(k)} e^{\pm\left(\frac{|x|^2}{4ia^2t}+\frac{\pi n}{4}i\right)}$, $k=1, 2, \ldots$, is a fundamental solution of the operator $\left(\frac{\partial}{\partial t}\pm ia^2\nabla^2\right)^k$ in R^n. *Hint.* Employ the result of Problem 11.11.

11.15. Prove that

1. $\mathscr{E}_1(x, t)=\frac{1}{2a}\theta(at-|x|)$ is a fundamental solution of the one-dimensional wave operator $\Box_a$. Establish the physical meaning of this solution.

2. $\mathscr{E}_2(x, t)=\frac{\theta(at-|x|)}{2\pi a\sqrt{a^2t^2-|x|^2}}$ is a fundamental solution of the two-dimensional wave operator $\Box_a$, $x=(x_1, x_2)$. Establish the physical meaning of this solution. *Hint.* Employ the result of Problem 9.26.

11.16. Prove that

1. $\mathscr{E}_3(x, t)=\frac{\theta(t)}{4\pi a^2 t}\delta_{S_{at}}(x)=\frac{\theta(t)}{2\pi a}\delta(a^2t^2-|x|^2)$, where S_{at}: $|x|=at$, is a fundamental solution of the three-dimensional wave operator $\Box_a$, $x=(x_1, x_2, x_3)$. Establish the physical meaning of this solution. *Hint.* Employ the result of Problem 9.27.

2. $\frac{1}{8\pi a^5}\theta(at-|x|)$ is a fundamental solution of the operator $\Box_a^2$ in R^4.

3. $\frac{1}{\pi 2^{2k-1}a^{2k+1}\Gamma(k)\Gamma(k-1)}(a^2t^2-|x|^2)^{k-2}\theta(at-|x|)$ is a fundamental solution of the operator $\Box_a^k$ in R^n.

4. The fundamental solution of the operator $\Box_a$ in R^4 can be represented in the form

$$\mathscr{E}_3(x, t)=\frac{1}{8\pi a^3}\Box_a\theta(at-|x|).$$

11.17. Prove that

$$\mathscr{E}_n(x, t)=\begin{cases}\dfrac{1}{(2a)^{n-2}\pi^{\frac{n-1}{2}}\Gamma\left(\frac{n-1}{2}\right)}\Box_a^{\frac{n-3}{2}}[\theta(t)\,\delta(a^2t^2-|x|^2)], & n \text{ odd and no less than } 3\\[2ex] \dfrac{1}{(2a)^{n-1}\pi^{\frac{n}{2}}\Gamma\left(\frac{n}{2}\right)}\Box_a^{\frac{n-2}{2}}\left[\dfrac{\theta(at-|x|)}{\sqrt{a^2t^2-|x|^2}}\right], & n \text{ even,}\end{cases}$$

is a fundamental solution of the wave operator $\Box_a$. *Hint.* For odd n's emloy the formula

$$\mathscr{E}_n(x, t) = \theta(t) F_\xi^{-1}\left[\frac{\sin|\xi|t}{|\xi|}\right]$$

and the result of Problem 9.27; for even n's apply the method of descent in the variable x_{n+1}.

11.18. Prove that $\mathscr{E}(x, t) = \frac{1}{2a}\theta(at - |x|)e^{\frac{b(at-x)}{2a^2}}$ is a fundamental solution of the operator

$$\Box_a - b\frac{\partial}{\partial x} - \frac{b}{a}\frac{\partial}{\partial t}, \text{ where } a, b > 0.$$

Hint. Employ the formula

$$\frac{1}{2\pi i}\int_{\alpha-i\infty}^{\alpha+i\infty}\frac{e^{z\tau}}{z}dz = \theta(\tau), \quad \alpha > 0.$$

11.19. Prove that

1. $\mathscr{E}(x, t) = -\theta(t)\theta(-x)e^{at+bx}$ is a fundamental solution of the operator

$$\frac{\partial^2}{\partial x\,\partial t} - a\frac{\partial}{\partial x} - b\frac{\partial}{\partial t} + ab, \text{ where } b > 0$$

(see the hint to Problem 11.18).

2. $\mathscr{E}(x, t) = \theta(t)\theta(x)I_0(2m\sqrt{xy})$ is a fundamental solution of the operator $\frac{\partial^2}{\partial x\,\partial t} - m^2$ in R^2.

11.20. Prove that the function

$$\mathscr{E}(x, t) = \frac{\theta(at - |x|)}{2a}I_0\left(\frac{m}{a}\sqrt{a^2t^2 - x^2}\right)$$

is a fundamental solution of the operator $\Box_a - m^2$.

11.21.* Prove that the following functions are fundamental solutions of the Klein-Gordon-Fock operator $\Box_a + m^2$:

$$\mathscr{E}(x, t) = \frac{\theta(at - |x|)}{2a}J_0\left(\frac{m}{a}\sqrt{a^2t^2 - x^2}\right), \quad n = 1;$$

$$\mathscr{E}(x, t) = \frac{1}{2\pi a^2}\theta(at - |x|)\frac{\cos\left(\frac{m}{a}\sqrt{a^2t^2 - |x|^2}\right)}{\sqrt{a^2t^2 - |x|^2}}, \quad n = 2;$$

$$\mathscr{E}(x, t) = \frac{\theta(t)}{2\pi a}\delta(a^2t^2 - |x|^2) - \frac{m}{4\pi a^2}\theta(at - |x|)\frac{J_1\left(\frac{m}{a}\sqrt{a^2t^2 - |x|^2}\right)}{\sqrt{a^2t^2 - |x|^2}}, \quad n = 3,$$

where J_0 and J_1 are Bessel functions.

11.22. Prove that the following functions are fundamental solutions of the telegrapher's equation $\Box_a + 2m\frac{\partial}{\partial t}$

$$\mathscr{E}(x,\ t)=\frac{1}{2a}e^{-mt}\theta(at-|x|)\,I_0\left(m\sqrt{t^2-\frac{x^2}{a^2}}\right),\quad n=1;$$

$$\mathscr{E}(x,\ t)=\frac{e^{-mt}\theta(at-|x|)\cosh\left(m\sqrt{t^2-\frac{|x|^2}{a^2}}\right)}{2\pi a^2\sqrt{t^2-\frac{|x|^2}{a^2}}},\quad n=2;$$

$$\mathscr{E}(x,\ t)=\frac{\theta(at)}{2\pi a}e^{-mt}\delta(a^2t^2-|x|^2)-\frac{me^{-mt}\theta(at-|x|)\,I_1\left(m\sqrt{t^2-\frac{|x|^2}{a^2}}\right)}{4a\pi^3\sqrt{t^2-\frac{|x|^2}{a^2}}},\quad n=3,$$

where $I_0(\xi) = J_0(i\xi)$ and $I_1(\xi) = -iJ_1(i\xi)$ are modified Bessel functions. *Hint.* Employ the result of Problem 11.21*.

11.23. 1. Prove that $\mathscr{E}(x,\ t) = v\theta(t)e^{-\alpha vt}\,\delta(x - vts)$, with $(\theta(t)\,e^{-\alpha vt}\,\delta(x - vts),\ \varphi(x,\ t)) = \int_0^\infty e^{-\alpha vt}\,\varphi(vts,\ t)\,dt$, is a fundamental solution of the transport operator $\frac{1}{v}\frac{\partial\mathscr{E}}{\partial t} + (s,\ \operatorname{grad}\mathscr{E}) + \alpha\mathscr{E} = \delta(x,\ t)$, $|s| = 1$, $v > 0$, $\alpha \geqslant 0$; $n = 3$.

2. Prove that

$$\mathscr{E}^0(x)=\left(\frac{e^{-\alpha|x|}}{|x|^2}\delta\left(s-\frac{x}{|x|}\right),\ \varphi\right)=\int_0^\infty e^{-\alpha\rho}\varphi(\rho s)\,d\rho$$

is a fundamental solution of the steady state transport operator

$$(s,\ \operatorname{grad}\mathscr{E}^0) = \alpha\mathscr{E}^0 = \delta(x),\ n = 3.$$

11.24. Find the fundamental solution of the equation $Z*\mathscr{E} = \delta$, with Z defined in Problem 8.30.

11.25. Prove that if $\mathscr{E}(x,\ t)$ is a fundamental solution of the transport operator

$$L(D)=a_1\frac{\partial}{\partial x_1}+\ldots+a_n\frac{\partial}{\partial x_n}+\alpha,\quad |a|\neq 0,$$

then

$$\frac{1}{\Gamma(k)\,|a|^{2(k-1)}}(\bar{a}_1x_1+\ldots+\bar{a}_nx_n)^{k-1}\mathscr{E}(x,\ t)$$

is a fundamental solution of the operator $L^k(D)$. *Hint.* Use induction in k.

Suppose $f(x,\ t) \in \mathscr{D}'(R^{n+1})$ and $\varphi(x) \in \mathscr{D}(R^n)$. Let us introduce a generalized function $(f(x,\ t),\ \varphi(x)) \in \mathscr{D}'(R^1)$ that operates on test functions $\psi \in \mathscr{D}(R^1)$ according to the rule

$$((f(x,\ t)\ \varphi(x)),\ \ \psi(t)) = (f,\ \varphi\psi).$$

From this definition it follows that

$$\left(\frac{\partial^k f(x, t)}{\partial t^k}, \varphi(x)\right) = \frac{d^k}{dt^k}(f(x, t), \varphi(x)), \quad k = 1, 2, \ldots .$$

We say that the generalized function $f(x, t)$ *belongs to class* C^p *in the variable* t *in the interval* (a, b) if for every $\varphi \in \mathscr{D}(R^n)$ the generalized function $(f(x, t), \varphi(x))$ belongs to $C^p(a, b)$.

11.26. Prove the validity of the following propositions for the fundamental solutions $\mathscr{E}_n(x, t)$, $n = 1, 2, 3$, of the wave operator considered in Problems 11.15-11.16:

1. $\mathscr{E}_n(x, t) \in C^\infty$ in $t \in [0, \infty)$.
2. $\mathscr{E}_n(x, t) \to 0, \quad \dfrac{\partial \mathscr{E}_n(x, t)}{\partial t} \to \delta(x), \quad \dfrac{\partial^2 \mathscr{E}_n(x, t)}{\partial t^2} \to 0$

at $t \to +0$ in $\mathscr{D}'(R^n)$.

11.27. For the fundamental solution $\mathscr{E}(x, t)$ of the heat conduction operator (see Problem 11.12) prove that

$$\mathscr{E}(x, t) \to \delta(x) \text{ as } t \to +0 \text{ in } \mathscr{D}'(R^n).$$

11.28. For the fundamental solution of Schrödinger's operator (see Problem 11.14) prove that

$$\mathscr{E}_1(x, t) \to -i\delta(x) \text{ as } t \to +0 \text{ in } \mathscr{D}'(R^1).$$

11.29. For the fundamental solution in Problem 11.18 prove that

1. $\mathscr{E}(x, t) \in C^\infty$ in $t \in [0, \infty)$.
2. $\mathscr{E}(x, t) \to 0, \quad \dfrac{\partial \mathscr{E}(x, t)}{\partial t} \to \delta(x),$

$$\frac{\partial^2 \mathscr{E}(x, t)}{\partial t^2} \to -\frac{b}{a}\delta(x) \text{ as } t \to +0 \text{ in } \mathscr{D}'(R^1).$$

11.30. For the fundamental solution in Problem 11.13 prove that

$$\mathscr{E}(x, t) \to \delta(x) \text{ as } t \to +0 \text{ in } \mathscr{D}'(R^1).$$

Answers to Problems of Sec. 11

11.1. *Uniqueness.* Obviously $\mathscr{E}(x) \in \mathscr{D}'_+$. For $u = \mathscr{E} - \mathscr{E}^*$, with $\mathscr{E}^* \in \mathscr{D}'_+$ another fundamental solution, we have $L(D)u = 0$. The convolution $u * \mathscr{E}$ exists (see formula (8.8)). We have $u = u * \delta = u * L(D)\mathscr{E} = L(D)u * \mathscr{E} = 0$. Hence, $\mathscr{E}^* = \mathscr{E}$.

11.3. 1. $\theta(x)\dfrac{1-e^{-4x}}{4}$. 2. $\theta(x)xe^x$. 3. $\theta(x)(e^{-x}-e^{-2x})$. 4. $\theta(x)e^{2x} \times \sin x$. 5. $\dfrac{\theta(x)}{3a^2}\left[e^{ax} - e^{\frac{-ax}{2}}\left(\cos\dfrac{a\sqrt{3}}{2}x + \sqrt{3}\sin\dfrac{a\sqrt{3}}{2}x\right)\right]$. 6. $\dfrac{\theta(x)}{2}(1-e^x)^2$. 7. $\dfrac{\theta(x)}{2a^3}(\sinh ax - \sin ax)$. 8. $\dfrac{\theta(x)}{2}(x\cosh x - \sinh x)$.

11.12. *Solution.* Applying the Fourier transform F_x to $\dfrac{\partial \mathscr{E}}{\partial t} - a^2\nabla^2\mathscr{E} = \delta(x, t)$ and employing the results of Problems 9.21.1

and 9.21.2 and formulas of Sec. 9, we obtain $\frac{\partial \tilde{\mathscr{E}}}{\partial t} + a^2 |\xi|^2 \tilde{\mathscr{E}} = 1(\xi)\cdot\delta(t)$, where $\tilde{\mathscr{E}}(\xi, t) = F_x[\mathscr{E}(x, t)]$. Substituting $a^2|\xi|^2$ for a in the formula for $\mathscr{E}(t)$ of Problem 11.2.1, we conclude that $\tilde{\mathscr{E}}(\xi, t) = \theta(t)\, e^{-a^2|\xi|^2 t}$. This, in view of the result of Problem 9.24, yields $\mathscr{E}(x, t) = F_\xi^{-1}[\tilde{\mathscr{E}}(\xi, t)] = \frac{\theta(t)}{(2a\sqrt{\pi t})^n} e^{-|x|^2/(4a^2 t)}$.

11.15. *Hint.* See the solution of Problem 11.12. For the unknown function $Z(t) \in C^2$ we arrive at the following boundary value problem: $Z'' + a^2\xi^2 Z = 0$, $Z(0) = 0$, $Z'(0) = 1$. This yields $Z(t) = \frac{\sin a\xi t}{a\xi}$ and, hence, $\tilde{\mathscr{E}}_1(\xi, t) = \theta(t)\frac{\sin a\xi t}{a\xi}$. Then employ the result of Problem 9.25.

11.24. $\frac{\theta(t)}{L} e^{-Rt/(2L)}\left(\cos \omega t - \frac{R}{2L\omega}\sin \omega t\right)$ if $4L - CR^2 > 0$, with $\omega = \frac{\sqrt{\frac{4L}{C} - R^2}}{2L}$.

Chapter IV

The Cauchy Problem

12 The Cauchy Problem for Second-order Equations of Hyperbolic Type

We start with the Cauchy problem in the plane. The Cauchy problem for the equation

$$a(x, y)\, u_{xx} + 2b(x, y)\, u_{xy} + c(x, y) u_{yy} + d(x, y)\, u_x + e(x, y)\, u_y + f(x, y)\, u = F(x, y) \tag{12.1}$$

with the boundary conditions

$$u\Big|_\Gamma = u_0(x, y), \quad \frac{\partial u}{\partial l}\Big|_\Gamma = u_1(x, y) \tag{12.2}$$

consists in the following. Suppose that in a region D we have an equation (12.1) of the hyperbolic type ($b^2 - ac > 0$) and that on a curve Γ that belongs to D or constitutes a section of the boundary of D two functions, $u_0(x, y)$ and $u_1(x, y)$, and the direction vector $\boldsymbol{l}(x, y)$ are given. We must find a function $u(x, y)$ that is a solution of Eq. (12.1) in D and on Γ satisfies the boundary conditions (12.2).

If at each point of curve Γ the direction vector $\boldsymbol{l}$ does not lie on the tangent to Γ and if the tangents to curve Γ do not coincide with the characteristics, then in D, which is bounded by characteristics passing through the ends of Γ, there is only one solution of the Cauchy problem (12.1), (12.2), provided the coefficients of Eq. (12.1) and the data in (12.2) are sufficiently smooth.

12.1. Suppose that in the interval (a, b) we have the functions $\varphi \in C^2$, $\varphi' \neq 0$, $u_0 \in C^2$, and $u_1 \in C^1$.

Prove that the Cauchy problem

$$u_{xy} = 0, \quad a < x < b, \quad c < y < d;$$
$$u\,|_{y=\varphi(x)} = u_0(x), \quad u_y\,|_{y=\varphi(x)} = u_1(x)$$

has a unique solution

$$u(x, y) = u_0(x) + \int_x^{\varphi^{-1}(y)} u_1(\xi)\, \varphi'(\xi)\, d\xi,$$

with $c = \inf \varphi(x)$ and $d = \sup \varphi(x)$, and $\varphi^{-1}(y)$ is the inverse of the function $\varphi(x)$.

12.2. Suppose that on the interval $(-1, 1)$ we have the functions $u_0 \in C^2$ and $u_1 \in C^1$. Prove that the Cauchy problem

$$u_{xx} - u_{yy} = 0;$$
$$u|_{y=0} = u_0(x), \quad u_y|_{y=0} = u_1(x)$$

has a unique solution in the square $\{|x - y| < 1, |x + y| < 1\}$. Prove that this square is the greatest region in which the solution of the problem is unique.

12.3. Prove that the Cauchy problem

$$u_{xy} = 0, \quad -\infty < x, \ y < \infty;$$
$$u|_{y=0} = u_0(x), \quad u_y|_{y=0} = u_1(x)$$

has a solution if and only if $u_0(x) \in C^2(R^1)$ and $u_1(x) \equiv \text{const}$. Show that the solution is not unique under the specified conditions and that all solutions are given by the formula

$$u(x, y) = u_0(x) + f(y) - f(0) + y[u_1(0) - f'(0)],$$

where $f(y)$ is any function of the class $C^2(R^1)$.

12.4. Prove that the solution of the Cauchy problem

$$u_{xy} = 0, \quad |x| < 1, \quad 0 < y < 1;$$
$$u|_{y=x^2} = 0, \quad u_y|_{y=x^2} = u_1(x)$$

is unique if and only if $u_1(x) \in C(-1, 1)$ and $xu_1(x) \in C^1(-1, 1)$, with $u_1(x)$ an even function. Show that under these conditions the solution of the problem is unique and is $u(x, y) = 2\int_x^{\sqrt{y}} \xi u_1(\xi) d\xi$.

12.5. Prove that the solution of the Cauchy problem

$$u_{xy} = 0, \quad |x| < 1, \quad |y| < 1;$$
$$u|_{y=x^3} = |x|^\alpha, \quad u_x|_{y=x^3} = 0$$

exists if and only if $\alpha = 0$ or $\alpha \geqslant 6$. Show that under these conditions the solution is unique and $u(x, y) = |y|^{\alpha/3}$.

12.6. Prove that the Cauchy problem

$$u_{xx} - u_{yy} = 6(x + y), \quad -\infty < x, \ y < \infty;$$
$$u|_{y=x} = 0, \quad u_x|_{y=x} = u_1(x)$$

has a solution if and only if $u_1(x) - 3x^2 \equiv \text{const}$. Show that under these conditions the solution is not unique and that all the solutions are given by the formula

$$u(x, y) = x^3 - y^3 + f(x - y) - f(0) + (x - y)[u_1(0) - f'(0)],$$

where $f(x)$ is any function of the class $C^2(R^1)$.

In Problems 12.7-12.19 the reader must find the greatest region in which the solution of the specified Cauchy problem is unique and also find the solution.

12.7. $u_{xy}=0;$

$$u|_{y=x^2}=0, \quad u_y|_{y=x^2}=\sqrt{|x|}, \quad |x|<1.$$

12.8. $u_{xy}+u_x=0;$

$$u|_{y=x}=\sin x, \quad u_x|_{y=x}=1, \quad |x<\infty.$$

12.9. $u_{xx}-u_{yy}+2u_x+2u_y=0;$

$$u|_{y=0}=x, \quad u_y|_{y=0}=0, \quad |x|<\infty.$$

12.10. $u_{xx}-u_{yy}-2u_x-2u_y=4;$

$$u|_{x=0}=-y, \quad u_x|_{x=0}=y-1, \quad |y|<\infty,$$

12.11. $u_{xx}+2u_{xy}-3u_{yy}=2;$

$$u|_{y=0}=0, \quad u_y|_{y=0}=x+\cos x, \quad |x|<\infty.$$

12.12. $u_{xy}+yu_x+xu_y+xyu=0;$

$$u|_{y=3x}=0, \quad u_y|_{y=3x}=e^{-5x^2}, \quad x<1.$$

12.13. 1. $xu_{xx}-u_{yy}+\frac{1}{2}u_x=0;$

$$u|_{y=0}=x, \quad u_y|_{y=0}=0, \quad x>0.$$

2. $xu_{xy}-yu_{yy}-u_y=2x^3; \quad u|_{y=x}=\sin x, \quad u_x|_{y=x}=\cos x, \quad x>0.$

12.14. $xu_{xx}+(x+y)u_{xy}+yu_{yy}=0;$

$$u\Big|_{y=\frac{1}{x}}=x^3, \quad u_x\Big|_{y=\frac{1}{x}}=2x^2, \quad x>0.$$

12.15. $u_{xx}+2(1+2x)u_{xy}+4x(1+x)u_{yy}+2u_y=0;$

$$u|_{x=0}=y, \quad u_x|_{x=0}=2, \quad |y|<\infty.$$

12.16. 1. $x^2u_{xx}-y^2u_{yy}-2yu_y=0;$

$$u|_{x=1}=y, \quad u_x|_{x=1}=y, \quad y<0.$$

2. $u_{xx}-4x^2u_{yy}-\frac{1}{x}u_x=0; \quad u|_{x=1}=y^2+1, \quad u_x|_{x=1}=4, \quad |y|<\infty.$

12.17. $x^2u_{xx}-2xyu_{xy}-3y^2u_{yy}=0;$

$$u|_{y=1}=0, \quad u_y|_{y=1}=\sqrt{x^7}, \quad x>0.$$

12.18. $yu_{xx}+x(2y-1)u_{xy}-2x^2u_{yy}-\frac{y}{x}u_x+\frac{2x}{1+2y}(u_x+2xu_y)=0;$

$$u|_{y=0}=x^2, \quad u_y|_{y=0}=1, \quad x>0.$$

12.19. $yu_{xx}-(x+y)u_{xy}+xu_{yy}-\frac{x+y}{x-y}(u_x-u_y)=0;$

$$u|_{y=0}=x^2, \quad u_y|_{y=0}=x, \quad x>0.$$

Solve Problems 12.20-12.24 by Riemann's method.

12.20. $u_{xy}+2u_x+u_y+2u=1, \quad 0<x, \quad y<1;$

$$u|_{x+y=1}=x, \quad u_x|_{x+y=1}=x.$$

12.21. $xyu_{xy}+xu_x-yu_y-u=2y,\quad 0<x,\ y<\infty.$

$$u|_{xy=1}=1-y,\quad u_y|_{xy=1}=x-1.$$

12.22. $u_{xy}+\frac{1}{x+y}(u_x+u_y)=2,\quad 0<x,\quad y<\infty;$

$$u|_{y=x}=x^2,\quad u_x|_{y=x}=1+x.$$

12.23. $u_{xx}-u_{yy}+\frac{2}{x}u_x-\frac{2}{y}u_y=0,\ |x-y|<1,\ |x+y-2|<1;$

$$u|_{y=1}=u_0(x),\quad u_y|_{y=1}=u_1(x),\quad u_0\in C^2(0,\ 2),\quad u_1\in C^1(0,\ 2).$$

12.24. $2u_{xy}-e^{-x}u_{yy}=4x,\quad -\infty<x,\ y<\infty;$

$$u|_{y=x}=x^5\cos x,\quad u_y|_{y=x}=x^2+1.$$

The *classical Cauchy problem for the wave equation* is the problem of finding a function $u(x, t)$ of the class $C^2(t>0)\cap C^1(t\geqslant 0)$ that satisfies the equation

$$u_{tt}=a^2\nabla^2 u+f(x,\ t) \tag{12.3}$$

at $t>0$ and the initial conditions

$$u|_{t=0}=u_0(x),\quad u_t|_{t=0}=u_1(x) \tag{12.4}$$

where f, u_0, and u_1 are known functions

If

$$\begin{aligned} &f\in C^1(t\geqslant 0)\quad u_0\in C^2(R^1),\quad u_1\in C^1(R^1),\quad n=1,\\ &f\in C^2(t\geqslant 0)\quad u_0\in C^3(R^n),\quad u_1\in C^2(R^n),\quad n=2,\ 3 \end{aligned} \tag{12.5}$$

then the Cauchy problem (12.3),(12.4) has a solution that is unique and is expressed by

D'Alembert's formula when $n=1$,

$$u(x,\ t)=\frac{1}{2}[u_0(x+at)+u_0(x-at)]$$
$$+\frac{1}{2a}\int_{x-at}^{x+at}u_1(\xi)\,d\xi+\frac{1}{2a}\int_0^t\int_{x-a(t-\tau)}^{x+a(t-\tau)}f(\xi,\ \tau)\,d\xi\,d\tau; \tag{12.6}$$

Poisson's formula when $n=2$,

$$u(x,\ t)=\frac{1}{2\pi a}\int_0^t\int_{|\xi-x|<a(t-\tau)}\frac{f(\xi,\ \tau)\,d\xi\,d\tau}{\sqrt{a^2(t-\tau)^2-|\xi-x|^2}}$$
$$+\frac{1}{2\pi a}\int_{|\xi-x|<at}\frac{u_1(\xi)\,d\xi}{\sqrt{a^2t^2-|\xi-x|^2}}$$
$$+\frac{1}{2\pi a}\frac{\partial}{\partial t}\int_{|\xi-x|<at}\frac{u_0(\xi)\,d\xi}{\sqrt{a^2t^2-|\xi-x|^2}}; \tag{12.7}$$

Kirchhoff's formula when $n=3$,

$$u(x, t)=\frac{1}{4\pi a^2}\int\limits_{|\xi-x|<at}\frac{1}{|\xi-x|}f\left(\xi, t-\frac{|\xi-x|}{a}\right)d\xi$$
$$+\frac{1}{4\pi a^2 t}\int\limits_{|\xi-x|=at}u_1(\xi)\,dS+\frac{1}{4\pi a^2}\frac{\partial}{\partial t}\left[\frac{1}{t}\int\limits_{|\xi-x|=at}u_0(\xi)\,dS\right]. \quad (12.8)$$

12.25. Suppose that $u(x, t)$ is a solution of the Cauchy problem

$$u_{tt}=a^2u_{xx};\quad u|_{t=0}=u_0(x),\quad u_t|_{t=0}=u_1(x).$$

Prove that for every $T>0$ the following Cauchy problem has a solution:

$$v_{tt}=a^2v_{xx},\quad t<T,\quad x\in R^1\quad v|_{t=T}=u|_{t=T},\quad v_t|_{t=T}=u_t|_{t=T}.$$

Show that $u(x, t)\equiv v(x, t)$ at $0\leqslant t\leqslant T$.

12.26. Prove that if the Cauchy problem

$$u_{tt}=a^2u_{xx};\quad u|_{t=0}=u_0(x);\ u_t|_{t=0}=u_1(x)$$

has a solution, then $u\in C^2(t\geqslant 0)$, $u_0\in C^2(R^1)$, $u_1\in C^1(R^1)$.

12.27. Suppose $u(x, t)$ is a solution of the Cauchy problem

$$u_{tt}=a^2\nabla^2u\quad u|_{t=0}=\varphi(x),\quad u_t|_{t=0}=0.$$

Prove that the function $v(x, t)=\int\limits_0^t u(x, \tau)\,d\tau$ is a solution of the Cauchy problem

$$v_{tt}=a^2\nabla^2v,\quad v|_{t=0}=0,\quad v_t|_{t=0}=\varphi(x).$$

12.28. Suppose $u(x, t, t_0)$ is a solution to the Cauchy problem

$$u_{tt}=a^2\nabla^2u\quad u|_{t=t_0}=0,\quad u_t|_{t=t_0}=f(x, t_0)$$

for each nonnegative value of t_0. Prove that the function $v(x, t, t_0)=\int\limits_{t_0}^t u(x, t, \tau)\,d\tau$ is a solution of the Cauchy problem

$$v_{tt}=a^2\nabla^2v+f(x, t),\quad v|_{t=t_0}=0,\quad v_t|_{t=t_0}=0.$$

12.29. Prove that if the functions $f(x)$, $u_0(x)$ and $u_1(x)$ are harmonic in R^n and $g(t)$ belongs to $C^1(t\geqslant 0)$, then the solution of the Cauchy problem

$$u_{tt}=a^2\nabla^2u+g(t)f(x);\quad u|_{t=0}=u_0(x),\quad u_t|_{t=0}=u_1(x)$$

is given by the formula

$$u(x, t)=u_0(x)+tu_1(x)+f(x)\int\limits_0^t(t-\tau)g(\tau)\,d\tau.$$

12.30. Find the solution of the Cauchy problem

$$u_{tt} = a^2\nabla^2 u + f(x); \quad u|_{t=0} = u_0(x), \quad u_t|_{t=0} = u_1(x),$$

if $(\nabla^2)^N f = 0$, $(\nabla^2)^N u_0 = 0$, and $(\nabla^2)^N u_1 = 0$.

12.31. Prove that for the Cauchy problem

$$u_{tt} = a^2\nabla^2 u, \quad x \in R^2; \quad u|_{t=0} = f(x_1) + g(x_2) \quad u_t|_{t=0} = F(x_1) + G(x_2)$$

to have a solution, it is sufficient that the functions $f(x_1)$ and $g(x_2)$ belong to class $C^2(R^1)$ and the functions $F(x_1)$ and $G(x_2)$ to class $C^1(R^1)$. Find the solution.

12.32. Prove that for the Cauchy problem

$$u_{tt} = a^2\nabla^2 u \quad x \in R^3;$$

$$u|_{t=0} = f(x_1)\, g(x_2, x_3), \quad u_t|_{t=0} = 0$$

to have a solution, it is sufficient that the function $g(x_2, x_3)$ be harmonic and $f(x_1)$ belong to $C^2(R^1)$. Find the solution.

12.33. Prove that for the Cauchy problem

$$u_{tt} = a^2\nabla^2 u \quad x \in R^3;$$

$$u|_{t=0} = \alpha(|x|), \quad u_t|_{t=0} = \beta(|x|)$$

to have a solution, it is sufficient that $\alpha(r) \in C^3 (r \geqslant 0)$, $\beta(r) \in C^2 (r \geqslant 0)$, and $\alpha'(0) = 0$. Find the solution.

12.34. Prove that the Cauchy problem

$$u_{tt} = \nabla^2 u, \quad x \in R^3;$$

$$u|_{t=0} = \theta(1 - |x|)\, |x|^{\alpha} (1 - |x|)^{\beta}, \quad u_t|_{t=0} = 0$$

has a solution if and only if $\alpha \geqslant 2$ and $\beta \geqslant 3$. Find this solution. Compare the result with the sufficient conditions (9.5) for $2 < \alpha < 3$, $\beta \geqslant 3$ and $\alpha = 2$, $2 < \beta < 3$.

12.35. Solve Cauchy problem

$$u_{tt} = u_{xx};$$

$$u|_{t=0} = \theta(1 - |x|)(x^2 - 1)^3, \quad u_t|_{t=0} = 0.$$

Construct the graph of the functions $u(x, 0)$, $u(x, 1/2)$, $u(x, 1)$, and $u(x, 2)$.

The solutions of Problems 12.36-12.38 can be found via formulas (9.6)-(9.8), but in some cases it is expedient to use the Fourier's method of variable separation or the results of Problems 12.27-12.32.

12.36. Solve the following Cauchy problems ($n = 1$):

1. $u_{tt} = u_{xx} + 6$; $u|_{t=0} = x^2$, $u_t|_{t=0} = 4x$.
2. $u_{tt} = 4u_{xx} + xt$; $u|_{t=0} = x^2$, $u_t|_{t=0} = x$.
3. $u_{tt} = u_{xx} + \sin x$; $u|_{t=0} = \sin x$, $u_t|_{t=0} = 0$.
4. $u_{tt} = u_{xx} + e^x$; $u|_{t=0} = \sin x$, $u_t|_{t=0} = x + \cos x$.

5. $u_{tt}=9u_{xx}+\sin x;\quad u\,|_{t=0}=1,\quad u_t\,|_{t=0}=1.$
6. $u_{tt}=a^2u_{xx}+\sin \omega x;\quad u\,|_{t=0}=0,\quad u_t\,|_{t=0}=0.$
7. $u_{tt}+a^2u_{xx}+\sin \omega t;\quad u\,|_{t=0}=0,\quad u_t\,|_{t=0}=0.$

12.37. Solve the following Cauchy problems ($n = 2$):

1. $u_{tt}=\nabla^2u+2;\quad u\,|_{t=0}=x,\quad u_t\,|_{t=0}=y.$
2. $u_{tt}=\nabla^2u+6xyt;\quad u\,|_{t=0}=x^2-y^2,\quad u_t\,|_{t=0}=xy.$
3. $u_{tt}=\nabla^2u+x^3-3xy^2;\quad u\,|_{t=0}=e^x\cos y,\quad u_t\,|_{t=0}=e^y\sin x.$
4. $u_{tt}=\nabla^2u+t\sin y;\quad u\,|_{t=0}=x^2,\quad u_t\,|_{t=0}-\sin y.$
5. $u_{tt}=2\nabla^2u;\quad u\,|_{t=0}=2x^2-y^2,\quad u_t\,|_{t=0}=2x^2+y^2.$
6. $u_{tt}=3\nabla^2u+x^3+y^3;\quad u\,|_{t=0}=x^2,\quad u_t\,|_{t=0}=y^2.$
7. $u_{tt}=\nabla^2u+e^{3x+4y};\quad u\,|_{t=0}=u_t\,|_{t=0}=e^{3x+4y},$
8. $u_{tt}=a^2\nabla^2u;\quad u\,|_{t=0}=\cos(bx+cy),\quad u_t\,|_{t=0}=\sin(bx+cy).$
9. $u_{tt}=a^2\nabla^2u;\quad u_{t=0}=r^4,\quad u_t\,|_{t=0}=r^4.$
10. $u_{tt}=a^2\nabla^2u+r^2e^t;\quad u\,|_{t=0}=0;\quad u_t\,|_{t=0}=0.$

12.38. Solve the following Cauchy problems ($n=3$)

1. $u_{tt}=\nabla^2u+2xyz;\quad u\,|_{t=0}=x^2+y^2-2z^2,\quad u_t\,|_{t=0}=1.$
2. $u_{tt}=8\nabla^2u+t^2x^2;\quad u\,|_{t=0}=y^2,\quad u_t\,|_{t=0}=z^2.$
3. $u_{tt}=3\nabla^2u+6r^2;\quad u\,|_{t=0}=x^2y^2z^2,\quad u_t\,|_{t=0}=xyz.$
4. $u_{tt}=\nabla^2u+6te^{x\sqrt{2}}\sin y\cos z;$
$$u\,|_{t=0}=e^{x+y}\cos z\sqrt{2},\quad u_t\,|_{t=0}=e^{3y+4z}\sin 5x.$$
5. $u_{tt}=a^2\nabla^2u,\quad u\,|_{t=0}=u_t\,|_{t=0}=r^4.$
6. $u_{tt}=a^2\nabla^2u+r^2e^t;\quad u\,|_{t=0}=u_t\,|_{t=0}=0.$
7. $u_{tt}=a^2\nabla^2u+\cos x\sin ye^z;$
$$u\,|_{t=0}=x^2e^{y+z},\quad u_t\,|_{t=0}=\sin xe^{y+z}.$$
8. $u_{tt}=a^2\nabla^2u+xe^t\cos(3y+4z);$
$$u\,|_{t=0}=xy\cos z,\quad u_t\,|_{t=0}=yze^x.$$
9. $u_{tt}=a^2\nabla^2u,\quad u\,|_{t=0}=u_t\,|_{t=0}=\cos r.$

12.39. Suppose that the sufficient conditions (9.5) for the existence of a solution to the Cauchy problem

$$u_{tt}=a^2\nabla^2u;\quad u\,|_{t=0}=u_0(x),\quad u_t\,|_{t=0}=u_1(x)$$

are met and suppose that at $|x|\geqslant\delta>0$,

$$m\,|x|^\alpha\leqslant u_0(x)\leqslant M\,|x|^\alpha,\quad m\,|x|^{\alpha-1}\leqslant u_1(x)\leqslant M\,|x|^{\alpha-1},$$

where $\alpha>0$, $0<m<M$. Prove that for each point x_0 there are positive numbers t_0, C_1, and C_2 such that for all $t\geqslant t_0$ the following estimate holds:

$$C_1t^\alpha\leqslant u(x_0,t)\leqslant C_2t^\alpha.$$

12.40. Suppose that the sufficient conditions (9.5) for the existence of a solution to the Cauchy problem

$$u_{tt} = a^2\nabla^2 u; \quad u|_{t=0} = u_0(x), \quad u_t|_{t=0} = u_1(x)$$

are met and suppose that at $\alpha > 0$,

$$\lim_{|x|\to\infty} \frac{u_0(x)}{|x|^{\alpha}} = A, \qquad \lim_{|x|\to\infty} \frac{u_1(x)}{|x|^{\alpha-1}} = B.$$

Prove that $\lim\limits_{t\to+\infty} \frac{u(x, t)}{t^{\alpha}} = C_n$ and find the C_n, $n = 1, 2, 3$.

If the solution $u(x, t)$ of the classical Cauchy problem for the wave equation (namely, (12.3) and (12.4)) and the function $f(x, t) \in C(t \geqslant 0)$ are continued to zero at $t < 0$, then in R^{n+1} the function $u(x, t)$ satisfies (in the generalized sense) the equation

$$u_{tt} = a^2\nabla^2 u + f(x, t) + u_0(x)\cdot\delta'(t) + u_1(x)\cdot\delta(t).$$

The *generalized Cauchy problem* for the wave equation with the source function $F \in \mathscr{D}'(R^{n+1})$, $F(x, t) = 0$ at $t < 0$, is the problem of finding a generalized function $u \in \mathscr{D}'(R^{n+1})$ that satisfies the wave equation

$$u_{tt} = a^2\nabla^2 u + F(x, t) \tag{12.9}$$

and vanishes at $t < 0$.

A solution of the generalized Cauchy problem (12.9) exists, is unique, and is given by the formula

$$u = \mathscr{E}_n * F, \tag{12.10}$$

where $\mathscr{E}_n(x, t)$ is the fundamental solution of the wave operator:

$$\mathscr{E}_1(x, t) = \frac{1}{2a}\,\theta(at - |x|),$$

$$\mathscr{E}_2(x, t) = \frac{\theta(at - |x|)}{2\pi a\sqrt{a^2t^2 - |x|^2}},$$

$$\mathscr{E}_3(x, t) = \frac{\theta(t)}{4\pi a^2 t}\,\delta_{S_{at}}(x).$$

The convolution $V_n = \mathscr{E}_n * F$ is called the *generalized wave (retarded) potential with a density F*. In particular, if $F = u_1(x)\cdot\delta(t)$ or $F = u_0(x)\cdot\delta'(t)$, then the convolutions

$$V_n^{(0)} = \mathscr{E}_n(x, t) * [u_1(x)\cdot\delta(t)] = \mathscr{E}_n(x, t) * u_1(x),$$

$$V_n^{(1)} = \mathscr{E}_n(x, t) * [u_0(x)\cdot\delta'(t)] = (\mathscr{E}_n(x, t) * u_0(x))_t$$

are called the *generalized surface wave (retarded) potentials (of the simple and double layers with densities u_1 and u_0, respectively)*.

The wave (retarded) potential V_n satisfies Eq. (12.9).

12.41. Prove that if $F(x, t) \in \mathscr{D}'(R^{n+1})$, $F = 0$ at $t < 0$, then $\mathscr{E}_n * F$ exists in $\mathscr{D}'(R^{n+1})$.

12.42. Prove that the generalized Cauchy problem (12.9) has a unique solution in the class of generalized functions belonging to $\mathscr{D}'(R^{n+1})$ that vanish at $t < 0$.

12.43. Prove that
1. $V_n^{(0)}$ and $V_n^{(1)}$ belong to the class C^∞ in $t \in (0, \infty)$.
2. $V_n^{(0)}$ and $V_n^{(1)}$ satisfy the initial conditions (as $t \to +0$)

$$V_n^{(0)}(x, t) \to 0, \quad \frac{\partial V_n^{(0)}(x, t)}{\partial t} \to u_1(x) \text{ in } \mathscr{D}'(R^n),$$

$$V_n^{(1)}(x, t) \to u_0(x), \quad \frac{\partial V_n^{(1)}(x, t)}{\partial x} \to 0 \text{ in } \mathscr{D}'(R^n).$$

12.44. Solve the generalized Cauchy problem (12.9) ($x \in R^1$) with the following source functions $F(x, t)$:
1. $\delta(t)\cdot\delta(x)$. 2. $\delta(t - t_0)\cdot\delta(x - x_0)$, $t_0 \geqslant 0$.
3. $\delta(t)\cdot\delta'(x)$. 4. $\delta'(t)\cdot\delta(x)$.
5. $\delta'(t - t_0)\cdot\delta(x)$. 6. $\delta(t)\cdot\delta'(x_0 - x)$.
7. $\delta''(t)\cdot\delta(x)$. 8. $\delta(t)\cdot\delta''(x)$.
9. $\delta(t)\cdot\alpha(x)\,\delta(x)$, with $\alpha(x) \in C$ and $\alpha(0) = 0$.
10. $\delta(t)\cdot\beta(x)\,\delta(x)$, with $\beta(x) \in C$ and $\beta(0) = 1$.

In stating the generalized Cauchy problems below we will assume that the source function $F(x, t)$ has the form $f(x, t) + u_0(x)\cdot\delta'(t) + u_1(x)\cdot\delta(t)$, with $f = 0$ at $t < 0$.

12.45. Solve the generalized Cauchy problem with the following source functions ($x \in R^1$):
1. $f = \omega(t)\cdot\delta(x)$, where $\omega(t) \in C$ ($t \geqslant 0$), $\omega(t) = 0$ at $t < 0$, and $u_0 = u_1 = \delta(x)$.
2. $f = \theta(t)\cdot\delta(x)$, $u_0 = \delta(x - x_0)$, $u_1 = x\delta(x)$.
3. $f = \theta(t)\,t\cdot\delta(x)$, $u_0 = \delta(2 - x)$, $u_1 = \delta(3 - x)$; $a = 1$.
4. $f = \theta(t)\sin t\cdot\delta(x - x_0)$, $u_0 = 0$, $u_1 = x\delta'(x)$.
5. $f = \theta(t)\cos t\cdot\delta(x)$, $u_0 = 0$, $u_1 = x^2\delta''(x)$.
6. $f = \theta(t)\,e^{\alpha t}\cdot\delta(x)$, $u_0 = \delta(1 - |x|)$, $u_1 = 0$.
7. $f = \frac{\theta(t)}{\sqrt{t}}\cdot\delta(2 - x)$, $u_0 = 0$, $u_1 = \delta(R - |x|)$; $a = 1$.
8. $f = \theta(t)\,t^2\cdot\delta(x)$, $u_0 = C = \text{const}$, $u_1 = \theta'(R - |x|)$; $a = 1$.
9. $f = \theta(t)\ln t\cdot\delta(x)$, $u_0 = \frac{1}{1+x^2}\delta(x)$, $u_1 = 0$.
10. $f = \frac{\theta(t-1)}{1+t^2}\cdot\delta(x)$, $u_0 = \theta'(2 - |x|)$, $u_1 = 0$; $a = 1$.
11. $f = 0$, $u_0 = 0$, $u_1 = \theta''(2 - |x|)$; $a = 1$.
12. $f = \frac{\theta(t)}{1+t}\cdot\delta(x - 1)$, $u_0 = 0$, $u_1 = \sin x\delta'(x - \pi)$.
13. $f = \theta(at - |x|)$, $u_0 = u_1 = 0$.
14. $f = \theta(t)(\alpha t + \beta)\cdot x\delta'(x)$, $u_0 = 0$, $u_1 = x\delta''(x)$; $a = 1$.

12.46. Prove that if $u_1(x)$ is a locally integrable function in R^1, then $V_1^{(0)}(x, t)$ is a continuous function in R^2 and is given by the formula

$$V_1^{(0)}(x, t) = \frac{\theta(t)}{2a} \int_{x-at}^{x+at} u_1(\xi)\, d\xi. \qquad (12.11)$$

12.47. Prove that if $u_0(x)$ is a locally integrable function in R^1, then $V_1^{(1)}(x, t)$ is a continuous function in R^2 and is given by the formula

$$V_1^{(1)}(x, t) = \frac{\theta(t)}{2} [u_0(x+at) + u_0(x-at)]. \qquad (12.12)$$

Hint. Employ the fact that $V_1^{(1)} = \frac{\partial}{\partial t}[\mathcal{E}_1 * u_0(x)]$, in view of the results of Problems 8.35 and 12.46.

12.48. Prove that if $f(x, t)$ is a locally integrable function in R^2 that vanishes at $t < 0$, then the potential $V_1(x, t)$ belongs to $C(R^2)$ and is expressed by the formula

$$V_1(x, t) = \frac{1}{2a} \int_0^t \int_{x-a(t-\tau)}^{x+a(t-\tau)} f(\xi, \tau)\, d\xi\, d\tau. \qquad (12.13)$$

12.49. Solve the following generalized problems

1. $u_{tt} = a^2 u_{xx} + \theta(x) \cdot \delta'(t) + \theta(x) \cdot \delta(t)$.
2. $u_{tt} = a^2 u_{xx} + \theta(t)(x-1) + x \cdot \delta'(t) + \operatorname{sign} x \cdot \delta(t)$.
3. $u_{tt} = a^2 u_{xx} + \theta(t)\, tx + \frac{\theta(x)}{\sqrt{x}} \cdot \delta(t)$.
4. $u_{tt} = u_{xx} + \frac{\theta(t)}{t+1} + \theta(-x) \cdot \delta(t)$.
5. $u_{tt} = u_{xx} + \theta(t-2) \ln t + |x| \cdot \delta'(t)$.
6. $u_{tt} = a^2 u_{xx} + \theta(t)\, t^m + \theta(2-|x|) \cdot \delta'(t), \quad m = 1, 2, \ldots$.
7. $u_{tt} = u_{xx} + \theta(t)\, e^{x+t} + \theta(x)\, e^{-x} \cdot \delta(t)$.
8. $u_{tt} = 9u_{xx} + \theta(t-\pi) \cos t + \theta(x-3) \cdot \delta'(t) + 1(x) \cdot \delta(t)$.
9. $u_{tt} = u_{xx} + \theta(t)\, \theta(x)$.
10. $u_{tt} = u_{xx} + 2\theta(t)\, \theta(x)\, x + e^{\alpha x} \cdot \delta(t), \quad \alpha \neq 0$.
11. $u_{tt} = u_{xx} + \theta(t-1)(x+t) + |x| \cdot \delta(t)$.
12. $u_{tt} = u_{xx} + \theta(t-2)\, t + \theta(x-1) \ln x \cdot \delta'(t)$.
13. $u_{tt} = u_{xx} + \theta(x)\, x^m \cdot \delta'(t) + \theta(x)\, x^m \cdot \delta(t), \quad m = 1, 2, \ldots$.
14. $u_{tt} = u_{xx} + \frac{\theta(t)}{\sqrt{t}} + \theta(x) \cos x \cdot \delta(t)$.
15. $u_{tt} = u_{xx} + \theta(t) \sqrt{t}\, x + \theta(-x) \cdot \delta'(t) + \theta(-x)\, x \cdot \delta(t)$.
16. $u_{tt} = u_{xx} + \theta(x)\, e^{-\sqrt{x}} \cdot \delta'(t) + x^2 \cdot \delta(t)$.
17. $u_{tt} = u_{xx} + \theta(t) \sin(x+t) + \sin x \cdot \delta(t)$.
18. $u_{tt} = u_{xx} + \theta(1-|x|) \cdot \delta(t)$.

12.50. Prove that

1. If $u_0 \in C^2(R^1)$ and $u_1 \in C^1(R^1)$, then the potentials $V_1^{(0)}$ and $V_1^{(1)}$ belong to the class $C^2(t \geqslant 0)$ and satisfy the equation $\Box_a u = 0$

at $t > 0$ and the initial conditions

$$V_1^{(0)}\,|_{t=+0}=0, \quad (V_1^{(0)})_t\,|_{t=+0}=u_1(x),$$
$$V_1^{(1)}\,|_{t=+0}=u_0(x), \quad (V_1^{(1)})_t\,|_{t=+0}=0.$$

Hint. The required properties follow directly from (12.11) and (12.12).

2. If $f \in C^1 (t \geqslant 0)$, then the potential $V_1 \in C^2 (R^2)$ satisfies the equation $\Box_a u = f(x, t)$ at $t > 0$ and the initial conditions

$$V_1\,|_{t=+0}=0, \quad (V_1)_t\,|_{t=+0}=0.$$

Hint. The required properties follow directly from (12.13).

12.51. Suppose that in the (generalized) Cauchy problem

$$u_{tt} = a^2 u_{xx} + u_0(x)\cdot\delta'(t) + u_1(x)\cdot\delta(t)$$

the function u_0 belongs to C^2 and the function u_1 to C^1 for all x's except $x = x_0$, where both functions (or their derivatives) have jump discontinuities. Show that the solution of this problem is classical everywhere in the half-plane $t > 0$ except at points that lie on the characteristics that pass through the point $x = x_0$, $t = 0$ (the decay of a discontinuity) for the following cases:

1. $u_0 = \theta(x)\,\omega(x)$, where $\omega \in C^2(R^1)$, $\omega(0) \neq 0$, and $u_1 = 0$.
2. $u_0 = 0$, $u_1 = \theta(x - x_0)\,\omega(x)$, where $\omega \in C^1(R^1)$ and $\omega(x_0) \neq 0$.
3. $u_0 = \theta(x - 1)$, $u_1 = \theta(x - 2)$.

12.52. For the Cauchy problem (12.9) prove that

1. the source of perturbation

$$F = u_0(x)\cdot\delta'(t) = \theta(x_0 - |x|)f(x)\cdot\delta'(t), \quad x_0 > 0,$$

$f \in C^2(R^1)$, generates two waves that at each moment $t > 0$ have a *front edge* at points $x = \pm(at + x_0)$, respectively, and at each moment $t > x_0/a$ have a *rear edge* at points $x = \pm(at - x_0)$ (*Huygens's principle*);

2. the source of perturbation

$$F = u_1(x)\cdot\delta(t) = \theta(x_0 - |x|)\,f(x)\cdot\delta(t), \quad x_0 > 0,$$

$f \in C^1(R^1)$, generates two waves that at each moment $t > 0$ have a *front edge* at points $x = \pm(at + x_0)$, but do not have a *rear edge* (*wave diffusion*). *Hint.* Employ (12.11) and (12.12).

12.53. Solve the following generalized problems and prove that the solutions are those of the classical Cauchy problem (12.3), (12.4):

1. $u_{tt} = a^2 u_{xx} + \theta(t)(x + t) + e^{\alpha x}\cdot\delta'(t)$.
2. $u_{tt} = a^2 u_{xx} + \theta(t)\,t\,\ln t + 3^x\cdot\delta'(t)$.
3. $u_{tt} = a^2 u_{xx} + \theta(t)(x^2 + t^2) + x^m\cdot\delta'(t)$, $m = 1, 2, \ldots$.
4. $u_{tt} = u_{xx} + \theta(t)\,x^2 + \cos x\cdot\delta'(t) + \cos x\cdot\delta(t)$.
5. $u_{tt} = a^2 u_{xx} + x^2 \ln|x|\cdot\delta(t)$.
6. $u_{tt} = u_{xx} + \theta(t)\cos(x + t) + 2^x\cdot\delta(t)$.
7. $u_{tt} = u_{xx} + \theta(t)\sin t + \frac{1}{1+x^2}\cdot\delta(t)$.

8. $u_{tt}=a^2u_{xx}+\theta(t)e^t+\frac{1}{1+x^2}\cdot\delta'(t)$.
9. $u_{tt}=u_{xx}+(\alpha x^2+\beta)\cdot\delta'(t)+x^{4/3}\cdot\delta(t)$.
10. $u_{tt}=u_{xx}+\ln(1+e^x)\cdot\delta'(t)+e^{-x^2}\cdot\delta(t)$.
11. $u_{tt}=u_{xx}+\theta(t)\, t^m x+\sin(x)\cdot\delta'(t)+x^m\delta(t)$, $m=1, 2, \ldots$.
12. $u_{tt}=u_{xx}+\theta(t)\arctan t+\ln(1+x^2)\cdot\delta'(t)$.
13. $u_{tt}=4u_{xx}+\theta(t)\cos x+\sqrt{1+x^2}\cdot\delta'(t)$.
14. $u_{tt}=u_{xx}+\theta(t)\, x\sin t+x^2e^{-|x|}\cdot\delta'(t)$.
15. $u_{tt}=4u_{xx}+e^{-x^2}\cdot\delta'(t)+e^{-x}\sin x\cdot\delta(t)$.
16. $u_{tt}=u_{xx}+\sin^2 x\cdot\delta'(t)+xe^{-|x|}\cdot\delta(t)$.
17. $u_{tt}=u_{xx}+\theta(t)\frac{x}{1+t^2}+\frac{1}{2-\cos x}\cdot\delta'(t)$.
18. $u_{tt}=u_{xx}+\theta(t)(xe^t+te^x)+\frac{1}{\sqrt{1+x^2}}\cdot\delta(t)$.

12.54. Solve the generalized Cauchy problem for each wave equation ($x\in R^2$):
1. $u_{tt}=a^2\nabla^2u+\theta(t)\cdot\delta(x)+\delta(x)\cdot\delta'(t)+\delta(x)\cdot\delta(t)$.
2. $u_{tt}=a^2\nabla^2u+\theta(t)\, t^2\cdot\delta(x)+|x|^m\delta(x)\cdot\delta'(t)+$ $+\delta(x-x^0)\cdot\delta(t)$, $m=1, 2, \ldots$.
3. $u_{tt}=a^2\nabla^2u+\omega(t)\cdot\delta(x)+e^{|x|}\delta(x)\cdot\delta(t)$, where $\omega\in C(t\geqslant 0)$ and $\omega=0$ at $t<0$.
4. $u_{tt}=a^2\nabla^2u+\theta(t)(\alpha t+\beta)\cdot\delta(x)+\delta(x-x_0)\cdot\delta(t)$.

12.55. Solve the generalized Cauchy problem for each wave equation ($x\in R^3$):
1. $u_{tt}=a^2\nabla^2u+\theta(t)\cdot\delta(x)+\delta(x)\cdot\delta'(t)+\delta(x)\cdot\delta(t)$.
2. $u_{tt}=a^2\nabla^2u+\theta(t-t_0)\cdot\delta(x-x^0)+\delta(x-x')\cdot\delta(t)$, $t_0\geqslant 0$.
3. $$u_{tt}=a^2\nabla^2u+\omega(t)\cdot\delta(x)+|x|^2\frac{\partial^2\delta(x)}{\partial x_k^2}\cdot\delta'(t)+\frac{\partial\delta(x)}{\partial x_k}\cdot\delta(t),\quad k=1, 2, 3,$$

where $\omega\in C(t\geqslant 0)$ and $\omega=0$ at $t<0$.
4. $u_{tt}=a^2\nabla^2u+\theta(t)\sin t\cdot\delta(x)+e^{-|x|^2}\frac{\partial\delta(x)}{\partial x_k}\cdot\delta'(t)$.

12.56. Prove that if $u_1(x)$ is a locally integrable function in R^n, $n=2, 3$, then $V_n^{(0)}$ is a locally integrable function in R^{n+1} and is given by the formulas

$$V_2^{(0)}(x, t)=\frac{\theta(t)}{2\pi a}\int\limits_{|x-\xi|<at}\frac{u_1(\xi)\,d\xi}{\sqrt{a^2t^2-|x-\xi|^2}},\qquad (12.14_1)$$

$$V_3^{(0)}(x, t)=\frac{\theta(t)}{4\pi a^2t}\int\limits_{|x-\xi|=at}u_1(\xi)\,ds.\qquad (12.14_2)$$

Remark. Since $V_n^{(1)}=\frac{\partial}{\partial t}(\mathscr{E}_n(x, t)*u_0(x))$, by substituting u_0 for u_1 in (12.14_1) and (12.14_2) and differentiating with respect to t

we find that

$$V_2^{(1)}(x,\ t)=\frac{\partial}{\partial t}\left(\frac{\theta(t)}{2\pi a}\int\limits_{|x-\xi|<at}\frac{u_0(\xi)\,d\xi}{\sqrt{a^2t^2-|x-\xi|^2}}\right),\quad (12.14_3)$$

$$V_3^{(1)}(x,\ t)=\frac{\partial}{\partial t}\left(\frac{\theta(t)}{4\pi a^2 t}\int\limits_{|x-\xi|=at}u_0(\xi)\,ds\right).\quad (12.14_4)$$

12.57. Prove that if $f(x, t)$ is a locally integrable function in R^{n+1}, $n=2, 3$, that vanishes at $t<0$, then V_2 is continuous and V_3 locally integrable in R^{n+1} and are expressed through the following formulas:

$$V_2(x,\ t)=\frac{1}{2\pi a}\int\limits_0^t\int\limits_{|x-\xi|<a(t-\tau)}\frac{f(\xi,\tau)\,d\xi\,d\tau}{\sqrt{a^2(t-\tau)^2-|x-\xi|^2}},\quad (12.15_1)$$

$$V_3(x,\ t)=\frac{1}{4\pi a^2}\int\limits_{|x-\xi|<at}\frac{f\left(\xi,\ t-\frac{|x-\xi|}{a}\right)}{|x-\xi|}\,d\xi.\quad (12.15_2)$$

12.58. Prove that

1. If $u_0\in C^3(R^n)$ and $u_1\in C^2(R^n)$ at $n=2, 3$, then $V_n^{(0)}$ and $V_n^{(1)}$, $n=2, 3$, belong to the class $C^2(t\geqslant 0)$ and satisfy the equation $\Box_a u=0$ at $t>0$ and the initial conditions

$$V_n^{(0)}|_{t=+0}=0,\quad \left.\frac{\partial V_n^{(0)}}{\partial t}\right|_{t=+0}=u_1(x),$$

$$V_n^{(1)}|_{t=+0}=u_0(x),\quad \left.\frac{\partial V_n^{(1)}}{\partial t}\right|_{t=+0}=0.$$

2. If $f\in C^2(t>0)$, then V_n belong to $C^2(t\geqslant 0)$, $n=2, 3$, satisfy the equation $\Box_a u=f(x, t)$ at $t>0$ and the initial conditions

$$V|_{t=+0}=0,\quad \left.\frac{\partial V_n}{\partial t}\right|_{t=+0}=0.$$

Hint. The required properties follow directly from (12.14) and (12.15) if we substitute $at\eta$ for $\xi-x$ and $a(t-\tau)\eta$ for $\xi-x$ in the equations, respectively.

12.59. Solve the generalized Cauchy problem for the wave equation ($x\in R^2$) and check whether the solutions are those of the classical Cauchy problem (12.3), (12.4):

1. $f=\theta(t)$, $u_0=C$, $u_1=C$, $C=\text{const}$.
2. $f=\theta(t)|x|^2$, $u_0=|x|^2$, $u_1=|x|^2$.
3. $f=\theta(t)t^2$, $u_0=0$, $u_1=1+|x|^2$.
4. $f=\theta(t)e^{-t}|x|^2$, $u_0=1+|x|^2$, $u_1=0$.

12.60. Solve the Cauchy problem for the wave equation ($x\in R^3$) with the following data:

1. $f=\theta(t)|x|^2$, $u_0=0$, $u_1=|x|^2$.
2. $f=\theta(t)t^2|x|^2$, $u_0=1$, $u_1=1$.
3. $f=\omega(t)$, where $\omega\in C^2(t\geqslant 0)$ and $\omega=0$ at $t<0$, $u_0=0$, $u_1=\alpha|x|^2+\beta$.

4. $f = \theta(t) \ln |x|$, $u_0 = 0$, $u_1 = 0$; $a = 1$.
5. $f = \theta(t)$, $u_0 = \frac{1}{1+|x|^2}$, $u_1 = 0$.
6. $f = 0$, $u_0 = \sin |x|^2$, $u_1 = \sinh |x|^2$; $a = 1$.
7. $f = \theta(t)\, t$, $u_0 = |x|^2$, $u_1 = \frac{1}{1+|x|^2}$.
8. $f = \theta(t) e^{-ikt}\omega(x)$, where $\omega \in C^2$, $u_0 = \sqrt{1+|x|^2}$, $u_1 = 0$; $a = 1$.
9. $f = \theta(t) e^{-|x|^2}$, $u_0 = 0$, $u_1 = \cos |x|^2$; $a = 1$.
10. $f = 0$, $u_0 = \ln(1+|x|^2)$, $u_1 = e^{-|x|^2}$; $a = 1$.
11. $f = 0$, $u_0 = e^{-|x|^2}$, $u_1 = \ln |x|$; $a = 1$.
12. $f = \theta(t) \sin t$, $u_0 = \cos |x|^2$, $u_1 = 0$.
13. $f = 0$, $u_0 = C\theta(R - |x|)$, $u_1 = 0$.
14. $f = \theta(at - |x|)$, $u_0 = u_1 = 0$.

The Cauchy problems for the equations in Problems 12.61-12.63 are the same as for the wave equation.

12.61. Solve the generalized Cauchy problem for the hyperbolic-type equation

$$\Box_a u = bu_x + \frac{b}{a} u_t + F(x, t), \quad a > 0, \ b > 0,$$

where

$$F(x, t) = f(x, t) + u_0(x) \cdot \delta'(t) + \left[u_1(x) - \frac{b}{a} u_0(x)\right] \cdot \delta(t),$$

with the following data:
1. $f = \theta(t) \cdot \delta(x)$, $u_0 = \delta(x)$, $u_1 = \delta(x)$.
2. $f = \theta(t)\, x$, $u_0 = 0$, $u_1 = \theta(x)$: $a = b = 1$.
3. $f = \theta(t)\, t$, $u_0 = 1$, $u_1 = x$; $a = b = 1$.
4. $f = \theta(t) e^t$, $u_0 = e^x$, $u_1 = e^x$; $b = 1$.
5. $f = \theta(t) e^x$, $u_0 = \alpha x + \beta$, $u_1 = 0$.

12.62. Solve the generalized Cauchy problem for the Klein-Gordon-Fock equation

$$\Box_a u + m^2 u = f(x, t) + u_0(x) \cdot \delta'(t) + u_1(x) \cdot \delta(t)$$

with the following data:
1. $f = 0$, $u_0 = \delta(x)$, $u_1 = \delta(x)$; $a = m = 1$.
2. $f = \omega(t) \cdot \delta(x)$, where $\omega \in C(t \geqslant 0)$ and $\omega = 0$ at $t < 0$, $u_0 = 0$, $u_1 = x$; $a = m = 1$.
3. $f = \theta(t)$, $u_0 = 1$; $u_1 = 1$; $a = m = 1$.
4. $f = 0$, $u_0 = \theta(x)$, $u_1 = \theta(x)$; $a = m = 1$.

12.63. Solve the generalized Cauchy problem for the telegrapher's equation

$$\Box_a u + 2mu_t = f(x, t) + u_0(x) \cdot \delta'(t) + u_1(x) \cdot \delta(t)$$

with the following data:
1. $f = 0$, $u_0 = \delta(x)$, $u_1 = \delta(x)$; $a = m = 1$.
2. $f = \omega(t) \cdot \delta(x)$, where $\omega \in C(t \geqslant 0)$ and $\omega = 0$ at $t < 0$, $u_0 = 0$, $u_1 = 0$; $a = m = 1$.
3. $f = 0$, $u_0 = 1$, $u_1 = \theta(x)$; $a = m = 1$.

Answers to Problems of Sec. 12

12.7. $\frac{4}{5}(y^{5/4} - |x|^{5/2})$; $|x| < 1,\ 0 < y < 1$.

12.8. $\sin\ y - 1 + e^{x-y}$; $-\infty < x,\ y < \infty$.

12.9. $x - y - \frac{1}{2} + \frac{1}{2}e^{2y}$; $-\infty < x,\ y < \infty$.

12.10. $\frac{1}{2}[1 - x - 3y + (x + y - 1)e^{2x}]$; $-\infty < x,\ y < \infty$.

12.11. $xy + \frac{3}{2}\sin\frac{2y}{3}\cos\left(x + \frac{y}{3}\right)$; $-\infty < x,\ y < \infty$.

12.12. $(y - 3x)e^{-(x^2+y^2)/2}$; $x < 1,\ y < 3$.

12.13. 1. $x + \frac{1}{4}y^2$; $x > 0,\ |y| < 2\sqrt{x}$.

2. $x^3y + \sin x - \frac{1}{2}x^4 - \frac{1}{2}x^2y^2$; $x > 0,\ y > 0$.

12.14. $\frac{x^2}{y}$; $x > 0,\ y > 0$. **12.15.** $2x + y - x^2$; $-\infty < x,\ y < \infty$.

12.16. 1. $\frac{y}{3x} + \frac{2x^2y}{3}$; $x > 0,\ y < 0$. 2. $x^4 + y^2$; $x > 0$.

12.17. $\frac{3}{4}\sqrt[4]{x^7y}\left(\sqrt[3]{y} - \frac{1}{y}\right)$; $x > 0,\ y > 0$.

12.18. $x^2 + 2y^2 + 1$; $x > 0, -\frac{1}{2} < y < x^2$.

12.19. $x^2 + xy + y^2$; $x > |y|$.

12.20. $\frac{1}{2} + (4 - 3y)e^{1-x-y} - \left(2x + \frac{3}{2}\right)e^{2(1-x-y)}$; $R = e^{x-\xi+2(y-\eta)}$.

12.21. $xy - y$; $R = \frac{\xi y}{x\eta}$; **12.22.** $x - y + xy$; $R = \frac{x+y}{\xi+\eta}$.

12.23.
$$\frac{1}{2xy}[(x+y-1)u_0(x+y-1) + (x-y+1)u_0(x-y+1)] + \frac{1}{2xy}\int_{x-y+1}^{x+y-1}[u_0(\xi) + u_1(\xi)]\xi\, d\xi.$$

12.24. $(y - x)(x^2 + 1) + x^5\cos x$.

12.30.
$$\sum_{k=0}^{N-1}\left[\frac{(at)^{2k}}{(2k)!}(\nabla^2)^k u_0(x) + \frac{a^{2k}t^{2k+1}}{(2k+1)!}(\nabla^2)^k u_1(x) + \frac{a^{2k}t^{2k+2}}{(2k+2)!}(\nabla^2)^k f(x)\right].$$

12.31. $\frac{1}{2}[f(x_1+at)+f(x_1-at)+g(x_2+at)+g(x_2-at)]+$

$$+\frac{1}{2a}\int_{x_1-at}^{x_1+at} F(\xi)\,d\xi+\frac{1}{2a}\int_{x_2-at}^{x_2+at} G(\eta)\,d\eta.$$

12.32. $\frac{1}{2}g(x_2, x_3)[f(x_1+at)+f(x_1-at)]$.

12.33. $\frac{1}{2|x|}[(|x|+at)\alpha(|x|+at)+$

$+(|x|-at)\alpha(||x|-at|)]+\frac{1}{2a|x|}\int_{||x|-at|}^{|x|+at} r\beta(r)\,dr$ at $|x|\neq 0$

and $u(0, t)=\alpha(at)+at\alpha'(at)$.

12.34. $\frac{1}{2|x|}[\theta(1-|x|-t)(|x|+t)^{\alpha+1}(1-|x|-t)^{\beta}+$
$+\theta(1-|x|-t|)\operatorname{sign}(|x|-t)||x|-t|^{\alpha+1}(1-||x|-t|)^{\beta}]$
at $|x|\neq 0$ and $u(0, t)=\theta(1-t)\,t^{\alpha}(1-t)^{\beta-1}[(\alpha+1)(1-t)-\beta t]$.

12.35. $\frac{1}{2}\theta(1-|x+t|)[(x+t)^2-1]^3+$

$$+\frac{1}{2}\theta(1-|x-t|)[(x-t)^2-1]^3.$$

12.36. 1. $(x+2t)^2$. 2. $x^2+xt+4t^2+\frac{1}{6}xt^3$. 3. $\sin x$. 4. $xt+$ $+\sin(x+t)-(1-\cosh t)e^x$. 5. $1+t+\frac{1}{9}(1-\cos 3t)\sin x$.
6. $\frac{1}{a^2\omega^2}(1-\cos a\omega t)\sin\omega x$. 7. $\frac{t}{\omega}-\frac{1}{\omega^2}\sin\omega t$.

12.37. 1. $x+ty+t^2$. 2. $xyt(1+t^2)+x^2-y^2$. 3. $\frac{1}{2}t^2(x^3-3xy^2)+$ $+e^x\cos y+te^y\sin x$. 4. $x^2+t^2+t\sin y$. 5. $2x^2-y^2+(2x^2+y^2)t+$ $+2t^2+2t^3$. 6. $x^2+ty^2+\frac{1}{2}t^2(6+x^3+y^3)+t^3+\frac{3}{4}t^4(x+y)$.
7. $e^{3x+4y}\left[\frac{26}{25}\cosh 5t-\frac{1}{25}+\frac{1}{5}\sinh 5t\right]$. 8. $\cos(bx+cy)\times$ $\times\cos(at\sqrt{b^2+c^2})+\frac{1}{a\sqrt{b^2+c^2}}\sin(bx+cy)\sin(at\sqrt{b^2+c^2})$.
9. $(x^2+y^2)^2(1+t)+8a^2t^2(x^2+y^2)\left(1+\frac{1}{3}t\right)+\frac{8}{3}a^4t^4\left(1+\frac{1}{5}t\right)$.
10. $(x^2+y^2+4a^2)(e^t-1-t)-2a^2t^2\left(1+\frac{1}{3}t\right)$.

12.38. 1. $x^2+y^2-2z^2+t+t^2xyz$. 2. $y^2+tz^2+8t^2+\frac{8}{3}t^3+\frac{1}{12}t^4x^2+$ $+\frac{2}{45}t^6$. 3. $x^2y^2z^2+txy+3t^2(x^2+y^2+z^2+x^2y^2+x^2z^2+y^2z^2)+$ $+3t^4\left(\frac{3}{2}+x^2+y^2+z^2\right)+\frac{9}{5}t^6$. 4. $e^{x+y}\cos(z\sqrt{2})+te^{3y+4z}\sin 5x+$

$+t^3e^x\sqrt{2}\sin y\cos z$. 5. $(1+t)(x^2+y^2+z^2)^2+10a^2t^2\left(1+\frac{1}{3}t\right)(x^2+$ $+y^2+z^2)+a^4t^4(5+t)$. 6. $(x^2+y^2+z^2+6a^2)(e^t-1-t)-$ $-a^2t^2(3+t)$. 7. $\frac{1}{a^2}(1-\cos at)e^z\cos x\sin y+e^{y+z}\times$ $\times\left[\frac{1}{a}\sinh at\sin x+\frac{at}{\sqrt{2}}\sinh(at\sqrt{2})+x^2\cosh(at\sqrt{2})\right]$.
8. $xy\cos z\cos at+\frac{1}{a}yze^x\sinh at+\frac{x}{1+25a^2}\cos(3y+4z)\left(e^t-\cos 5at-\right.$ $\left.-\frac{1}{5a}\sin 5at\right)$. 9. $\left(\cos at+\frac{1}{a}\sin at\right)\cos\sqrt{x^2+y^2+z^2}+$ $+\frac{1}{\sqrt{x^2+y^2+z^2}}\sin\sqrt{x^2+y^2+z^2}\left(t\cos at-at\sin at-\frac{1}{a}\sin at\right)$.

12.40. $C_1=a^\alpha\left(A+\frac{B}{a\alpha}\right)$, $C_2=a^\alpha\left[A(a+1)\int_0^{\pi/2}\sin^{\alpha+1}\varphi\,d\varphi+\right.$ $\left.+\frac{B}{a}\int_0^{\pi/2}\sin^\alpha\varphi\,d\varphi\right]$, $C_3=a^\alpha\left[A(a+1)+\frac{B}{a}\right]$.

12.41. *Solution.* In view of the result of Problem 8.34, the convolution $\mathscr{E}_n*F$ exists and is given by the formula of Problem 8.34, where $g=\mathscr{E}_n$ and $f=F$, since $F(x,t)=0$ at $t<0$ and $\operatorname{supp}\mathscr{E}_n(x,t)\subset\overline{\Gamma}^+$, in view of the results of Problems 11.15-11.17.

12.42. *Solution.* For $w=u-u^*$, where $u^*(x,t)\in\mathscr{D}'(R^{n+1})$ (with $u^*=0$ at $t<0$) is another solution of the Cauchy problem (12.9), we have $w\in\mathscr{D}'(R^{n+1})$ (with $w=0$ at $t<0$) and $w_{tt}=$ $=a^2\nabla^2w$. In view of the result of Problem 12.41, the convolution $\mathscr{E}_n*w$ exists. Then $w=\delta*w=((\mathscr{E}_n)_{tt}-a^2\nabla^2\mathscr{E}_n)*w=\mathscr{E}_n*$ $*(w_{tt}-a^2\nabla^2w)=0$. Hence $u=u^*$.

12.43. *Solution.* 1. $\mathscr{E}_n(x,t)\in C^\infty$ in $t\in[0,\infty)$, in view of the result of Problem 11.26. For each positive value of t, $\operatorname{supp}\mathscr{E}_n$ lies in the ball $|x|\leqslant at$ and, hence, remains uniformly bounded in R^n as $t\to t_0\geqslant 0$. Whence, in view of the continuity of the convolution in $\mathscr{D}'$, we have

$$\left(\frac{\partial^k\mathscr{E}_n(x,t)}{\partial t^k}*u_1(x),\varphi(x)\right)\in C[0,\infty),\ k=0,1,\ldots,\tag{12.16}$$

for all $\varphi\in\mathscr{D}(R^n)$ (for the definition of the generalized function $(u(x,t),\varphi(x))\in\mathscr{D}'(R^n)$ see the material following Problem 11.25). In view of the result of Problem 8.35 and Eq. (12.16),

$$\frac{\partial^k}{\partial t^k}(V_n^{(0)}(x,t),\varphi(x))=\left(\frac{\partial^k}{\partial t^k}(\mathscr{E}_n(x,t)*u_1(x)\cdot\delta(t)),\varphi\right)$$
$$=\left(\frac{\partial^k\mathscr{E}_n(x,t)}{\partial t^k}*u_1(x),\varphi\right)\in C[0,\infty).$$

Hence, $(V_n^{(0)}(x, t), \varphi(x)) \in C^\infty[0, \infty)$, i.e. $V_n^{(0)} \in C^\infty$ with respect to $t \in [0, \infty)$. A similar result can be obtained for $V_n^{(1)}$.

2. In view of the result of Problem 11.26, as $t \to +0$,

$V_n^{(0)}(x, t) = \mathscr{E}_n(x, t) * u_1(x) \to 0 * u_1(x) = 0$ in $\mathscr{D}'(R^n)$,

$$\frac{\partial V^{(0)}(x, t)}{\partial t} = \frac{\partial}{\partial t}[\mathscr{E}_n(x, t) * u_1(x)]$$

$$= \frac{\partial \mathscr{E}_n(x, t)}{\partial t} * u_1(x) \to \delta * u_1 - u_1(x) \text{ in } \mathscr{D}'(R^n).$$

12.44. *Hint.* Use formula (12.10), the result of Problem 11.15, formulas (8.3) and (8.3$_1$), and the results of Problems 8.31 and 8.8. 1. $u = \mathscr{E}_1(x, t) = \frac{1}{2a}\theta(at - |x|)$. 2. $\frac{1}{2a}\theta(a(t - t_0) - |x - x_0|)$. 3. $\frac{\partial \mathscr{E}_1}{\partial x} = \frac{1}{2a}\theta(t)\delta(at + x) - \frac{1}{2a}\theta(t)\delta(at - x)$. 4. $\frac{\partial \mathscr{E}_1}{\partial t} = \frac{1}{2}\delta(at - |x|)$. 5. $\frac{1}{2}\delta(a(t - t_0) - |x|)$. 6. $\frac{1}{2a}\theta(t)\delta(at + x - x_0) - \frac{1}{2a}\theta(t)\delta(at - x + x_0)$. 7. $\frac{\partial^2 \mathscr{E}_1}{\partial t^2} = \frac{1}{2}\frac{\partial}{\partial t}\delta(at - |x|)$. 8. $\frac{\partial^2 \mathscr{E}_1}{\partial x^2} = \frac{\theta(t)}{2a}[\delta'(at + x) - \delta'(at - x)]$. 9. 0. 10. $\frac{1}{2a}\theta(at - |x|)$.

12.45. See the hint to Problem 12.44. 1. *Solution.* Equation (12.9) for the unknown function $u(x, t)$ has the form

$$\begin{aligned} u_{tt} &= a^2 u_{xx} + f(x, t) + u_0(x)\cdot\delta'(t) + u_1(x)\cdot\delta(t) \\ &= a^2 u_{xx} + \omega(t)\cdot\delta(x) + \delta(x)\cdot\delta'(t) + \delta(x)\cdot\delta(t). \end{aligned} \quad (12.17)$$

In view of formula (12.10),

$$\begin{aligned} u = V_1 + V_1^{(1)} + V_1^{(0)} &= \mathscr{E}_1 * [\omega(t)\cdot\delta(x)] \\ &+ \mathscr{E}_1 * [\delta(x)\cdot\delta'(t)] + \mathscr{E}_1 * [\delta(x)\cdot\delta(t)]. \end{aligned} \quad (12.18)$$

In view of the result of Problem 8.36.1,

$$V_1 = \frac{1}{2a}\theta(at - |x|)\int_0^{t - |x|/a} \omega(\tau)\, d\tau.$$

In view of the results of Problems 12.44.1 and 12.44.4, $V_1^{(0)}(x, t) = \frac{1}{2a}\theta(at - |x|)$ and $V_1^{(1)}(x, t) = \frac{1}{2}\delta(at - |x|)$. Substituting V_1, $V_1^{(1)}$, and $V_1^{(0)}$ into (12.18), we arrive at the solution of the generalized Cauchy problem (12.17). From the result of Problem 12.2 follows the uniqueness of the solution to the Cauchy problem (12.17). The result of Problem 12.43 yields the following limiting relationships:

1. $u(x, t) \to \delta(t)$ and $u_t(x, t) \to \delta(t)$ as $t \to 0$ in $\mathscr{D}'(R^n)$. 2. $\frac{1}{2a^2}\theta(at - |x|)(at - |x|) + \frac{1}{2}\delta(at - |x - x_0|)$. 3. $\frac{1}{4}\theta(t - |x|)(t - |x|)^2 +$

$+\frac{1}{2}\theta(t-|x-3|)+\frac{1}{2}\delta(t-|x-2|)$. 4. $\frac{1}{2a}\theta(at-|x-x_0|)\times\left[1-\cos\left(t-\frac{|x-x_0|}{a}\right)\right]-\frac{1}{2a}\theta(at-|x|)$. *Hint.* $x\delta'(x)=-\delta(x)$. 5. $\frac{1}{2a}\theta(at-|x|)\left[2+\sin\left(t-\frac{|x|}{a}\right)\right]$. *Hint.* $x^2\delta''(x)=2\delta(x)$. 6. $\frac{1}{2a\alpha}\theta(at-|x|)(e^{\alpha(t-|x|/a)}-1)+\frac{1}{2}\delta(at-|x+1|)+\frac{1}{2}\delta(at-|x-1|)$. 7. $\theta(t-|x-2|)\sqrt{1-|x-2|}+\frac{1}{2}\theta(t-|x+R|)+\frac{1}{2}\theta(t-|x-R|)$. 8. $\frac{1}{6}\theta(t-|x|)(t-|x|)^3+c\theta(t)+\frac{1}{2}\theta(t-|x+R|)-\frac{1}{2}\theta(t-|x-R|)$. *Hint.* See the result of Problem 7.14.1. 9. $\frac{1}{2a}\theta(at-|x|)\left(t-\frac{|x|}{a}\right)\ln\left[e^{-1}\left(t-\frac{|x|}{a}\right)\right]+\frac{1}{2}\delta(at-|x|)$. 10. $\frac{1}{2}\theta(t-1-|x|)\left(\arctan(t-|x|)-\frac{\pi}{4}\right)+\frac{1}{2}\delta(t-|x+R|)-\frac{1}{2}\delta(t-|x-R|)$. 11. $\frac{1}{2}\theta(t)\delta(t+x+2)-\frac{1}{2}\theta(t)\delta(t-x-2)-\frac{1}{2}\theta(t)\delta(t+x-2)+\frac{1}{2}\theta(t)\delta(t-x+2)$. *Hint.* See Problems 7.14.1 and 8.8.2. 12. $\frac{1}{2a}\theta(at-|x-1|)\times\ln\left(1+t-\frac{|x-1|}{a}\right)+\frac{1}{2a}\theta(at-|x-\pi|)$. 13. $\theta(at-|x|)\frac{a^2t^2-x^2}{4a^2}$. 14. $-\frac{1}{4}\theta(t-|x|)[\alpha(t-|x|)^2+2\beta(t-|x|)]-\theta(t)\delta(t+x)+\theta(t)\delta(t-x)$. *Hint.* Employ the fact that $x\delta''(x)=-2\delta'(x)$ and the result of Problem 8.8.2.

12.49. *Hint.* Employ formulas (12.10)-(12.13). 1. *Solution.* $u=V_1+V_1^{(1)}+V_1^{(0)}$; $V_1=0$. In view of formula (12.11), $V_1^{(0)}=\frac{\theta(t)}{2a}\int\limits_{x-at}^{x+at}\theta(\xi)\,d\xi=\frac{\theta(t)}{2a}\left[\int\limits_0^{x+at}\theta(\xi)\,d\xi-\int\limits_0^{x-at}\theta(\xi)\,d\xi\right]=\frac{\theta(t)}{2a}\times[\theta(x+at)(x+at)-\theta(x-at)(x-at)]$. $V_1^{(1)}=\frac{\partial}{\partial t}\times\left[\frac{\theta(t)}{2a}\int\limits_{x-at}^{x+at}\theta(\xi)\,d\xi\right]=\frac{\theta(t)}{2}(\theta(x+at)+\theta(x-at))$. 2. $\theta(t)\times\left(x+\frac{t^2}{2}(x-1)+\frac{|x+at|-|x-at|}{2a}\right)$. 3. $\theta(t)\left[\frac{xt^3}{6}+\frac{1}{a}\theta(x+at)\sqrt{x+at}-\frac{1}{a}\theta(x-at)\sqrt{x-at}\right]$. 4. $\theta(t)\left[(t+1)\ln(t+1)-t+\frac{1}{2}\theta(t-x)(t-x)+\frac{1}{2}\theta(-t-x)(t+x)\right]$. *Hint.* $V_1=\frac{1}{2}\theta\times(t-|x|)*\frac{\theta(t)}{t+1}\cdot 1(x)$. 5. $\theta(t-2)\left(t^2\ln\sqrt{t}+(1-t)\ln 4-(1-\right.$

$-t)^2+\frac{t^2}{4}\Big)+\frac{\theta(t)}{2}(|x+t|+|x-t|)$. 6. $\frac{\theta(t)}{2}\Big[\frac{2t^{m+2}}{(m+1)(m+2)}+$ $+\theta(2-|x+at|)+\theta(2-|x-at|)\Big]$. 7. $\frac{\theta(t)}{2}\Big[te^{x+t}-e^x\sinh t+$ $+\theta(x+t)(1-e^{-x-t})-\theta(x-t)(1-e^{t-x})\Big]$. 8. $-\theta(t-\pi)\times$ $\times(1+\cos t)+\frac{\theta(t)}{2}[\theta(x+3t-3)+\theta(x-3t-3)+2t]$. 9. $\frac{\theta(t)}{4}\times$ $\times[\theta(x+t)(x+t)^2+\theta(x-t)(x-t)^2-2\theta(x)x^2]$. 10. $\frac{\theta(t)}{6}\times$ $\times\Big[\theta(x+t)(x+t)^3+\theta(x-t)(x-t)^3-2\theta(x)x^3+\frac{6}{\alpha}e^{\alpha x}\sinh\alpha t\Big]$, $\alpha\neq\theta$. 11. $\frac{\theta(t-1)}{6}(3x+t+2)(t-1)^2+\frac{\theta(t)}{4}[\operatorname{sign}(x+t)(x+t)^2-$ $-\operatorname{sign}(x-t)(x-t)^2]$. 12. $\frac{\theta(t-2)}{6}(t+4)(t-2)^2+\frac{\theta(t)}{2}[\theta(x-$ $-1+t)\ln(x+t)+\theta(x-1-t)\ln(x-t)]$. 13. $\frac{\theta(t)}{2}\Big[\theta(x+t)\times$ $\times(x+t)^m\Big(1+\frac{x+t}{m+1}\Big)+\theta(x-t)(x-t)^m\Big(1-\frac{x-t}{m+1}\Big)\Big]$. 14. $\frac{\theta(t)}{6}\times$ $\times[8t^{3/2}+3\theta(x+t)\sin(x+t)-3\theta(x-t)\sin(x-t)]$. 15. $\theta(t)\times$ $\times\Big[\frac{4}{15}xt^{5/2}+\theta(-x-t)\Big(\frac{1}{2}+\frac{(x+t)^2}{4}\Big)+\theta(-x+t)\Big(\frac{1}{2}-$ $-\frac{(-x+t)^2}{4}\Big)\Big]$. 16. $\frac{\theta(t)}{6}[6x^2t+2t^3+3\theta(x+t)e^{-\sqrt{x+t}}+3\theta(x-$ $-t)e^{-\sqrt{x-t}}]$. 17. $\frac{\theta(t)}{2}[\cos x\sin t+2\sin x\sin t-t\cos(x+t)]$. 18. $\frac{\theta(t)}{2}[\theta(1+x+t)(1+x+t)-\theta(1+x-t)(1+x-t)+\theta(-1+$ $+x-t)(-1+x-t)-\theta(-1+x+t)(-1+x+t)]$.

12.53. *Hint.* Employ formulas (12.10)-(12.13), the result of Problem 12.50, and the solution of Problem 12.45.1.

1. $\theta(t)\Big(\frac{t^3}{6}+\frac{xt^2}{2}+e^{\alpha x}\cosh\alpha at\Big)$. 2. $\theta(t)\Big[t^3\Big(\frac{1}{6}\ln t-\frac{5}{36}\Big)+$ $+3^x\cosh at\Big]$. *Hint.* $V_1=\frac{1}{2a}\theta(at-|x|)*\theta(t)\,t\ln t\cdot 1(x)$.
3. $\frac{\theta(t)}{12}[(1-a^2)t^4+6t^2x^2+6(x+at)^m+6(x-at)^m]$.
4. $\frac{\theta(t)}{12}[t^4+6t^2x^2+12\cos x(\sin t+\cos t)]$. 5. $\frac{\theta(t)}{18a}[3(x+at)^3\ln\times$ $\times|x+at|-3(x-at)^3\ln|x-at|-6ax^2t-2a^3t^3]$. 6. $\frac{\theta(t)}{2}\times$ $\times\Big[t\sin(x+t)-\sin x\sin t+\frac{2^{x+t}-2^{x-t}}{\ln 2}\Big]$. 7. $\theta(t)\Big[t-\sin t+$ $+\frac{1}{2}\arctan\frac{2t}{1+x^2-t^2}\Big]$. 8. $\theta(t)\Big\{e^t-t-1+\frac{1}{2[1+(x+at)^2]}+$ $+\frac{1}{2[1+(x-at)^2]}\Big\}$. 9. $\theta(t)\Big[\alpha(x^2+t^2)+\beta+\frac{3}{14}((x+t)^{7/3}-$ $-(x-t)^{7/3})\Big]$. 10. $\frac{\theta(t)}{2}\Big[\ln(1+e^{2x}+2e^x\cosh t)+\int_{x-t}^{x+t}e^{-z^2}dz\Big]$.

11. $\theta(t)\left[\frac{xt^{m+2}}{(m+1)(m+2)}+\sin x\cos t+\frac{(x+t)^{m+1}-(x-t)^{m+1}}{2(m+1)}\right]$.
12. $\frac{\theta(t)}{2}\{(t^2-1)\arctan t+t-t\ln(t^2+1)+\ln[(t^2+x^2+1)^2-$
$-4t^2x^2]\}$. 13. $\frac{\theta(t)}{2}(\cos x\sin^2 t+\sqrt{1+(x+2t)^2}+\sqrt{1+(x-2t)^2})$.
14. $\frac{\theta(t)}{2}[2x(t-\sin t)+(x+t)^2e^{-|x+t|}+(x-t)^2e^{-|x-t|}]$. 15. $\theta(t)\times$
$\times\left\{e^{-x^2-4t^2}\cosh 4xt+\frac{\sqrt{2}}{8}e^{-x}\left[e^{2t}\cos\left(x-2t-\frac{\pi}{4}\right)-e^{-2t}\cos\times\right.\right.$
$\left.\left.\times\left(x+2t-\frac{\pi}{4}\right)\right]\right\}$. 16. $\theta(t)\left[\sin^2 x\cos^2 t+\cos^2 x\sin^2 t+\right.$
$\left.+\frac{1}{2}e^{-|x-t|}(1+|x-t|)-\frac{1}{2}e^{-|x+t|}(1+|x+t|)\right]$. 17. $\theta(t)\times$
$\times\left[x(t\arctan t-\ln\sqrt{1+t^2})+\frac{1}{4-2\cos(x+t)}+\frac{1}{4-2\cos(x-t)}\right]$.
18. $\theta(t)\left[e^x\sinh t+x(e^t-1)-xt-te^x+\frac{1}{2}\ln\frac{x+t+\sqrt{1+(x+t)^2}}{x-t+\sqrt{1+(x-t)^2}}\right]$.

12.54. *Hint.* Employ formula (12.10) and the result of Problem 11.15.2.
1. $\frac{\theta(at-|x|)}{2\pi a^2}\left(\ln\frac{at+\sqrt{a^2t^2-|x|^2}}{|x|}+\frac{a}{\sqrt{a^2t^2-|x|^2}}\right)+\frac{\partial\mathscr{E}_2(x,t)}{\partial t}$.
2. $\frac{\theta(at-|x|)}{4\pi a^3}\left[\left(2at^2+\frac{|x|^2}{a}\right)\ln\frac{at+\sqrt{(at)^2-|x|^2}}{|x|}-3t\sqrt{(at)^2-|x|^2}\right]$
$+\mathscr{E}_2(x-x_0,t)$ 3. $\frac{\theta(at-|x|)}{2\pi a}\int\limits_0^{t-|x|/a}\frac{\omega(\tau)\,d\tau}{\sqrt{a^2(t-\tau)^2-|x|^2}}+\mathscr{E}_2(x,t)$.
4. $\frac{\theta(at-|x|)}{2\pi a^2}\left[(\alpha t+\beta)\ln\frac{at+\sqrt{(at)^2-|x|^2}}{|x|}-\frac{\alpha}{a}\sqrt{(at)^2-|x|^2}\right]+$
$+\mathscr{E}_2(x-x_0,t)$.

12.55. *Hint.* Employ formula (12.10) and the result of Problem 11.16.1. 1. $\frac{\theta(at-|x|)}{4\pi a^2|x|}+\mathscr{E}_3(x,t)+\frac{\partial\mathscr{E}_3(x,t)}{\partial t}$, where $\mathscr{E}_3=$
$=\frac{\theta(t)}{4\pi a^2t}\delta_{S_{at}}(x)$ 2. $\frac{\theta(a(t-t_0)-|x-x_0|)}{4\pi a^2|x-x_0|}+\mathscr{E}_3(x-x',t)$. 3. *Solution.* $u=V_3+V_3^{(0)}+V_3^{(1)}$. In view of the result of Problem 8.35.1, $(V_3,\varphi)=(\mathscr{E}_3*\omega(t),\varphi)\left(\frac{\theta(t)}{4\pi a^2t}\delta_{S_{at}}(x)\cdot\omega(\tau),\ \eta\ (a^2t^2-|x|^2)\varphi\times\right.$

$$\left.\times(x,t+\tau)\right)=\int\limits_{-\infty}^{\infty}\omega(\tau)\left\{\int\limits_0^{\infty}\int\limits_{|x|=at}\frac{\varphi(x,t+\tau)}{4\pi a^2t}\,dS_x\,dt\right\}d\tau.$$

Since $dS_x d(at)=dx$ is the volume element in R^3, then

$$(V_3,\varphi)=\int\limits_{-\infty}^{\infty}\omega(\tau)\left[\int\limits_{R^3}\frac{\varphi(x,\tau+|x|/a)}{4\pi a^2|x|}\,dx\right]d\tau$$

$$=\int\limits_{-\infty}^{\infty}\int\limits_{R^3}\frac{\omega(t-|x|/a)}{4\pi a^2|x|}\varphi(x,t)\,dx\,dt.$$

Hence, $V_3 = \dfrac{\omega\left(t-\frac{|x|}{a}\right)}{4\pi a^2 |x|}$, $V_3^{(0)} = \dfrac{\partial \mathscr{E}_3(x, t)}{\partial x_k}$,

$$V_3^{(1)} = 2\frac{\partial \mathscr{E}_3(x, t)}{\partial t} \text{ since } |x|^2 \frac{\partial^2 \delta(x)}{\partial x_k^2} = 2\delta(x).$$

4. $\theta(at-|x|)\dfrac{\sin\left(t-\frac{|x|}{a}\right)}{4\pi a^2|x|} + \dfrac{\partial^2 \mathscr{E}_3(x, t)}{\partial x_k \partial t}$, since

$$e^{-|x|^2}\frac{\partial \delta(x)}{\partial x_k} = \frac{\partial \delta(x)}{\partial x_k}.$$

12.59. *Hint.* Employ the result of Problem 11.15 and formulas (12.10), (12.14_1), (12.14_3), and (12.15_1). 1. $\theta(t)\left(\frac{t^2}{2}+Ct+C\right)$.

2. $\theta(t)\left[\frac{a^2t^4}{6}+\frac{2}{3}a^2t^3+2a^2t^2+|x|^2\left(\frac{t^2}{2}+t+1\right)\right]$.

3. $\theta(t)\left[\frac{t^4}{12}+\frac{2}{3}a^2t^3+t(1+|x|^2)\right]$.

4. $\theta(t)\left[\frac{2}{3}a^2t^3+(4a^2+|x|^2)(t-1+e^{-t})+1+|x|^2\right]$.

12.60. *Hint.* Employ the result of Problem 11.16 and formulas (12.10), (12.14_2), (12.14_4), and (12.15_2). 1. $\theta(t)\left(\frac{|x|^2t^4}{2}+\frac{a^2t^4}{4}+\right.$ $\left.+|x|^2t+a^2t^3\right)$. 2. $\theta(t)\left(\frac{a^2t^6}{60}+\frac{|x|^2t^4}{12}+t+1\right)$. 3. $\int_0^t \omega(\tau)\,\times$ $\times(t-\tau)\,d\tau+\theta(t)(\alpha a^2t^3+\alpha|x|^2t+\beta t)$. 4. $\frac{\theta(t)}{12}\left[\frac{(|x|+t)^3}{|x|}\,\times\right.$ $\left.\times\ln(|x|+t)+\frac{(|x|-t)^3}{|x|}\ln(|x|-t)-2|x|^2\ln|x|-3t^2\right]$.

5. $\frac{\theta(t)}{2|x|}\left(|x|\,t^2+\frac{|x|+at}{1+(|x|+at)^2}+\frac{|x|-at}{1+(|x|-at)^2}\right)$.

6. $\frac{\theta(t)}{2|x|}\left[(|x|+t)\sin(|x|+t)^2+(|x|-t)\sin(|x|-t)^2+\right.$ $\left.+\frac{1}{2}\cosh(|x|+t)^2-\frac{1}{2}\cosh(|x|-t)^2\right]$. 7. $\frac{\theta(t)}{12}\left(2t^3+\right.$

$$\left.+12|x|^2+36a^2t^2+\frac{3}{a|x|}\ln\frac{1+(|x|+at)^2}{1+(|x|-at)^2}\right).$$

8. $\theta(t)\left(\frac{e^{-ikt}}{4\pi}\int_{|z|<t}\frac{e^{ik|z|}\,\omega(x-z)}{|z|}\,dz+\right.$

$$\left.+\frac{|x|+t}{2|x|}\sqrt{1+(|x|+t)^2}+\frac{|x|-t}{2|x|}\sqrt{1+(|x|-t)^2}\right].$$

9. $\frac{\theta(t)}{4|x|}\left[2e^{-|x|^2}\int_0^t e^{-\rho^2}\sin 2\rho|x|\,d\rho+\sin(|x|+t)^2-\sin(|x|-t)^2\right]$.

10. $\frac{\theta(t)}{2|x|}\,[(|x|+t)\ln(1+(|x|+t)^2)+$

$+(|x|-t)\ln(1+(|x|-t)^2)+e^{t^2-|x|^2}\sinh 2t|x|]$. 11. $\frac{\theta(t)}{8|x|}\times$ $\times\{8e^{-(|x|^2+t^2)}(|x|\cosh 2t|x|-t\sinh 2t|x|)+[(|x|+t)^2\times$ $\times\ln(|x|+t)^2-(|x|-t)^2\ln(|x|-t)^2-4t|x|]\}$. 12. $\theta(t)\Big(t-\sin t+$ $+\frac{|x|+at}{2|x|}\cos(|x|+at)^2+\frac{|x|-at}{2|x|}\cos(|x|-at)^2\Big)$.
13. $\frac{\theta(t)c}{2|x|}[(|x|-at)\,\theta(R-||x|-at|)+(|x|+at)\,\theta(R-|x|-at)]$. *Hint*. The solution depends only on $|x|$ and t. By the substitution $u_1(r,t)=ru(r,t)$ the problem can be reduced to the Cauchy problem for the equation of string vibrations. Then employ formula (12.12). 14. $\theta(at-|x|)\frac{a^2t^2-|x|^2}{8a^2}$.

12.61. *Hint*. Employ formula (12.10) and the result of Problem 11.18. 1. *Solution*. $u_1=V_1+V_1^{(0)}+V_1^{(1)}$, where

$$V_1=\mathscr{E}*f=\mathscr{E}*\theta(t)=\frac{\theta(at-|x|)}{b}\left(e^{-\frac{b(x-at)}{2a^2}}-e^{-\frac{b(x-|x|)}{2a^2}}\right);$$

$$V_1^{(0)}=\mathscr{E}*\left[u_1(x)\cdot\delta(t)-\frac{b}{a}u_0(x)\cdot\delta(t)\right]=\left(1-\frac{b}{a}\right)\mathscr{E};$$

$$V_1^{(1)}=\mathscr{E}*[u_0(x)\cdot\delta'(t)]=\frac{\partial\mathscr{E}}{\partial t}*[\delta(x)\cdot\delta(t)]=\frac{\partial\mathscr{E}}{\partial t}=\frac{b}{2a}\mathscr{E}+$$
$$+\frac{\delta(at-|x|)}{2}e^{-\frac{b(x-at)}{2a^2}},$$

with $\mathscr{E}(x,t)$ given in Problem 11.18. 2. $\theta(t)\Big[(e^t-1)(x+t-3)+$ $+3t-xt+\frac{t^2}{2}+\theta(x+t)e^t-\theta(x-t)-\theta(t-|x|)e^{(t-x)/2}]$.

3. $\theta(t)\left[e^t(-1+x+t)-x-\frac{t^2}{2}+2\right]$. 4. $\theta(t)\Big[\frac{a}{a-1}(e^t-ae^{t/a}+$ $+a-1)+\frac{a^2+a}{2a^2+1}e^{x+at+t/a}+\frac{a^2-a+1}{2a^2+1}e^{x-at}\Big]$, $a\neq 1$.

5. $\theta(t)\Big\{\frac{ex}{b+2a^2}\Big[\frac{a^2}{b+a^2}(e^{(b+a^2)t/a}-1)+e^{-at}-1\Big]+\beta+\alpha(x-at)+$ $+\frac{\alpha a^2}{b}(e^{bt/a}-1)\Big\}$.

12.62. *Hint*. Employ the result of Problem 11.21*. 1. $\frac{\theta(t-|x|)}{2}\times$ $\times\left[J_0(\sqrt{t^2-x^2})-\frac{t}{\sqrt{t^2-x^2}}J_1(\sqrt{t^2-x^2})\right]+\frac{1}{2}\delta(t-|x|)$.

2. $\frac{\theta(t-|x|)}{2}\int\limits_0^{t-|x|}\omega(\tau)\,J_0(\sqrt{(t-\tau)^2-x^2})\,d\tau+\frac{\theta(t)}{2}\int\limits_{x-t}^{x+t}\xi J_0\times$ $\times(\sqrt{t^2-(x-\xi)^2})\,d\xi$. 3. $\theta(t)\Big[2-\sin t-\cos t+\int\limits_0^t J_0(\sqrt{t^2-\xi^2})\,d\xi-$

$-t\int_0^t \frac{J_1(\sqrt{t^2-\xi^2})}{\sqrt{t^2-\xi^2}}\,d\xi\Big]$. 4. $\frac{\theta(t)}{2}\Big[\theta(x+t)+\theta(x-t)+$

$+\int_{-t}^{t}\theta(x-\xi)\Big(J_0(\sqrt{t^2-\xi^2})-t\,\frac{J_1\sqrt{t^2-\xi^2})}{\sqrt{t^2-\xi^2}}\Big)\,d\xi\Big]$.

12.63. *Hint*. Employ the result of Problem 11.22. 1. $\frac{1}{2}e^{-t}\delta\times$ $\times(t-|x|)+e^{-t}\theta(t-|x|)\Big[I_0(\sqrt{t^2-x^2})+\frac{t}{2}\,\frac{I_1(\sqrt{t^2-x^2}}{\sqrt{t^2-x^2}}\Big]$.

2. $\frac{1}{2}\theta(t-|x|)\,e^{-t}\int_0^{t-|x|}\omega(\tau)\,e^{\tau}J_0(i\sqrt{(t-\tau)^2-x^2})\,d\tau$.

3. $\theta(t)\Big(1+\frac{1}{2}\int_{-t}^{t}\theta(x-\xi)\,e^{-t}J_0(i\sqrt{t^2-\xi^2})\,d\xi\Big)$.

13 The Cauchy Problem for the Heat Conduction Equation

The *classical Cauchy problem* for the heat conduction equation is the problem of finding a function $u(x, t)$ of class $C^2(t>0)\cap C(t\geqslant 0)$ that satisfies the equation

$$u_t = a^2\nabla^2 u + f(x, t) \tag{13.1}$$

for $x\in R^n$, $t>0$ and the initial conditions

$$u|_{t=0} = u_0(x), \tag{13.2}$$

where f and u_0 are known functions.

If the function $f\in C^2(t\geqslant 0)$ and all its derivatives up to second order inclusive are bounded in every strip $0\leqslant t\leqslant T$, while the function u_0 belongs to the class $C(R^n)$ and is bounded, then the solution of the Cauchy problem (13.1), (13.2) in the class of functions $u(x, t)$ that are bounded in every strip $0\leqslant t\leqslant T$ exists, is unique, and is represented by Poisson's formula

$$u(x, t)=\frac{1}{(2a\sqrt{\pi t})^n}\int_{R^n} u_0(\xi)\,e^{-\frac{|x-\xi|^2}{4a^2t}}\,d\xi$$
$$+\int_0^t\int_{R^n}\frac{f(\xi,\tau)}{[2a\sqrt{\pi(t-\tau)}]^n}\,e^{-\frac{|x-\xi|^2}{4a^2(t-\tau)}}\,d\xi\,d\tau. \tag{13.3}$$

13.1. Let the function $u(x, t, t_0)$ belong to class C^2 for $x\in R^n$, $t\geqslant t_0\geqslant 0$. Prove that for each nonnegative value of t_0 the function $u(x, t, t_0)$ is a solution of the Cauchy problem

$$u_t = a^2\nabla^2 u,\quad u|_{t=t_0} = f(x, t_0)$$

if and only if for every nonnegative value of t_0 the function

$$v(x, t, t_0) = \int_{t_0}^{t} u(x, t, \tau)\, d\tau$$

is a solution of the Cauchy problem

$$v_t = a^2\nabla^2 v + f(x, t), \quad v|_{t=t_0} = 0.$$

13.2. Suppose $u_k(x_k, t)$ is a solution of the Cauchy problem

$$u_t = a^2\nabla^2 u, \quad u|_{t=0} = f_k(x_k), \quad k = 1, 2, \ldots, n.$$

Prove that the function $u(x, t) = \prod_{k=1}^{n} u_k(x_k, t)$ is a solution of the Cauchy problem

$$u_t = a^2\nabla^2 u, \quad u|_{t=0} = \prod_{k=1}^{n} f_k(x_k).$$

13.3. Suppose the function $f(x, t) \in C^2$ $(t \geqslant 0)$ is harmonic in x for each fixed nonnegative value of t. Prove that the function $u(x, t) = \int_0^t f(x, \tau)\, d\tau$ is a solution of the Cauchy problem

$$u_t = a^2\nabla^2 u + f(x, t), \quad u|_{t=0} = 0.$$

13.4. Suppose u_0 belongs to $C^\infty(R^n)$, while the series $\sum_{k=0}^{\infty} \frac{\delta^k}{k!}(\nabla^2)^k u_0(x)$, $\delta > 0$, and all the series obtained through term-by-term differentiation of the initial series up to second order inclusive are uniformly convergent in each finite region. Prove that the function

$$u(x, t) = \sum_{k=0}^{\infty} \frac{a^{2k} t^k}{k!} (\nabla^2)^k u_0(x)$$

is a solution of the Cauchy problem

$$u_t = a^2\nabla^2 u, \quad 0 < t < \frac{\delta}{a^2}; \quad u|_{t=0} = u_0(x).$$

The solutions to Problems 13.5-13.8 can be found via Poisson's formulas, but sometimes it is expedient to use Fourier's method of variable separation or the results of Problems 13.1-13.4.

13.5. Solve the following Cauchy problems ($n = 1$):

1. $u_t = 4u_{xx} + t + e^t, \quad u|_{t=0} = 2.$
2. $u_t = u_{xx} + 3t^2, \quad u|_{t=0} = \sin x.$
3. $u_t = u_{xx} + e^{-t}\cos x, \quad u|_{t=0} = \cos x.$
4. $u_t = u_{xx} + e^t \sin x, \quad u|_{t=0} = \sin x.$
5. $u_t = u_{xx} + \sin t, \quad u|_{t=0} = e^{-x^2}.$

6. $4u_t = u_{xx}$, $\quad u|_{t=0} = e^{2x-x^2}$.
7. $u_t = u_{xx}$, $\quad u|_{t=0} = xe^{-x^2}$.
8. $4u_t = u_{xx}$, $\quad u|_{t=0} = \sin xe^{-x^2}$.

13.6. Solve the following Cauchy problems ($n=2$):

1. $u_t = \nabla^2 u + e^t$; $\quad u|_{t=0} = \cos x \sin y$.
2. $u_t = \nabla^2 u + \sin t \sin x \sin y$; $\quad u|_{t=0} = 1$.
3. $u_t = \nabla^2 u + \cos t$; $\quad u|_{t=0} = xye^{-x^2-y^2}$.
4. $8u_t = \nabla^2 u + 1$; $\quad u|_{t=0} = e^{-(x-y)^2}$.
5. $2u_t = \nabla^2 u$; $\quad u|_{t=0} = \cos xy$.

13.7. Solve the following Cauchy problems ($n=3$):

1. $u_t = 2\nabla^2 u + t \cos x$; $u|_{t=0} = \cos y \cos z$.
2. $u_t = 3\nabla^2 u + e^t$; $u|_{t=0} = \sin(x-y-z)$.
3. $4u_t = \nabla^2 u + \sin 2z$; $u|_{t=0} = \frac{1}{4} \sin 2z + e^{-x^2} \cos 2y$.
4. $u_t = \nabla^2 u + \cos(x-y+z)$; $u|_{t=0} = e^{-(x+y-z)^2}$.
5. $u_t = \nabla^2 u$; $u|_{t=0} = \cos(xy) \sin z$.

13.8. Solve the Cauchy problem

$$u_t = \nabla^2 u, \quad u|_{t=0} = u_0(x), \quad x \in R^n,$$

for the following u_0's:

1. $u_0 = \cos \sum_{k=1}^{n} x_k$. 2. $u_0 = e^{-|x|^2}$. 3. $u_0 = \left(\sum_{k=1}^{n} x_k\right) e^{-|x|^2}$.

4. $u_0 = \left(\sin \sum_{k=1}^{n} x_k\right) e^{-|x|^2}$. 5. $u_0 = e^{-\left(\sum_{k=1}^{n} x_k\right)^2}$.

If the solution $u(x, t)$ of the classical Cauchy problem (13.1), (13.2) and the function $f(x, t) \in C$ are continued to zero at $t < 0$, then $u(x, t)$ satisfies (in the generalized sense) the equation

$$u_t = a^2 \nabla^2 u + f(x, t) + u_0(x) \cdot \delta(t) \tag{13.4}$$

in R^{n+1}.

The *generalized Cauchy problem* for the heat conduction equation with a source function $F(x, t) \in \mathscr{D}'(R^{n+1})$ (with $F = 0$ at $t < 0$) is the problem of finding a generalized function $u \in \mathscr{D}'$ that vanishes at $t < 0$ and satisfies in R^{n+1} the heat conduction equation

$$u_t = a^2 \nabla^2 u + F(x, t). \tag{13.5}$$

If there exists the convolution $\mathscr{E} * F$, where

$$\mathscr{E}(x, t) = \frac{\theta(t)}{(2a\sqrt{\pi t})^n} e^{-\frac{|x|^2}{4a^2 t}}$$

is the fundamental solution of the heat conduction operator, then

$$u = \mathscr{E} * F$$

is a solution of the generalized Cauchy problem (13.5). This solution is unique in the class of generalized functions $u(x, t)$ for which there is the convolution $\mathscr{E} * u$.

The convolution $V = \mathscr{E} * F$ is called the *generalized heat potential with a density* F. In particular, if $F = u_0(x)\cdot\delta(t)$, where $u_0 \in \mathscr{D}'(R^n)$ the convolution

$$V^{(0)} = \mathscr{E}(x, t) * u_0(x)\cdot\delta(t) = \mathscr{E}(x, t) * u_0(x)$$

(provided it exists) is called the *generalized surface heat potential with a density* u_0. The heat potential V satisfies Eq. (13.5).

We denote by M the class of all functions that are locally integrable in R^{n+1}, vanish at $t < 0$, and are bounded in each strip $0 \leqslant t \leqslant T$, $x \in R^n$.

13.9. Find the solution of the generalized Cauchy problem (13.5) for the following source functions F:

1. $\delta(t)\cdot\delta(x)$. 2. $\delta(t-t_0)\cdot\delta(x-x_0)$, $t_0 \geqslant 0$.

3. $\delta(t)\cdot\dfrac{\partial\delta(x)}{\partial x_k}$. 4. $\delta'(t)\cdot\delta(x)$.

5. $\delta(t-t_0)\cdot\dfrac{\partial^2\delta(x)}{\partial x_k^2}$, $t_0 \geqslant 0$. 6. $\delta'(t)\cdot\dfrac{\partial\delta(x)}{\partial x_k}$.

7. $\theta(t)\cdot\delta(x)$. 8. $\theta(t-t_0)\cdot\delta(x-x_0)$, $t_0 \geqslant 0$.

9. $\delta'(t)\,\delta(x-x_0)$. 10. $\omega(t)\cdot\delta(x)$, where $\omega \in C(t \geqslant 0)$, $\omega = 0$ at $t < 0$.

13.10. Suppose $f(x, t)$ belongs to M. Show that the convolution $V = \mathscr{E} * f$

(1) exists in M and is given by the formula

$$V(x, t) = \int_0^t \int_{R^n} \frac{f(\xi, \tau)}{[2a\sqrt{\pi(t-\tau)}]^n} e^{-\frac{|x-\xi|^2}{4a^2(t-\tau)}}\, d\xi\, d\tau, \tag{13.6}$$

(2) satisfies the estimate

$$|V(x, t)| \leqslant t \sup_{\substack{\xi \\ 0\leqslant\tau\leqslant t}} |f(\xi, \tau)|, \quad t > 0,$$

(3) is the only (generalized) solution of the equation $V_t = a^2\nabla^2 V + f(x, t)$ in class M.

13.11. Suppose the function $u_0(x)$ is bounded in R^n. Prove that the convolution

$$V^{(0)} = \mathscr{E}(x, t) * u_0(x)\cdot\delta(t) = \mathscr{E}(x, t) * u_0(x)$$

(1) exists in M and is given by the formula

$$V^{(0)}(x, t) = \frac{\theta(t)}{(2a\sqrt{\pi t})^n} \int_{R^n} u_0(\xi)\, e^{-\frac{|x-\xi|^2}{4a^2t}}\, d\xi, \tag{13.7}$$

(2) satisfies the estimate

$$|V^{(0)}(x, t)| \leqslant \sup_{\xi} |u_0(\xi)|, \quad t > 0,$$

(3) is the only (generalized) solution of the equation $V_t^{(0)} = a^2\nabla^2 V^{(0)} + u_0(x)\cdot\delta(t)$ in class M.

13.12. Prove that the solution of the generalized Cauchy problem

$$u_t = a^2\nabla^2 u + f(x, t) + u_0(x)\cdot\delta(t) \qquad (13.8)$$

is expressed by the classical Poisson formula

$$u(x, t) = \frac{\theta(t)}{(2a\sqrt{\pi t})^n}\int_{R^n} u_0(\xi)\, e^{-\frac{|x-\xi|^2}{4a^2t}}\,d\xi$$

$$+\int_0^t\int_{R^n}\frac{f(\xi, \tau)}{[2a\sqrt{\pi(t-\tau)}]^n}\, e^{-\frac{|x-\xi|^2}{4a^2(t-\tau)}}\,d\xi\,d\tau \qquad (13.9)$$

if the function f is locally integrable in R^{n+1} and vanishes at $t < 0$, if u_0 is locally integrable in R^n, and if both terms on the right-hand side of (13.9) are locally integrable in R^{n+1}.

13.13. Prove that

(1) If the function $f \in C^2$ $(t \geqslant 0)$ and all of its derivatives up to second order inclusive belong to class M, then $V = \mathscr{E} * f \in C^2$ $(t > 0) \cap C^1$ $(t \geqslant 0)$ satisfy the equation $V_t = a^2\nabla^2 V + f(x, t)$ at $t > 0$ and the initial condition $V\,|_{t=+0} = 0$.

(2) If $u_0(x)$ is continuous and bounded, then

$$V^{(0)} = \mathscr{E} * u_0 = C^\infty(t > 0) \cap C(t \geqslant 0)$$

satisfies the equation $V_t^{(0)} = a^2\nabla^2 V^{(0)}$ and the initial condition $V^{(0)}\,|_{t=+0} = u_0(x)$.

(3) If conditions (1) and (2) are met, then the function $u = V + V^{(0)}$, where V and $V^{(0)}$ are defined in (13.6) and (13.7), is the classical solution of the Cauchy problem (13.1), (13.2). *Hint.* The required properties follow directly from Eq. (13.8).

13.14. Find the solution of the generalized Cauchy problem

$$u_t = u_{xx} + u_0(x)\cdot\delta(t)$$

for the following u_0's:

1. $\theta(x)$. 2. $\theta(1-x)$. 3. $\theta(1-|x|)$.
4. $\theta(x)\,e^{-x}$. 5. $\theta(x)(x+1)$. 6. $\theta(x-1)\,x$.

Show that the solutions $u(x, t)$ at $t > 0$ belong to class C^∞ and satisfy the equation $u_t = u_{xx}$, while, as $t \to +0$, they remain continuous at all points at which the functions $u_0(x)$ are continuous and satisfy the initial condition $u\,|_{t=+0} = u_0(x)$ at these points.

13.15. Solve the generalized Cauchy problem

$$u_t = u_{xx} + f(x, t)$$

for the following f's:

1. $\theta(t-1)\,e^t$. 2. $\theta(t-\pi)\cos t$. 3. $\theta(t-1)\,x$.
4. $\theta(t-2)\,e^x$. 5. $\theta(t)\,\theta(x)$. 6. $\theta(t)\cdot\theta(1-|x|)$.

Show that the solutions $u(x, t)$ belong to class $C(R^2)$, satisfy the initial conditions $u|_{t=0} = 0$, and belong to class C^2 at the points where the $f(x, t)$ are continuous.

13.16. Solve the generalized Cauchy problem (13.8) for the heat conduction equation ($x \in R^1$) with the data given below and check whether the solutions are those of the classical Cauchy problem (13.1), (13.2):

1. $f = \theta(t)\,x$, $u_0 = x$. 2. $f = \theta(t)\,x^2$, $u_0 = x^2$.
3. $f = \theta(t)\,2xt$, $u_0 = x^3 + x^4$, $a = 1$.
4. $f = \theta(t)\,3x^2t^2$, $u_0 = e^x$, $a = 1$.
5. $f = \theta(t)\sqrt{t}$, $u_0 = \sinh x$. 6. $f = \frac{\theta(t)}{\sqrt{t}}$, $u_0 = xe^x$.
7. $f = \theta(t)\ln t$, $u_0 = x \sin x$, $a = 1$.
8. $f = \theta(t)\,x \cos x$, $u_0 = x \cos x$, $a = 1$.
9. $f = \theta(t)\,e^x$, $u_0 = \theta(x)\,x$, $a = 1$.
10. $f = \theta(t)\,xe^x$, $u_0 = \theta(x)\,x^2$, $a = 1$.

13.17. Solve the generalized Cauchy problem (13.8) for the heat conduction equation ($x \in R^2$) with the data given below and check whether the solutions are those of the classical Cauchy problem (13.1), (13.2):

1. $f = \theta(t)\,xye^t$, $u_0 = x^2 - y^2$.
2. $f = \theta(t)\,(x^2 + y^2)$, $u_0 = x^2 + y^2$.
3. $f = \theta(t)\,4xy$, $u_0 = x^2y^2$, $a = 1$.
4. $f = \theta(t)\,e^x \cos y$, $u_0 = e^{x+y}$.
5. $f = 0$, $u_0 = x \cos y$.
6. $f = \theta(t)\,xy$, $u_0 = \cos y$.

13.18. Solve the generalized Cauchy problem (13.8) for the heat conduction equation ($x \in R^3$) with the data given below and check whether the solutions are those of the classical Cauchy problem (13.1), (13.2):

1. $f = \theta(t)\,xye^z$, $u_0 = xe^y \cos z$.
2. $f = \theta(t)\,xy \cos z$, $u_0 = (x^2 + y^2) \cos z$, $a = 1$.
3. $f = \theta(t)\,xyz \cos t$, $u_0 = xy^2z^3$.
4. $f = \theta(t)\,(x^2 - 2y^2 + z^2)\,e^t$, $u_0 = x + y^2 + z^3$.
5. $f = \theta(t) \cos t \cdot \sin 3x \cos 4ye^{5z}$, $u_0 = \sin 3x \cdot \cos 4ye^{4z}$, $a = 1$.

13.19. Solve the generalized Cauchy problem (13.8) for the heat conduction equation ($x \in R^n$) with the data given below and check whether the solutions are those of the classical Cauchy problem (13.1), (13.2):

1. $f = \theta(t)\,|x|^2$, $u_0 = |x|^2$.
2. $f = \theta(t) \sum\limits_{k=1}^{n} x_k^3$, $u_0 = \sum\limits_{k=1}^{n} x_k^3$.
3. $f = \theta(t)\,e^t$, $u_0 = e^{\sum\limits_{k=1}^{n} x_k}$.

4. $f=0,\ u_0=\sum_{k=1}^{n} x_k e^{\sum_{k=1}^{n} x_k}$.

5. $f=0,\ u_0=\left(\cos \sum_{k=1}^{n} x_k\right) e^{\sum_{k=1}^{n} x_k}$.

The equation $u_t = a^2u_{xx} - bu_x - cu = f(x, t)$, where a, b and c are constants, can be reduced to the heat conduction equation by the substitution

$$v(y, t) = e^{-ct}u(y - bt, t).$$

13.20. Solve the problem

$$u_t - a^2u_{xx} - bu_x - cu = f(x, t),\ u\mid_{t=0} = u_0(x)$$

with the following data:

1. $f = 1,\ u_0 = 1,\ c = 1$.
2. $f = e^t,\ u_0 = \cos x,\ a = c = 1,\ b = 0$.
3. $f = e^t,\ u_0 = \cos x,\ a = \sqrt{2},\ c = 2,\ b = 0$.
4. $f = t \sin x,\ u_0 = 1,\ a = c = 1,\ b = 0$.
5. $f = 0,\ u_0 = e^{-x^2}$.
6. $f = w(t) \in C^1 (t \geqslant 0)$, with u_0 belonging to C and bounded.

13.21. Solve the generalized Cauchy problem

$$u_t - a^2u_{xx} - bu_x - cu = f(x, t) + u_0(x)\cdot\delta(t)$$

with the following data:

1. $f = \theta(t - 1),\ u_0 = \theta(x),\ c \neq 0$.
2. $f = \theta(t - 1),\ u_0 = \theta(1 - x),\ c = 0$.
3. $f = \theta(t - 1)e^t,\ u_0 = \theta(1 - |x|),\ c \neq 1$.
4. $f = \theta(t - 1)e^t,\ u_0 = \theta(x)e^x,\ c = 1$.
5. $f = \theta(t - 1)e^x,\ u_0 = x\theta(x),\ a = 2,\ b = c = -2$.
6. $f = \theta(t)\theta(x),\ u_0 = x$.

Investigate the smoothness of the solutions, just as in Problems 13.14 and 13.15.

13.22. Solve the generalized Cauchy problem

$$u_t - a^2u_{xx} - bu_x - cu = f(x, t) + u_0(x)\cdot\delta(t)$$

with the data given below, and check whether the solutions are those of the classical Cauchy problem $u_t - a^2u_{xx} - bu_x - cu = f(x, t)$, $u\mid_{t=0} = u_0(x)$:

1. $f=\theta(t)x^2,\ u_0=x^2,\ a=b=c=1$.
2. $f=\dfrac{\theta(t)}{\sqrt{t}},\ u_0=e^x$.
3. $f=\theta(t)te^x,\ u_0=xe^x,\ a=2,\ b=-1,\ c=-2$.
4. $f=\theta(t)xe^x,\ u_0=xe^x+\sinh x,\ a=c=1,\ b=-2$.

5. $f=\theta(t)\, e^x \cos t\cdot\sin x,\ u_0=e^x\cos x,\ a=1,\ b=-2,\ c=2.$
6. $f=\theta(t)\, x,\ u_0=x\sin x,\ a=b=c=1.$

13.23. Suppose $u(x, t)$ is a solution of the Cauchy problem

$$u_t=a^2\nabla^2 u, \quad u|_{t=0}=u_0(x),$$

where u_0 belongs to $C(R^n)$ and $|u_0(x)|\leqslant Me^{-\delta|x|^2}$, $\delta\geqslant 0$. Prove that

$$|u(t, x)|\leqslant M(1+4a^2\delta t)^{-n/2}e^{-\frac{\delta|x|^2}{1+4a^2\delta t}}$$

for all $t\geqslant 0$, $x\in R^n$.

13.24. Suppose $u(x, t)$ is a solution of the Cauchy problem

$$u_t=a^2\nabla^2 u, \quad u|_{t=0}=u_0(x),$$

where $u_0(x)$ is finite and continuous. Prove that for all $T>0$, $\delta<1/4a^2T$ there is a positive M such that

$$|u(x, t)|\leqslant Me^{-\delta|x|^2},\ x\in R^n,\ 0\leqslant t\leqslant T.$$

13.25. Suppose u_0 belongs to $C(R^n)$ and $|u_0(x)|\leqslant M_\delta e^{\delta|x|^2}$, $\delta>0$. Prove that at $0<t<1/4a^2\delta$, $x\in R^n$, the function

$$u(x, t)=\frac{1}{(2a\sqrt{\pi t})^n}\int_{R^n} u_0(\xi)\, e^{-\frac{|x-\xi|^2}{4a^2t}}\, d\xi \qquad (13.10)$$

belongs to class C^∞ and is a solution of the Cauchy problem

$$u_t=a^2\nabla^2 u, \quad 0<t<\frac{1}{4a^2\delta}; \quad u|_{t=0}=u_0(x).$$

13.26. Prove that if the hypothesis of Problem 13.25 is true for all positive δ's, then function (13.10) belongs to class C^∞ for $t>0$, $x\in R^n$, and is a solution of the classical Cauchy problem

$$u_t=a^2\nabla^2 u, \quad t>0; \ u|_{t=0}=u_0(x).$$

13.27. Employ the method of generalized functions to solve the problem

$$u_t=a^2u_{xx},\ t>0,\ x>0;\ u|_{t=0}=u_0(x),$$
$$u|_{x=0}=0, \text{ where } u_0(x)\in C\ (x\geqslant 0).$$

Answers to Problems of Sec. 13

13.5. 1. $1+e^t+\frac{1}{2}t^2$. 2. $t^3+e^{-t}\sin x$. 3. $(1+t)\,e^{-t}\cos x$. 4. $\cosh t\sin x$. 5. $1-\cos t+(1+4t)^{-1/2}e^{-\frac{x^2}{1+4t}}$. 6. $(1+t)^{-\frac{1}{2}}e^{\frac{2x-x^2+t}{1+t}}$. 7. $x(1+4t)^{-3/2}e^{-\frac{x^2}{1+4t}}$. 8. $(1+t)^{-\frac{1}{2}}\sin\frac{x}{1+t}\,e^{-\frac{4x^2+t}{4(1+t)}}$.

13.6. 1. $e^t-1+e^{-2t}\cos x\sin y$. 2. $1+\frac{1}{5}\sin x\sin y\,(2\sin t-\cos t+$ $+e^{-2t})$. 3. $\sin t+\frac{xy}{(1+4t)^3}e^{-\frac{x^2+y^2}{1+4t}}$. 4. $\frac{t}{8}+\frac{1}{\sqrt{1+t}}e^{-\frac{(x-y)^2}{1+t}}$.

5. $\frac{1}{\sqrt{1+t^2}}\cos\frac{xy}{1+t^2}e^{-\frac{t(x^2+y^2)}{2(1+t^2)}}$.

13.7. 1. $\frac{1}{4}\cos x\,(e^{-2t}-1+2t)+\cos y\cos z e^{-4t}$. 2. $e^t-1+\sin(x-$ $-y-z)\,e^{-9t}$. 3. $\frac{1}{4}\sin 2z+\frac{\cos 2y}{\sqrt{1+t}}e^{-t-\frac{x^2}{1+t}}$. 4. $\frac{1}{3}\cos(x-y+z)\times$ $\times(1-e^{-3t})+\frac{1}{\sqrt{1+12t}}e^{-\frac{(x+y-z)^2}{1+12t}}$. 5. $\frac{\sin z}{\sqrt{1+4t^2}}\cos\frac{xy}{1+4t^2}e^{-t-\frac{t(x^2+y^2)}{1+4t^2}}$.

13.8. 1. $e^{-nt}\cos\sum_{k=1}^{n}x_k$. 2. $(1+4t)^{-n/2}e^{-\frac{|x|^2}{1+4t}}$. 3. $(1+4t)^{-\frac{n+2}{2}}\times$ $\times e^{-\frac{|x|^2}{1+4t}}\left(\sum_{k=1}^{n}x_k\right)$. 4. $(1+4t)^{-n/2}\sin\frac{\sum_{k=1}^{n}x_k}{1+4t}e^{-\frac{nt+|x|^2}{1+4t}}$.

5. $\frac{1}{\sqrt{1+4nt}}e^{-\frac{1}{1+4nt}\left(\sum_{k=1}^{n}x_k\right)^2}$.

13.9. 1. $\mathscr{E}(x,t)$. 2. $\mathscr{E}(x-x_0,\ t-t_0)$. 3. $-\frac{x_k}{2a^2t}\mathscr{E}(x,t)$.

4. $\left(\frac{|x|^2}{4a^2t^2}-\frac{n}{2t}\right)\mathscr{E}(x,t)+\delta(x,t)$. 5. $\frac{x_k^2-2a^2(t-t_0)}{4a^4(t-t_0)^2}\mathscr{E}(x,\ t-t_0)$.

6. $\frac{x_k}{4a^2t^2}\left(n+2-\frac{|x|^2}{2a^2t}\right)\mathscr{E}(x,t)+\frac{\partial\mathscr{E}(x,t)}{\partial x_k}$. 7. $\int_0^t\mathscr{E}(x,\tau)\,d\tau$.

8. $\int_0^{t-t_0}\mathscr{E}(x-x_0,\tau)\,d\tau$. 9. $\left[\frac{|x-x_0|^2}{4a^2t^2}-\frac{n}{2t}\right]\mathscr{E}(x-x_0,t)+\delta(x-x_0,t)$.

10. $\int_0^t\omega(\tau)\mathscr{E}(x,\ t-\tau)\,d\tau$.

13.14. 1. $\theta(t)\,\Phi\left(\frac{x}{\sqrt{2t}}\right)$. 2. $\theta(t)\,\Phi\left(\frac{1-x}{\sqrt{2t}}\right)$. 3. $\theta(t)\left[\Phi\left(\frac{x+1}{\sqrt{2t}}\right)-\right.$ $\left.-\Phi\left(\frac{x-1}{\sqrt{2t}}\right)\right]$. 4. $\theta(t)\,e^{t-x}\Phi\left(\frac{x-2t}{\sqrt{2t}}\right)$. 5. $\theta(t)\left[\sqrt{\frac{t}{\pi}}e^{-\frac{x^2}{4t}}+\right.$ $\left.+(x+1)\,\Phi\left(\frac{x}{\sqrt{2t}}\right)\right]$. 6. $\theta(t)\left[\sqrt{\frac{t}{\pi}}e^{-\frac{(x-1)^2}{4t}}+x\Phi\left(\frac{x-1}{\sqrt{2t}}\right)\right]$.

13.15. 1. $\theta(t-1)(e^t-e)$. 2. $\theta(t-\pi)\sin t$. 3. $\theta(t-1)(t-1)x$. 4. $\theta(t-2)(e^{t-2}-1)e^x$. 5. $\theta(t)\int\limits_0^t \Phi\left(\frac{x}{\sqrt{2\tau}}\right)d\tau$. 6. $\theta(t)\times$ $\times\int\limits_0^t\left[\Phi\left(\frac{x+1}{2\sqrt{\tau}}\right)-\Phi\left(\frac{x-1}{2\sqrt{\tau}}\right)\right]d\tau$.

13.16. *Hint.* For the proof see Problems 13.12 and 13.13 and for the solution see the text above Problem 13.5. 1. $\theta(t)(t+1)x$. 2. $\theta(t)(x^2+x^2t+2a^2t+a^2t^2)$. 3. $\theta(t)[x^3+x^4+6t(x+2x^2)+t^2(12+$ $+x)]$. 4. $\theta(t)\left(x^2t^3+\frac{1}{2}t^4+e^{x+t}\right)$. 5. $\theta(t)\left(\frac{2}{3}t^3t^2+e^{a^2t}\sinh x\right)$. 6. $\theta(t)(2\sqrt{t}+(x+2a^2t)e^{x+a^2t})$. 7. $\theta(t)[t\ln t-t+(x\sin x+$ $+2t\cos x)e^{-t}]$. 8. $\theta(t)[x\cos x+2\sin x(e^{-t}-1)]$. 9. $\theta(t)\left[e^x(e^t-\right.$ $\left.-1)+\sqrt{\frac{t}{\pi}}e^{-\frac{x^2}{4t}}+x\Phi\left(\frac{x}{\sqrt{2t}}\right)\right]$. 10. $\theta(t)\left[(2-x)e^x+(x+2t-\right.$ $\left.-2)e^{x+t}+x\sqrt{\frac{t}{\pi}}e^{-\frac{x^2}{4t}}+(x^2+2t)\Phi\left(\frac{x}{\sqrt{2t}}\right)\right]$.

13.17. *Hint.* See the hint in the answer to Problem 13.16. 1. $\theta(t)[x^2-y^2+xy(e^t-1)]$. 2. $\theta(t)[(x^2+y^2)(t+1)+$ $+4a^2t+2a^2t^2]$. 3. $\theta(t)(x^2y^2+2t(x+y)^2+4t^2)$. 4. $\theta(t)(te^x\cos y+$ $+e^{x+y+2a^2t})$. 5. $\theta(t)\,xe^y\cos z$. 6. $\theta(t)(xyt+\cos ye^{-a^2t})$.

13.18. *Hint.* See the hint in the answer to Problem 13.16. 1. $\theta(t)\times$ $\times[xe^y\cos z+e^{-2}xye^z(e^{a^2t}-1)]$.
2. $\theta(t)\cos z[xy(1-e^{-t})+(x^2+y^2+4t)e^{-t}]$.
3. $\theta(t)[xyz\sin t+x(y^2+2a^2t)(z^3+6a^2tz)]$.
4. $\theta(t)[x+y^2+z^3+2a^2t(1+3z)+(x^2-2y^2+z^2)\times$ $\times(e^t-1)]$. 5. $\theta(t)[\sin 3x\cos 4ye^{4z}(e^{-9t}+\sin te^z)]$.

13.19. *Hint.* See the hint in the answer to Problem 13.16. 1. $\theta(t)[(1+t)|x|^2+na^2t(2+t)]$. 2. $\theta(t)\left(\sum\limits_{k=1}^n[(1+t)x_k^3+3a^2t\times\right.$ $\left.\times(2+t)x_k]\right)$. 3. $\theta(t)\left[e^t-1+\exp\left(na^2t+\sum\limits_{k=1}^n x_k\right)\right]$. 4. $\theta(t)\left[\left(2na^2t+\right.\right.$ $\left.\left.+\sum\limits_{k=1}^n x_k\right)\exp\left(na^2t+\sum\limits_{k=1}^n x_k\right)\right]$. 5. $\theta(t)\left[\cos\left(2a^2nt+\sum\limits_{k=1}^n x_k\right)\exp\sum\limits_{k=1}^n x_k\right]$.

13.20. 1. $2e^t-1$. 2. $te^t+\cos x$. 3. $e^{2t}-e^t+e^{-2t}\cos x$. 4. e^t+ $+\frac{1}{2}t^2\sin x$. 5. $(1+4a^2t)^{-1/2}\exp\left[ct-\frac{(x+bt)^2}{1+4a^2t}\right]$. 6. $\int\limits_0^t\omega(t-\tau)\times$ $\times e^{c\tau}d\tau+\frac{1}{2a\sqrt{\pi t}}\int\limits_{-\infty}^{\infty}u_0(\xi)\,e^{ct-\frac{(x-\xi+bt)^2}{4a^2t}}d\xi$.

13.21. 1. $\frac{1}{c}\theta(t-1)(e^{ct-c}-1)+\theta(t)e^{ct}\Phi\left(\frac{x+bt}{a\sqrt{2t}}\right)$. 2. $\theta(t-1)\times$ $\times(t-1)+\theta(t)\Phi\left(\frac{1-x-bt}{a\sqrt{2t}}\right)$. 3. $\frac{\theta(t-1)}{1-c}(e^t-e^{ct-c+1})+\theta(t)e^{ct}\times$ $\times\left[\Phi\left(\frac{x+bt+1}{a\sqrt{2t}}\right)-\Phi\left(\frac{x+bt-1}{a\sqrt{2t}}\right)\right]$. 4. $\theta(t-1)(t-1)e^t+$ $+\theta(t)e^{x+t(1+b+a^2)}\Phi\left(\frac{x+bt+2a^2t}{a\sqrt{2t}}\right)$. 5. $\theta(t-1)(t-1)e^x+\theta(t)\times$ $\times e^{-2t}\left[2\sqrt{\frac{t}{\pi}}e^{-\frac{(x-2t)^2}{16t}}+(x-2t)\Phi\left(\frac{x-2t}{2\sqrt{2t}}\right)\right]$. 6. $\theta(t)\left[(x+\right.$ $\left.+bt)e^{ct}+e^{ct}\int_0^t\Phi\left(\frac{x+bt}{a\sqrt{2\tau}}\right)d\tau\right]$.

13.22. 1. $\theta(t)(2x-x^2+2[t-x+(x+t)^2]e^t)$. 2. $\theta(t)\left(e^{x+t(a^2+b+c)}+\right.$ $\left.+\int_0^t\frac{1}{\sqrt{t-\tau}}e^{c\tau}d\tau\right)$. 3. $\theta(t)[(1+x+7t)e^{x+t}-(1+t)e^x]$. 4. $\theta(t)\times$ $\times[x(t+1)e^x+e^{2t}\sinh(x-2t)]$. 5. $\theta(t)(\cos x+\sin t\sin x)e^x$. 6. $\theta(t)[1-x+(x+t-1)e^t+(x+t)\sin(x+t)+2t\cos(x+t)]$.

13.27. $\frac{1}{2a\sqrt{\pi t}}\int_0^\infty u_0(y)\left[e^{-\frac{(x-y)^2}{4a^2t}}-e^{-\frac{(x+y)^2}{4a^2t}}\right]dy.$

14 The Cauchy Problem for Other Equations and Goursat's Problem

We start with the Cauchy problem for Schrödinger's equation. The statement of the classical Cauchy problem

$$u_t+i\nabla^2u+f(x,\ t),\quad u|_{t=0}=u_0(x) \tag{14.1}$$

and the generalized Cauchy problem

$$u_t=i\nabla^2u+F(x,\ t) \tag{14.2}$$

is similar to the respective problems for the heat conduction equation (see pp. 157 and 159).

The fundamental solution of Schrödinger's equation is

$$\mathscr{E}(x,\ t)=\frac{\theta(t)}{(2\sqrt{\pi t})^n}e^{\frac{i|x|^2}{4t}-\frac{\pi ni}{4}}.$$

For the Cauchy problem (14.1) the results similar to those formulated in Problems 13.1-13.4 are valid.

We will say that a function $u(x,\ t)$ belongs to class $\mathscr{P}$ if the function satisfies the following estimate

$$|u(x,\ t)|\leqslant c(1+|x|)^\lambda,\quad x\in R^n,\ t\geqslant 0,$$

for certain c and λ.

14.1. Prove that if $u_0(x)$ belongs to $\mathscr{S}(R^n)$, then the function

$$u(x, t)=\frac{1}{(2\pi)^n}\int_{R^n} e^{-it|y|^2-i(x, y)}\int_{R^n} u_0(\xi)\, e^{i(\xi, y)}\, d\xi\, dy \qquad (14.3)$$

is a solution of the Cauchy problem

$$u_t = i\nabla^2 u, \quad u|_{t=0} = u_0(x); \qquad (14.4)$$

$u(x, t) \in C^\infty(t \geqslant 0)$; $u(x, t) \in \mathscr{S}(R^n)$ for every fixed $t>0$; and the functions $x^\beta D^\alpha u(x, t)$ are uniformly bounded in $x \in R^n$, $t \geqslant 0$, for any values of α and β.

14.2. Suppose $u(x, t)$ is a solution of the Cauchy problem (14.4). Prove that the function $v(x, t) = u(x, T-t)$ for any positive T is a solution of the Cauchy problem

$$v_t = -i\nabla^2 v, \quad 0<t<T; \quad v|_{t=T} = u_0(x).$$

14.3. Let $u(x, t)$ and $v(x, t)$ be solutions of the problems

$$u_t = iu_{xx}, \quad u|_{t=0} = u_0(x); \qquad (14.5)$$
$$v_t = -iv_{xx}, \quad 0<t<T, \quad v|_{t=T} = v_0(x),$$

where $u(x, t)$ belongs to $\mathscr{P}$, and $v(x, t)$ is defined via the formulas in Problems 14.1 and 14.2. Prove that

$$\int_{R^1} u_0(x)\, v(x, 0)\, dx = \int_{R^1} u(x, T)\, v_0(x)\, dx.$$

Hint. In

$$\int_0^T \int_{-\delta}^{\delta} v(x, t)\, \varphi_\delta(x)\, [u_t(x, t) - iu_{xx}(x, t)]\, dx\, dt = 0,$$

where the function $\varphi_\delta(x)$ is the same as in Problem 6.5, integration by parts eliminates the derivatives of $u(x, t)$. Then pass to the limit with $\delta \to \infty$.

14.4. Prove the uniqueness of the solution of the Cauchy problem (14.5) in $\mathscr{P}$. *Hint.* Employ the result of Problem 14.3.

The solution of the Cauchy problem (14.1) is unique in class $\mathscr{P}$ (for $n=1$ see Problem 14.4). The solutions of Problems 14.5-14.10 are taken only from this class. In this case the existence of u_{tt} is not required.

14.5. Suppose $u_0(x) \in C^{n+1}(R^n)$, $|x|^{n+3}\,|u_0(x)| \leqslant M$, and $|x|^{n+1}|D^\alpha u_0(x)| \leqslant M$ for all α, $|\alpha| \leqslant n+1$. Prove that the Cauchy problem (14.4) has a solution and that this solution is given by the formula (14.3), which can be written in the form

$$u(x, t)=\frac{1}{(2\sqrt{\pi t})^n}\, e^{-\frac{\pi n i}{4}} \int_{R^n} u_0(\xi)\, e^{\frac{i|x-\xi|^2}{4t}}\, d\xi.$$

14.6. Suppose $u_0(x) \in C^{\alpha}(R^1)$, $\alpha \geqslant 2$, $u_0(x) = 0$ for $|x| \geqslant 1$ and $|u_0^{(r)}(x)| \leqslant M$, $r \leqslant \alpha$. Prove that the solution to the Cauchy problem (14.5) belongs to class $C^{\infty}(t > 0)$ and

$$\left|\frac{\partial^r}{\partial x^r} u(x, t)\right| \leqslant CM(1+|x|)^{2+r-\alpha}, \quad r = 0, 1, \ldots, \alpha-2,$$

for all $x \in R^1$, $t \geqslant 0$.

14.7. Suppose $u_0(x) \in C^{\alpha}(R^1)$, $|u_0^{(r)}(x)| \leqslant C(1+|x|)^{\lambda}$, $r \leqslant \alpha$, $\alpha \geqslant 2$, $\lambda < \alpha - 5$, and suppose that $u_k(x, t)$ is the solution of the Cauchy problem

$$u_t = iu_{xx}, \quad u|_{t=0} = u_0(x) e(x-k),$$

where $e(x)$ is the same as in Problem 6.4. Prove that the Cauchy problem (14.5) has a solution given by the formula $u(x, t) = \sum_{k=-\infty}^{\infty} u_k(x, t)$ and $|u(x, t)| \leqslant C_1(1+|x|)^{\alpha-2}$ for all $x \in R^1$, $t \geqslant 0$. *Hint.* Employ the result of Problem 14.6 to show that

$$|u_k(x, t)| \leqslant \frac{C_1(2+|k|)^{\lambda}}{(1+|x-k|)^{\alpha-2}} \leqslant \frac{C_1(1+|x|)^{\alpha-2}(2+|k|)^{\lambda}}{(1+|k|)^{\alpha-2}}.$$

14.8. Suppose $u_0(x) \in C^1(R^1)$ and $\int_{R^1} |xu_0'(x)|\, dx < \infty$. Prove that the Cauchy problem (14.5) has a solution and that this solution is given by the formula

$$u(x, t) = \frac{1}{2}[u_0(+\infty) + u_0(-\infty)] + \frac{1}{\sqrt{\pi}} e^{-\frac{\pi i}{4}} \int_{R^1} u_0'(\xi) \int_0^{\frac{x-\xi}{2\sqrt{t}}} e^{iy^2}\, dy\, d\xi.$$

14.9. Let $u_0(x) = e^{ia|x|^2}$, where a is a real number, $x \in R^n$. Prove that the Cauchy problem (14.4) has a solution for $a \geqslant 0$, while for $a < 0$ a solution exists only if $0 \leqslant t \leqslant -1/4a$. Find the solution. Compare the result with that of Problem 14.7 at $n = 1$ for $a = 0, \pm 1$.

14.10. Solve the following Cauchy problems:

1. $u_t = iu_{xx} + tx^3$; $u|_{t=0} = x^4$.
2. $u_t = iu_{xx}$, $0 < t < \frac{1}{4}$; $u|_{t=0} = xe^{-ix^2}$.
3. $u_t = i\nabla^2 u + x\cos t - y^2 \sin t$; $u|_{t=0} = x^2 + y^2$.
4. $u_t = i\nabla^2 u + 6x + y^2 + iz^3$; $u|_{t=0} = i(x^3 + y^3 + z^3)$.
5. $u_t = i\nabla^2 u$; $u|_{t=0} = e^{-|x|^2}$, $x \in R^n$.

14.11. Find solutions of the generalized Cauchy problem (14.2) with the following functions $F \in \mathscr{D}'(R^{n+1})$;

1. $\delta(t)\cdot\delta(x)$. 2. $\delta(t)\cdot\frac{\partial\delta(x)}{\partial x_k}$.
3. $\theta(t)\cdot\delta(x+x_0)$, $n = 1$. 4. $\theta(t-t_0)\cdot\delta(x)$, $n = 1$, $t_0 \geqslant 0$.

14.12. Solve the generalized Cauchy problem

$$u_t = iu_{xx} + f(x, t) + u_0(x)\cdot\delta(t)$$

at $t > 0$ for the following f and u_0 ($f = 0$ at $t < 0$ and is specified only for $t > 0$):

1. $f = \theta(x)$, $u_0 = \theta(x)$. 2. $f = \theta(t-1)$, $u_0 = \theta(1-|x|)$.

3. $f=\theta(t-\pi)\sin t$, $u_0=x^2$. 4. $f=\frac{1}{\sqrt{t}}$, $u_0=\cos x$.

5. $f=\theta(t-1)(e^t-e)$, $u_0=x\sin x$.

Prove that the functions $u(x, t)$ found in Problems 14.12.3-14.12.5 are solutions of the classical Cauchy problem.

Problems 14.13-14.20 deal with the Cauchy problem for the equation $u_{tt} = -(\nabla^2)^2u + f(x, t)$.

14.13. Suppose $u(x, t) \in C^4$ ($t \geqslant 0$). Prove that the function $u(x, t)$ is a solution of the Cauchy problem

$$u_{tt} = -(\nabla^2)^2u; \quad u|_{t=0} = \varphi(x), \quad u_t|_{t=0} = 0$$

if and only if

$$w(x, t)=u(x, t)+i\int_0^t \nabla^2u(x, \tau)\,d\tau$$

is a solution of the Cauchy problem

$$w_t = i\nabla^2w; \quad w|_{t=0} = \varphi(x).$$

14.14. Suppose the function $w(x, t) \in C^4$ ($t \geqslant 0$) is a solution of the Cauchy problem

$$w_t = i\nabla^2w; \quad w|_{t=0} = \varphi(x),$$

where $\varphi(x)$ is a real-valued function. Prove that the function $u(x, t) = \mathrm{Re}\, w(x, t)$ is a solution of the Cauchy problem

$$u_{tt} = -(\nabla^2)^2u; \quad u|_{t=0} = \varphi(x), \quad u_t|_{t=0} = 0.$$

14.15. Suppose the function $f(x, t) \in C^4$ ($t \geqslant 0$) is biharmonic, or $(\nabla^2)^2f = 0$, for every nonnegative value of t. Find the solution to the Cauchy problem

$$u_{tt} = -(\nabla^2)^2u + f(x, t); \quad u|_{t=0} = 0, \quad u_t|_{t=0} = 0.$$

14.16. Let $u_0(x)$ and $u_1(x)$ be biharmonic functions. Find the solution to the Cauchy problem

$$u_{tt} = -(\nabla^2)^2u; \quad u|_{t=0} = u_0(x), \quad u_t|_{t=0} = u_1(x).$$

14.17. Suppose the function $w(x, t) \in C^4(t \geqslant 0)$ is a solution to the Cauchy problem

$$w_t = i(\nabla^2)w; \quad w|_{t=0} = \varphi(x),$$

where $\varphi(x)$ is a real-valued function. Find the solution to the Cauchy problem

$$u_{tt} = -(\nabla^2)^2u; \quad u|_{t=0} = 0, \quad u_t|_{t=0} = \varphi(x).$$

14.18. Suppose the function $w(x, t) \in C^4(t \geqslant 0)$ is a solution of the Cauchy problem

$$w_t = i\nabla^2 w; \quad w|_{t=0} = \varphi(x),$$

with $\varphi(x)$ a pure imaginary function. Find the solution to the Cauchy problem

$$u_{tt} = -(\nabla^2)^2 u; \quad u|_{t=0} = \varphi(x), \quad u_t|_{t=0} = 0.$$

14.19. Suppose $u_0(x) \in C^{n+3}(R^n)$, $|x|^{n+5}|u_0(x)| \leqslant M$, and $|x|^{n+1}|D^\alpha u_0(x)| \leqslant M$, $|\alpha| \leqslant n+3$. Prove that the Cauchy problem

$$u_{tt} = -(\nabla^2)^2 u; \quad u|_{t=0} = u_0(x), \quad u_t|_{t=0} = 0$$

has a solution and that this solution is given by the formula

$$u(x, t) = \frac{1}{(2\sqrt{\pi t})^n} \int u_0(\xi) \cos\left(\frac{|x-\xi|^2}{4t} - \frac{\pi n}{4}\right) d\xi.$$

Hint. Use the results of Problems 14.5 and 14.14.

14.20. Solve the following Cauchy problems

1. $u_{tt} = -\frac{\partial^4 u}{\partial x^4} + 6tx^3; \; u|_{t=0} = 0, \; u_t|_{t=0} = x^4.$
2. $u_{tt} = -(\nabla^2)^2 u + xye^t; \quad u|_{t=0} = x^2y^2, \quad u_t|_{t=0} = 0.$
3. $u_{tt} = -(\nabla^2)^2 u + 6x^2y^2z^2; \quad u|_{t=0} = 0, \quad u_t|_{t=0} = 0.$
4. $u_{tt} = -\frac{\partial^4 u}{\partial x^4}, \quad 0 < t < \frac{1}{4}; \; u|_{t=0} = \cos x^2, \; u_t|_{t=0} = 0.$

For the equation

$$\frac{\partial u}{\partial t} = P\left(i\frac{\partial}{\partial x}\right) u, \quad t > 0, \quad x \in R^1, \tag{14.6}$$

where $P(\sigma) = a_0\sigma^N + a_1\sigma^{N-1} + \ldots + a_N$, $a_0 \neq 0$, $N \geqslant 2$, with the initial condition

$$u|_{t=0} = u_0(x), \tag{14.7}$$

the classical Cauchy problem is stated in the class of functions $u(x, t) \in C(t \geqslant 0)$ that have continuous derivatives $\partial u/\partial t$ and $\partial^N u/\partial x^N$ at $t > 0$.

The Cauchy problem (14.6), (14.7) is said to be *well* or *properly* or *correctly posed* in the class $\mathcal{S}$ (for the definition of class $\mathcal{S}$ see Sec. 9) if to every function $u_0(x) \in \mathcal{S}$ there corresponds a unique solution to (14.6), (14.7) that for every $t > 0$ belongs to $\mathcal{S}$ and decreases, as $|x| \to \infty$, together with its derivatives in Eq. (14.6) faster than any power of $|x|^{-1}$ uniformly in t in each interval $0 < t < T < \infty$.

14.21. Suppose that the Cauchy problem (14.6), (14.7) is properly posed in $\mathcal{S}$ and that $v(\sigma, t) = F[u(x, t)] = \int_{-\infty}^{\infty} u(x, t) e^{ix\sigma} dx$, where $u(x, t)$ is the solution to the Cauchy problem (14.6), (14.7). Prove that for every $t \geqslant 0$ the function $v(\sigma, t)$ belongs to $\mathcal{S}$ and is the solution to the problem

$$\frac{dv}{dt} = P(\sigma)\, v, \quad v|_{t=0} = F[u_0(x)]. \tag{14.8}$$

14.22. Suppose $u_0(x) \in \mathcal{S}$ and

$$\operatorname{Re} P(\sigma) \leqslant C < \infty \tag{14.9}$$

for all real values of σ. Prove that the function

$$u(x, t) = \frac{1}{2\pi} \int_{-\infty}^{\infty} e^{tP(\sigma) - ix\sigma} \int_{-\infty}^{\infty} u_0(\xi)\, e^{i\sigma\xi}\, d\xi\, d\sigma \tag{14.10}$$

is the solution to problem (14.6), (14.7), belongs to class C^∞ $(t \geqslant 0)$, and decreases, as $|x| \to \infty$, together with its derivatives faster than any power of $|x|^{-1}$ uniformly in $t \geqslant 0$.

14.23. Prove that condition (14.9) is necessary and sufficient for the Cauchy problem (14.6), (14.7) to be properly posed in $\mathcal{S}$. *Hint.* To prove necessity, first show that if condition (14.9) is not met, there is a function $u_0(x) \in \mathcal{S}$ for which the solution of problem (14.8) does not belong to $\mathcal{S}$.

14.24. Suppose the Cauchy problem (14.6), (14.7) is correctly posed in $\mathcal{S}$. Prove that the solution to this problem is expressed via formula (14.10), which can be written in the form

$$u(x, t) = \int_{-\infty}^{\infty} u_0(\xi)\, G(x - \xi, t)\, d\xi, \tag{14.11}$$

$$G(x, t) = \frac{1}{2\pi} \int_{-\infty}^{\infty} e^{tP(\sigma) - ix\sigma}\, d\sigma. \tag{14.12}$$

Hint. Use the estimate $|G(x, t)| \leqslant Ct^{-1/N}$.

14.25. Suppose condition (14.9) is met, $u_0(x) \in C^{N+2}(R^1)$, and

$$\int_{-\infty}^{\infty} |u_0^{(k)}(x)|\, dx < \infty, \quad k = 0, 1, \ldots, N+2.$$

Prove that the Cauchy problem (14.6), (14.7) has a solution, that this solution is expressed through the formula (14.10) (or formulas (14.11) and (14.12)), and that $u(x, t)$ is bounded at $t \geqslant 0$ together with its derivatives in Eq. (14.6).

14.26. Solve the following Cauchy problems for first-order equations:

1. $u_t + 2u_x + 3u = 0$, $u|_{t=0} = x^2$.
2. $u_t + 2u_x + u = xt$, $u|_{t=0} = 2 - x$.
3. $2u_t = u_x + xu$, $u|_{t=0} = 1$. 4. $2u_t = u_x - xu$, $u|_{t=0} = 2xe^{x^2/2}$.

5. $u_t+(1+x^2)u_x-u=0, \quad u|_{t=0}=\arctan x.$
6. $u_t+(1+t^2)u_x+u=1, \quad u|_{t=0}=e^{-x}.$
7. $u_t=u_x+\frac{2x}{1+x^2}u, \quad u|_{t=0}=1.$
8. $2tu_t+xu_x-3x^2u=0, \quad u|_{t=1}=5x^2.$

Problems 14.27-14.55 deal with Goursat's problem. For the statement of this problem see Vladimirov [2], pp. 70 and 221.

14.27. Prove that Goursat's problem

$$u_{xy}=0, \quad 0<y<\alpha x, \quad x>0, \ y>0;$$
$$u|_{y=0}=f(x), \quad u|_{y=\alpha x}=g(x)$$

has a unique solution $u(x, y)=f(x)+g\left(\frac{y}{\alpha}\right)-f\left(\frac{y}{\alpha}\right)$, provided $f(x)$ and $g(x)$ belong to class $C^2(x>0)\cap C(x\geqslant 0)$ and $f(0)=g(0)$.

14.28. Prove that Goursat's problem

$$u_{xy}=0, \ x>0, \ y>0, \ u|_{y=0}=f(x), \ u|_{x=0}=g(y)$$

has a unique solution $u(x, y)=f(x)+g(y)-f(0)$ provided $f(x)$ and $g(x)$ belong to class $C^2(x>0)\cap C(x\geqslant 0)$ and $f(0)=g(0)$.

14.29. Prove that the solution of Goursat's problem

$$u_{xy}=0, \ y>\alpha x, \ x>0, \ \alpha<0;$$
$$u|_{y=\alpha x}=0, \quad u|_{x=0}=0$$

is not unique. Show that every solution of this problem is given by the formula $u(x, y)=f(x)-f(y/\alpha)$, where $f(x)$ is any function belonging to $C^2(R^1)$ that vanishes at $x\leqslant 0$.

14.30. Prove that Goursat's problem

$$u_{xy}=0, \ 0<y<\varphi(x), \ x>0;$$
$$u|_{y=0}=f(x), \quad u|_{y=\varphi(x)}=g(x)$$

has a unique solution

$$u(x, y)=f(x)+g(\varphi^{-1}(y))-f(\varphi^{-1}(y)),$$

provided the functions $f(x)$, $g(x)$, and $\varphi(x)$ belong to $C^2(x>0)\cap C(x\geqslant 0)$, $f(0)=g(0)$, $\varphi(0)=0$, $\varphi'(x)>0$; $\varphi^{-1}(y)$ is the inverse of $\varphi(x)$.

14.31. Suppose the functions $\varphi(x)$ and $\psi(x)$ belong to class $C^2(x>0)\cap C(x\geqslant 0)$ and $\varphi(0)=\psi(0)$. For what real values of a does Goursat's problem

$$au_{xx}+u_{yy}=0, \ x>0, \ y>0,$$
$$u|_{y=0}=\varphi(x), \ u|_{x=0}=\psi(y)$$

have a unique solution? Find the solution.

14.32. For what positive values of parameter b does the problem

$$u_{tt}=a^2u_{xx};\ 0<t<bx,\ x>0,$$
$$u|_{t=0}=0,\ u|_{t=bx}=0$$

have a unique solution?

In Problems 14.33-14.55 the reader must find the solution of Goursat's problem and prove its uniqueness.

14.33. $u_{xy}+u_x=x,\ x>0,\ y>0;$

$$u|_{x=0}=y^2,\ u|_{y=0}=x^2.$$

14.34. $u_{xy}+x^2yu_x=0,\ x>0,\ y>0;$

$$u|_{x=0}=0,\ u|_{y=0}=x.$$

14.35. $u_{xy}+u_y=1,\ x>0,\ y>0;$

$$u|_{x=0}=\varphi(y),\ u|_{y=0}=\psi(x),$$

where the functions $\varphi(x)$ and $\psi(x)$ belong to class $C^2(x>0)\cap\cap C(x\geqslant 0)$ and $\varphi(0)=\psi(0)$.

14.36. $u_{xy}+xu_x=0,\ x>0,\ y>0;$

$$u|_{x=0}=\varphi(y),\ u|_{y=0}=\psi(x),$$

where the functions $\varphi(x)$ and $\psi(x)$ belong to class $C^2(x>0)\cap\cap C(x\geqslant 0)$ and $\varphi(0)=\psi(0)$.

14.37. $2u_{xx}-2u_{yy}+u_x+u_y=0,\ y>|x|;$

$$u|_{y=x}=1,\ u|_{y=-x}=(x+1)e^x.$$

14.38. $2u_{xx}+u_{xy}-u_{yy}+u_x+u_y=0,\ -\frac{1}{2}x<y<x,\ x>0,$

$$u|_{y=x}=1+3x,\ u|_{y=-\frac{1}{2}x}=1.$$

14.39. $u_{xx}+6u_{xy}+5u_{yy}=0,\ x<y<5x,\ x>0;$

$$u|_{y=x}=\varphi(x),\ u|_{y=5x}=\psi(x),$$

where the functions $\varphi(x)$ and $\psi(x)$ belong to class $C^2(x>0)\cap\cap C(x\geqslant 0)$ and $\varphi(0)=\psi(0)$.

14.40. $u_{xx}+yu_{yy}+\frac{1}{2}u_y=0,\ -\frac{1}{4}x^2<y<0,\ x>0;$

$$u|_{y=0}=0,\ u|_{y=-\frac{1}{4}x^2}=x^2.$$

14.41. $u_{xy}-e^xu_{yy}=0,\ y>-e^x,\ x>0;$

$$u|_{x=0}=y^2,\ u|_{y=-e^x}=1+x^2.$$

14.42. $yu_{xx}+(x-y)u_{xy}-xu_{yy}-u_x+u_y=0,\ 0<y<x,\ x>0;$

$$u|_{y=0}=0,\ u|_{y=x}=4x^4$$

14.43. $xu_{xx}+(x-y)u_{xy}-yu_{yy}=0,\ 0<y<x,\ x>0;$

$$u|_{y=0}=0,\ u|_{y=x}=x.$$

14.44. $y^2u_{xx}+u_{xy}=0,\ y^3-8<3x<y^3,\ 0<y<2;$
$$u|_{y=2}=3x+8,\ u|_{3x=y^3}=2y^3.$$

14.45. $x^2u_{xx}-y^2u_{yy}=0,\ y>x,\ x>1;$
$$u|_{x=1}=1,\ u|_{y=x}=x.$$

14.46. $x^2u_{xx}-y^2u_{yy}+xu_x-yu_y=0,\ \frac{1}{x}<y<x,\ x>1;$
$$u|_{y=x}=x,\ u|_{y=\frac{1}{x}}=1+\ln x.$$

14.47. $3x^2u_{xx}+2xyu_{xy}-y^2u_{yy}=0,\ x<y<\frac{1}{\sqrt[3]{x}},\ 0<x<1;$
$$u|_{x=y}=y,\ u|_{xy^3=1}=y^2.$$

14.48. $3x^2u_{xx}+2xyu_{xy}-y^2u_{yy}=0,\ 1<y<x,\ x>1;$
$$u|_{y=x}=0,\ u|_{y=1}=\cos\frac{\pi x}{2}.$$

14.49.
$$u_{xx}-2\sin x u_{xy}-\cos^2 x u_{yy}-\cos x u_y=0,\ |y-\cos x|<x,\ x>0;$$
$$u|_{y=x+\cos x}=\cos x,\ u|_{y=-x+\cos x}=\cos x.$$

14.50. $u_{xy}-\frac{1}{x-y}(u_x-u_y)=1,\ y<-x,\ x>2;$
$$u|_{y=-x}=0,\ u|_{x=2}=2+2y+\frac{1}{2}y^2.$$

14.51. $u_{xx}-u_{yy}+\frac{2}{x}u_x=0,\ y>1+|x|;$
$$u|_{y=x+1}=1-x,\ u|_{y=1-x}=1+x.$$

14.52. $u_{xx}-u_{yy}+\frac{4}{x}u_x+\frac{2}{x^2}u=0,\ y>x,\ x>1;$
$$u|_{y=x}=1,\ u|_{x=1}=y.$$

14.53. $u_{xy}=1,\ \alpha x<y<\beta x,\ x>0,\ 0<\alpha<\beta;$
$$u|_{y=\alpha x}=0,\ u|_{y=\beta x}=0.$$

14.54. $u_{xy}=0,\ x^2<y<2x^2,\ x>0;$
$$u|_{y=x^2}=x^4,\ u|_{y=2x^2}=x^2.$$

14.55. $u_{xy}=0,\ x^4<y<x^2,\ 0<x<1;$
$$u|_{y=x^2}=0,\ u|_{y=x^4}=x(1-x).$$

Problems 14.56-14.63 deal with the Cauchy problem for quasi-linear and nonlinear equations.

14.56. Find the solution of the Cauchy problem
$$u_t+uu_x=0,\ t>0;\ u|_{t=0}=\operatorname{sign} x,$$

continuous for $t \geqslant 0$, $|x| + t \neq 0$, and continuously differentiable at $t \neq |x|$.

14.57. Find the solution of the Cauchy problem

$$u_t = uu_x = 0, \quad t > 0; \quad u|_{t=0} = \begin{cases} \alpha & \text{at } x < 0, \\ \beta & \text{at } x > 0 \end{cases}$$

(where α and β ($\geqslant \alpha$) are constants), continuous for $t \geqslant 0$, $|x| + t \neq 0$, and continuously differentiable outside the straight lines $t = x/\alpha$ and $t = x/\beta$.

14.58. Prove that the Cauchy problem for Burgers' equation

$$u_t + uu_x = a^2 u_{xx}$$

with the initial condition

$$u|_{t=0} = u_0(x)$$

is reduced, via the substitution $u = -2a^2 v_x/v$, to the Cauchy problem

$$v_t = a^2 v_{xx}, \quad v|_{t=0} = \exp\left[-\frac{1}{2a^2}\int_0^x u_0(\xi)\, d\xi\right].$$

14.59. Let u be the solution of the Cauchy problem

$$u_t + uu_x = \varepsilon u_{xx}, \quad u|_{t=0} = \operatorname{sign} x,$$

continuous at $t \geqslant 0$, $|x| + t \neq 0$, and continuously differentiable at $t > 0$. Prove that this solution tends, $\varepsilon \to +0$, to the solution of Problem 14.56 (E. Hopf's theorem).

14.60. See whether the solution of the Korteweg-de Vries differential equation

$$u_t + 6uu_x + u_{xxx} = 0$$

is the function

$$u(x, t) = \frac{a}{2\cosh^2\left[\frac{\sqrt{a}}{2}(x - x_0 - at)\right]}, \quad a > 0,$$

which describes a solitary wave. Such solutions are called *solitons*. Prove that a soliton has a finite energy,

$$\int_{-\infty}^{\infty} (u_t^2 + u_x^2)\, dx < \infty.$$

14.61. Verify the validity of the following propositions for Liouville's equation

$$u_{tt} - u_{xx} = ge^u, \quad g > 0.$$

1. The function

$$u(x, t) = \ln \frac{\alpha^2(1-a^2)}{2g\cosh^2\left[\frac{\alpha}{2}(x-x_0-at)\right]}, \quad 0 \leqslant a \leqslant 1,$$

is a solution for all values of x and t.

2. The function

$$u(x, t) = \ln \frac{8\varphi'(x+t)\psi'(x-t)}{g[\varphi(x+t)-\psi(x-t)]^2}$$

is a solution for all functions φ and ψ such that $\varphi, \psi \in C^3$ and $\varphi'\psi' > 0$.

3. The function

$$u(x, t) = \frac{1}{2}[u_0(x+t)+u_0(x-t)] - \ln\left\{\cos^2\left[\sqrt{\frac{g}{8}}\int_{x-t}^{x+t} e^{\frac{u_0(\xi)}{2}}\,d\xi\right]\right\}$$

is a solution of the Cauchy problem with the initial conditions

$$u|_{t=0} = u_0(x), \quad u_t|_{t=0} = 0,$$

provided

$$\left|\sqrt{\frac{g}{8}}\int_{x-t}^{x+t} e^{\frac{u_0(\xi)}{2}}\,d\xi\right| < \frac{\pi}{2}.$$

14.62. See whether the solution of the sine-Gordon equation

$$u_{tt} - u_{xx} = -g\sin u, \quad g > 0$$

is the function

$$u(x, t) = 4\arctan e^{\pm\sqrt{g}(x-x_0-at)/\sqrt{1-a^2}}, \quad 0 \leqslant a < 1,$$

which is also a soliton with a finite energy

$$\int_{-\infty}^{+\infty}(u_t^2+u_x^2)\,dx < \infty.$$

14.63. See whether the solution of the nonlinear Schrödinger equation

$$iu_t + u_{xx} + \nu|u|^2u = 0, \quad \nu > 0,$$

is the function

$$u(x, t) = \sqrt{\frac{2\alpha}{\nu}}\,\frac{\exp\left\{i\left[\frac{a}{2}x - \left(\frac{a^2}{4}-\alpha\right)t\right]\right\}}{\cosh\left[\sqrt{\alpha}(x-x_0-at)\right]}, \quad \alpha \geqslant 0.$$

Answers to Problems of Sec. 14

14.9. $(1+4at)^{-\frac{n}{2}} e^{\frac{ia|x'|^2}{1+4at}}$.

14.10. 1. $x^4 + t^2\left(\frac{1}{2}x^3 - 12\right) + itx(12x + t^2)$. 2. $\frac{x}{(1-4t)^{3/2}} e^{\frac{-ix^2}{1-4t}}$.
3. $x\sin t + x^2 + y^2\cos t + 2i(t+\sin t)$. 4. $i(x^3+y^3+z^3) - t(6y+$

$+6z-y^2-iz^3)+t^2(i-3z)$. 5. $(\sqrt{1+4it})^{-n}\,e^{\frac{-|x|^2}{1+4it}}$, $0\leqslant\arg\sqrt{1+4it}<\pi/2$.

14.11. 1. $\mathscr{E}(x,\ t)$. 2. $\frac{ix_k}{2t}\mathscr{E}(x,\ t)$. 3. $\int\limits_0^1 \mathscr{E}(x+x_0,\ \tau)\,d\tau$. 4. $\int\limits_0^{t-t_0}\mathscr{E}(x,\ \tau)\,d\tau$.

14.12. 1. $\frac{1}{\sqrt{\pi}}e^{-\pi i/4}\Bigg(\int\limits_{-\infty}^{\frac{x}{2\sqrt{t}}} e^{iy^2}\,dy+\int\limits_0^t\int\limits_{-\infty}^{\frac{x}{2\sqrt{\tau}}} e^{iy^2}\,dy\,d\tau\Bigg)$. 2. $\theta(t-1)\times\times(t-1)+\frac{1}{\sqrt{\pi}}e^{-\frac{\pi i}{4}}\int\limits_{\frac{x-1}{2\sqrt{t}}}^{\frac{x+1}{2\sqrt{t}}} e^{iy^2}\,dy$. 3. $x^2+2it-\theta(t-\pi)(1+\cos t)$. 4. $2\sqrt{t}+\cos x e^{-it}$. 5. $\theta(t-1)(e^t-e-te)+(x\sin x+2it\cos x)\,e^{-it}$.

14.15. $\int\limits_0^t (t-\tau)\,f(x,\ \tau)\,d\tau$. **14.16.** $u_0(x)+tu_1(x)$.

14.17. $\mathrm{Re}\int\limits_0^t w(x,\ \tau)\,d\tau$. **14.18.** $i\,\mathrm{Im}\,w(x,\ t)$.

14.20. 1. $tx^4+t^3(x^3-4)$. 2. $x^2y^2-4t^2+xy(e^t-1-t)$. 3. $3x^2y^2z^2t^2-\\-2(x^2+y^2+z^2)t^4$. 4. $\frac{1}{2}\left(\frac{1}{\sqrt{1+4t}}\cos\frac{x^2}{1+4t}+\frac{1}{\sqrt{1-4t}}\cos\frac{x^2}{\sqrt{1-4t}}\right)$.

14.26. 1. $(x-2t)^2e^{-3t}$. 2. $4-x-2t+xt-2e^{-t}$. 3. $e^{\frac{t}{8}(4x+t)}$. 4. $(2x+t)\,e^{x^2/2}$. 5. $(\arctan x-t)\,e^t$. 6. $1-e^{-t}+e^{-x+\frac{1}{3}t^3}$. 7. $\frac{1+(x+t)^2}{1+x^2}$. 8. $\frac{5x^2}{t}e^{\frac{3x^2(t-1)}{2t}}$.

14.31. $\varphi(x-ay)-\varphi(-ay)+\psi(y)$, $a\leqslant 0$.

14.32. $b\leqslant\frac{1}{a}$. **14.33.** $y^2+\frac{1}{2}x^2(1+e^{-y})$.

14.34. $\int\limits_0^x e^{-\frac{1}{2}\xi^2y^2}\,d\xi$.

14.35. $y+\psi(x)+[\varphi(y)-\varphi(0)-y]\,e^{-x}$.

14.36. $\varphi(y)+\int\limits_0^x \psi'(\xi)\,e^{-\xi y}\,d\xi$.

14.37. $\left(1+\frac{1}{2}x-\frac{1}{2}y\right)e^{\frac{1}{2}(x-y)}$ **14.38.** $1+(x+2y)e^{\frac{1}{3}(y-x)}$.

14.39. $\varphi\left(\frac{5x-y}{4}\right)+\psi\left(\frac{y-x}{4}\right)-\varphi(0)$.

14.40. $2x\sqrt{-y}$ **14.41.** $x^2+(y-1+e^x)^2$.

14.42. $xy(x+y)^2$. **14.43.** y. **14.44.** $3x+y^3$.

14.45. x. **14.46** $\sqrt{xy}\cdot\frac{1}{2}\ln\frac{x}{y}$ **14.47.** $\sqrt[4]{\frac{y^5}{x}}$.

14.48. $y\cos\frac{\pi x}{2y}$. **14.49.** $-1+2\cos\frac{x}{2}\cos\frac{y-\cos x}{2}$.

14.50. $\frac{1}{2}(x+y)^2$. *Hint*. Perform the substitution $u=\frac{1}{x-y}v$.

14.51. $2-y$. *Hint*. Perform the substitution $u=\frac{1}{x}v$.

14.52. $\frac{y}{x}$. *Hint*. Perform the substitution $u=\frac{1}{x^2}v$.

14.53. $\frac{1}{\alpha+\beta}(y-\alpha x)(\beta x-y)$. **14.54.** $\frac{4}{3}x^4-x^2+y-\frac{1}{3}y^2$.

14.55. $x-\sqrt{y}$.

14.56. -1 at $x\leqslant -t$, $+1$ at $x\geqslant t$, and x/t at $x\leqslant t$. ***Hint***. Look for the solution in the form $f(x/t)$.

14.57. α at $x\leqslant t\alpha$, β at $x\geqslant t\beta$, and x/t at $t\alpha\leqslant x\leqslant t\beta$.

Chapter V

Boundary Value Problems for Equations of Elliptic Type

Let S be the smooth surface that is the boundary of a region $G \in R^n$ and let $\mathbf{n}_x$ be the outward normal vector at point $x \in S$, i.e. normal to S at point x. It is said that a function u has a *regular normal derivative on S outward* if there exists, uniformly in all $x \in S$, a limit,

$$\lim_{\substack{x' \to x \\ x' \in G \cap (-\mathbf{n}_x)}} \frac{\partial u(x')}{\partial \mathbf{n}_x} = \frac{\partial u(x)}{\partial \mathbf{n}_x} = \frac{\partial u}{\partial \mathbf{n}}.$$

(i) The *interior Dirichlet problem for Laplace's equation*: to find a function $u \in C(\overline{G})$ that is harmonic in $G \subset R^3$ and assumes specified (continuous) values u_0^- on S (the boundary of G).

(ii) The *exterior Dirichlet problem*: to find a function $u \in C(\overline{G_1})$ that is harmonic in $G_1 = R^3 \setminus \overline{G}$, vanishes at infinity, and assumes specified (continuous) values u_0^+ on S.

(iii) The *interior Neumann problem*: to find a function $u \in C(\overline{G})$ that is harmonic in G and possesses a specified (continuous) regular normal derivative u_1^- on S.

(iv) The *exterior Neumann problem*: to find a function $u \in C(\overline{G_1})$ that is harmonic in G_1, vanishes at infinity, and possesses a specified (continuous) regular normal derivative u_1^+ on S.

Problems (i), (ii), and (iv) have unique solutions, but the solution of Problem (iii) is defined to within an arbitrary constant (the solvability condition for this problem is $\int_S u_1^- \, dS = 0$).

Problems (i)-(iv) in R^2 are stated in a similar manner, the only difference being that the exterior problems require only the boundedness of the solutions at $|x| \to \infty$. Problems (i) and (ii) have unique solutions, while the solutions of problems (iii) and (iv) are defined to within arbitrary constants (with $\int_S u_1^{\mp} \, dS = 0$ being the solvability condition).

15 The Sturm-Liouville Problem

Consider the boundary value problem

$$Lu \equiv -(p(x)\,y'(x))' + q(x)\,y(x) = f(x), \tag{15.1}$$

$$\left.\begin{aligned}\alpha_1 y(a) - \alpha_2 y'(a) = 0,\\ \beta_1 y(b) + \beta_2 y'(b) = 0,\end{aligned}\right\} \tag{15.2}$$

where $\alpha_1^2 + \alpha_2^2 \neq 0$, $\beta_1^2 + \beta_2^2 \neq 0$, $p \in C^1([a, b])$, $p(x) \neq 0$, $q \in C([a, b])$, $f \in C(a, b) \cap L_2(a, b)$.

Physical problems often obey the following conditions:

$$\alpha_1\alpha_2 \geqslant 0, \quad \beta_1\beta_2 \geqslant 0,$$
$$p(x) > 0, \quad q(x) \geqslant 0.$$

The domain of definition M_L of operator L consists of functions $y(x)$ of class $C^2(a, b) \cap C^1([a, b])$, $y'' \in L_2(a, b)$, that satisfy the boundary conditions (15.2).

The problem of finding the values of λ (called the *eigenvalues* of operator L) for which the equation $Ly = \lambda y$ has nonzero solutions $y(x)$ from M_L is known as the *Sturm-Liouville problem* (the solutions are called the *eigenfunctions* of L corresponding to the eigenvalues).

If $\lambda = 0$ is not an eigenvalue of L, the solution of the boundary value problem (15.1) is unique in M_L and is expressed through the formula

$$y(x) = \int_a^b G(x, \xi)\, f(\xi)\, d\xi,$$

where $G(x, \xi)$ is the Green's function of the boundary value problem (15.1), (15.2) or of operator L.

The Green's function $G(x, \xi)$ is represented in the form

$$G(x, \xi) = -\frac{1}{k}\begin{cases} y_1(x)\, y_2(\xi), & a \leqslant x \leqslant \xi, \\ y_1(\xi)\, y_2(x), & \xi \leqslant x \leqslant b, \end{cases}$$

where $y_1(x)$ and $y_2(x)$ are the nonzero solutions of the equation $Ly = 0$ that satisfy the first and second boundary conditions in (15.2), and

$$k = p(x)\,w(x) = p(a)\,w(a) \neq 0, \quad x \in [a, b],$$

$$w(x) = \begin{vmatrix} y_1(x) & y_2(x) \\ y_1'(x) & y_2'(x) \end{vmatrix}$$

is the *Wronskian*.

The boundary value problem

$$Ly = \lambda y + f,$$

where $f \in C(a, b) \cap L_2(a, b)$, provided $\lambda = 0$ is not an eigenvalue of L, is equivalent to the integral equation

$$y(x) = \lambda \int_a^b G(x, \xi)\, y(\xi)\, d\xi + \int_a^b G(x, \xi)\, f(\xi)\, d\xi.$$

This method can sometimes be applied to problems involving degeneracy, when $p(x)$ vanishes or becomes infinitely large or when $q(x)$ becomes infinitely large at an end point of segment $[a, b]$.

15.1. Find the Green's function of operator L in the interval $(0, 1)$ for the following cases:

1. $Ly = -y''$, $y(0) = y(1) = 0$.
2. $Ly = -y''$, $y'(0) = y(0)$, $y'(1) + y(1) = 0$.
3. $Ly = -y''$, $y(0) = hy'(0)$, $h \geqslant 0$, $y(1) = 0$.
4. $Ly = -y'' - y$, $y(0) = y(1) = 0$.
5. $Ly = -y'' - y$, $y(0) = y'(0)$, $y(1) = y'(1)$.
6. $Ly = -y'' + y$, $y(0) = y(1) = 0$.
7. $Ly = -y'' + y$, $y'(0) = y'(1) = 0$.

15.2. Find the Green's function of operator L in the interval $(1, 2)$ for the following cases:

1. $Ly = -x^2y'' - 2xy'$, $y'(1) = 0$, $y(2) = 0$.
2. $Ly = -xy'' - y'$, $y'(1) = 0$, $y(2) = 0$.
3. $Ly = -x^3y'' - 3x^2y - xy$, $y(1) = 0$, $y(2) + 2y'(2) = 0$.
4. $Ly = -x^4y'' - 4x^3y' - 2x^2y$, $y(1) + y'(1) = 0$, $y(2) + 3y'(2) = 0$.

15.3. Find the Green's function of operator L in the interval $(0, \pi/4)$ for the following cases:

1. $Ly = -(\cos^2 x \cdot y')'$, $y(0) = 0$, $y(\pi/4) = 0$.
2. $Ly = -(y'/\cos x)'$, $y(0) = 0$, $y(\pi/4) = 0$.
3. $Ly = -\cos^2 x \cdot y'' + \sin 2x \cdot y'$, $y(0) = y'(0)$, $y\left(\frac{\pi}{4}\right) + y'\left(\frac{\pi}{4}\right) = 0$.

15.4. Find the Green's function of operator L in the interval $(0, 1)$ for the following cases:

1. $Ly = -(1 + x^2)y'' - 2xy'$, $y(0) = y'(0)$, $y(1) = 0$.
2. $Ly = -(1 + x^2)y'' - 2xy'$, $y(0) = 0$, $y(1) + y'(1) = 0$.
3. $Ly = -(3 + x^2)y'' - 2xy'$, $y(0) = y'(0)$, $y(1) = 0$.
4. $Ly = -(x + 1)^2y'' - 2(x + 1)y' + 2y$, $y(0) = y(1) = 0$.
5. $Ly = -\left(\frac{1}{x-2} y'\right)' + \frac{3y}{(x-2)^3}$, $y(0) = 0$, $y(1) = 0$.
6. $Ly = -(4 - x^2)y'' + 2xy'$, $y(0) = y(1) = 0$.
7. $Ly = -(xy')' + \frac{4}{x} y$, $y(0) = y(1) = 0$.
8. $Ly = -\frac{1}{x^2} y'' + \frac{2}{x^3} y' - \frac{2}{x^4} y$, $y'(0) = y(1) = 0$.

15.5. Find the Green's function of operator L in the interval $(0, \pi/4)$ under the conditions $|y(0)| < \infty$ and $y(\pi/4) = 0$ for the following cases:

1. $Ly = -(\tan^2 x \cdot y')'$. 2. $Ly = -(\tan x \cdot y')'$.

15.6. Find the Green's function of operator L in the interval $(0, \pi/2)$ for the following cases:

1. $Ly = -\cos^2 x \cdot y'' + \sin 2x \cdot y'$, $y(0) = 0$, $|y(\pi/2)| < \infty$
2. $Ly = -\sin^2 x \cdot y'' - \sin 2x \cdot y'$, $|y(0)| < \infty$, $y(\pi/2) = 0$
3. $Ly = -\sin^2 x \cdot y'' - \sin 2x \cdot y'$, $|y(0)| < \infty$, $y\left(\frac{\pi}{2}\right) + y'\left(\frac{\pi}{2}\right) = 0$

15.7. Find the Green's function of operator L in the interval $(0, 1)$ under the condition $|y(0)| < \infty$ for the following cases:

1. $Ly = -x^2 y'' - 2xy + 6y$, $y'(1) + 3y(1) = 0$.
2. $Ly = -y'' + \frac{2}{x^2} y$, $y(1) = 0$.
3. $Ly = -x^2 y'' - 2xy' + 2y$, $y'(1) = 0$.
4. $Ly = -(xy')'$, $y(1) = 0$.
5. $Ly = -xy'' - y'$, $y'(1) + y(1) = 0$.
6. $Ly = -x^2 y'' - 2xy' + 2y$, $y(1) + y'(1) = 0$.
7. $Ly = -x^2 y'' - 2xy' + 2y$, $2y(1) + y'(1) = 0$.
8. $Ly = -y'' + \frac{a(a-1)}{x^2} y$, $a > 1$, $y(1) = 0$.
9. $Ly = -(xy')' + (1 + x) y$, $y(1) = 0$.

15.8. Find the Green's function of the operator $Ly = -x^4 y'' - 4x^3 y' - 2x^2 y$ in the interval $(1, 3)$ if $y(1) + y'(1) = 0$ and $2y(3) + 3y'(3) = 0$.

15.9. Find the Green's function of operator L in the interval $(0, 1)$ for the following cases:

1. $Ly = -(e^{-\frac{x^2}{2}} y')' + e^{-\frac{x^2}{2}} y$, $y(0) = y(1) = 0$.
2. $Ly = -e^{x^2} y'' - 2xe^{x^2} y'$, $y(0) = 2y'(0)$, $y(1) = 0$.
3. $Ly = -y'' + (1 + x^2) y$, $y(0) = y'(1) = 0$.

Hint. A particular solution of the equation $-y'' + (1 + x^2) y = 0$ may be sought in the form $y = e^{z(x)}$.

15.10. Find the Green's function of the operator $Ly = -(\sqrt{x} y')' + 3x^{-3/2} y$ in the interval $(0, 2)$ if $|y(0)| < \infty$ and $y(2) = 0$.

15.11. Find the Green's function of the operator $Ly = -(x + 1) y'' - y'$ if $|y(-1)| < \infty$ and $y(0) = 0$.

15.12. Find the Green's function of the operator $Ly = -x^2 y'' - xy' + n^2 y$ if $|y(0)| < \infty$ and $y(1) = 0$.

15.13. Find the Green's function of the operator $Ly = -[(x^2 - 1) y']' + 2y$ if $|y(1)| < \infty$ and $y(2) = 0$.

15.14. Reduce the Sturm-Liouville problem to an integral equation for the following cases:

1. $Ly \equiv -(1 + e^x) y'' - e^x y' = \lambda x^2 y$, $0 < x < 1$, $y(0) - 2y'(0) = 0$, $y'(1) = 0$.
2. $Ly \equiv -(x^2 + 1) y'' - 2xy' + 2y = \lambda y$, $0 < x < 1$, $y'(0) = 0$, $y(1) - y'(1) = 0$.

3. $Ly \equiv -\sqrt{1+e^{2x}}\, y'' - \frac{e^{2x}}{1+e^{2x}} y' = \lambda xy, \quad 0<x<1, \quad y(0) = \sqrt{2}\, y'(0),\ y'(1)=0.$

4. $Ly \equiv -(1-x^2)\, y'' + 2xy' - 2y = \lambda y,\ 0<x<1,\ y'(0) = 0,\ |y(1)| < \infty.$

5. $Ly \equiv -\cos^4 x \cdot y'' + 4\sin x \cos^3 x \cdot y' = \lambda xy, 0<x<\frac{\pi}{2}, 2y(0) - y'(0) = 0,\ |y(\pi/2)| < \infty.$

6. $Ly \equiv -x^2 y'' - 2xy' + (2\cos^2 x + 1)\, y = \lambda y \cos 2x,\ 1<x<2,\ y(1) = 0,\ y'(2) = 0.$

7. $Ly \equiv -y'' = \lambda y,\ 0<x<1,\ y'(0) = y'(1) = 0.$

15.15. Reduce the finding of solutions of the equation

$$-2xy'' - y' = 2\lambda \sqrt{x} y, \quad 0<x<1,$$

with the boundary conditions $\lim_{x\to 0} (\sqrt{x} y') = 0$ and $y(1) = 0$ to solving an integral equation.

15.16. Reduce the finding of solutions of the equation $-xy'' + y' = \lambda y$, $1<x<2$, with the boundary conditions $y(1) = y'(2) = 0$ to solving an integral equation.

15.17. Reduce the finding of solutions of each of the equations given below (with the specified boundary conditions) to solving an integral equation:

1. $-(1+x^2)\, y'' - 2xy' + \lambda y = 0,\ y(0) = y'(1) = 0.$
2. $-e^x y'' - e^x y' + \lambda y = 0,\ y(0) = 0,\ y(1) + y'(1) = 0.$
3. $-y'' + \lambda y = f(x),\ y(0) = hy'(0),\ h \geqslant 0,\ y(1) = 0.$
4. $-xy'' - y' + \lambda xy = 0,\ |y(0)| < \infty,\ y(1) = 0.$

15.18. Employ the Green's function in solving the following problems:

1. $-\frac{xy''}{1+x} - \frac{y'}{(1+x)^2} = f(x),\ 1<x<e,\ y(1)=0,\ y(e) - ey'(e) = 0$, where e is the base of the natural system of logarithms.

2. $-x^4 y'' - 4x^3 y' - 2x^2 y = f(x),\ 1<x<2,\ y(1) = 0,\ y(2) + y'(2) = 0.$

3. $-\frac{x}{1-x} y'' - \frac{1}{(1-x)^2} y' = f(x), \quad -1<x<0, \quad 2y(-1) + y'(-1) = 0,\ |y(0)| < \infty.$

4. $-(1+\cos x)\, y'' + \sin x \cdot y' = f(x),\ 0<x<\pi/2,\ y(0) - 2y'(0) = 0,\ y(\pi/2) = 0.$

5. $-y'' + \frac{2}{x^2} y = f(x),\ 1<x<2,\ 2y(1) = y'(1),\ y(2) + 2y'(2) = 0.$

15.19. Prove that the boundary value problem

$$-y'' + q(x)\, y = f(x),\ y'(a) - hy(a) = c_1,\ y'(b) + Hy(b) = c_2$$

is equivalent to the three Cauchy problems

(1) $g' + g^2 = q(x),\ g(a) = -h;$
(2) $Y' - g(x)Y = -f(x),\ Y(a) = c_1;$
(3) $y' + g(x)y = Y(x),\ y(b) = \frac{c_2 - Y(b)}{H - g(b)}.$

Hint. Factorize the operator $-\frac{d^2}{dx^2} + q$ as follows:

$$-\frac{d^2}{dx^2} + q = -\left(\frac{d}{dx} - g\right)\left(\frac{d}{dx} + g\right).$$

Answers to Problems of Sec. 15

15.1. 1. $\begin{cases} x(1-\xi), & 0 \leqslant x \leqslant \xi, \\ \xi(1-x), & \xi \leqslant x \leqslant 1. \end{cases}$ 2. $\frac{1}{3}\begin{cases} (x+1)(2-\xi), & 0 \leqslant x \leqslant \xi, \\ (\xi+1)(2-x), & \xi \leqslant x \leqslant 1. \end{cases}$

3. $-\frac{1}{h+1}\begin{cases} (x+h)(\xi-1), & 0 \leqslant x \leqslant \xi, \\ (\xi+h)(x-1), & \xi \leqslant x \leqslant 1. \end{cases}$

4. $\frac{1}{\sin 1}\begin{cases} \sin x \sin(1-\xi), & 0 \leqslant x \leqslant \xi, \\ \sin(1-x)\sin\xi, & \xi \leqslant x \leqslant 1. \end{cases}$

5. $\frac{1-\cot 1}{2}\begin{cases} (\sin x + \cos x)\left(\frac{\cot 1 + 1}{\cot 1 - 1}\sin\xi + \cos\xi\right), & 0 \leqslant x \leqslant \xi, \\ \left(\frac{\cot 1 + 1}{\cot 1 - 1}\sin x + \cos x\right)(\sin\xi + \cos\xi), & \xi \leqslant x \leqslant 1. \end{cases}$

6. $\frac{1}{\sinh 1}\begin{cases} \sinh x \sinh(1-\xi), & 0 \leqslant x \leqslant \xi, \\ \sinh\xi \sinh(1-x), & \xi \leqslant x \leqslant 1. \end{cases}$

7. $\frac{1}{2(e^2-1)}\begin{cases} (e^x + e^{-x})(e^{\xi} + e^{2-\xi}), & 0 \leqslant x \leqslant \xi, \\ (e^{\xi} + e^{-\xi})(e^x + e^{2-x}), & \xi \leqslant x \leqslant 1. \end{cases}$

15.2. 1. $\begin{cases} \frac{1}{2} - \frac{1}{\xi}, & 1 \leqslant x \leqslant \xi, \\ \frac{1}{2} - \frac{1}{x}, & \xi \leqslant x \leqslant 2. \end{cases}$ 2. $\begin{cases} \ln\frac{2}{\xi}, & 1 \leqslant x \leqslant \xi, \\ \ln\frac{2}{x}, & \xi \leqslant x \leqslant 2. \end{cases}$

3. $\begin{cases} \frac{-\ln x}{x\xi}, & 1 \leqslant \xi \leqslant 2, \\ \frac{-\ln\xi}{x\xi}, & \xi \leqslant x \leqslant 2. \end{cases}$ 4. $\begin{cases} \frac{1}{x}\left(\frac{1}{\xi^2} - \frac{2}{\xi}\right), & 1 \leqslant x \leqslant \xi, \\ \frac{1}{\xi}\left(\frac{1}{x^2} - \frac{2}{x}\right), & \xi \leqslant x \leqslant 2. \end{cases}$

15.3. 1. $\begin{cases} \tan x(1-\tan\xi), & 0 \leqslant x \leqslant \xi, \\ \tan\xi(1-\tan x), & \xi \leqslant x \leqslant \frac{\pi}{4}. \end{cases}$

2. $\begin{cases} -\sin x(\sqrt{2}\sin\xi - 1) & 0 \leqslant x \leqslant \xi, \\ -\sin\xi(\sqrt{2}\sin x - 1), & \xi \leqslant x \leqslant \frac{\pi}{4}. \end{cases}$

3. $\frac{1}{4}\begin{cases} (\tan x + 1)(\tan\xi - 3), & 0 \leqslant x \leqslant \xi, \\ (\tan\xi + 1)(\tan x - 3), & \xi \leqslant x \leqslant \frac{\pi}{4}. \end{cases}$

15.4. 1. $\begin{cases} \frac{4}{\pi+4}(1+\arctan x)\left(\arctan\xi-\frac{\pi}{4}\right), & 0\leqslant x\leqslant\xi, \\ \frac{4}{\pi+4}(1+\arctan\xi)\left(\arctan x-\frac{\pi}{4}\right), & \xi\leqslant x\leqslant 1. \end{cases}$

2. $\begin{cases} \arctan x\left(-\frac{4}{\pi+2}\arctan\xi+1\right), & 0\leqslant x\leqslant\xi, \\ \arctan\xi\left(-\frac{4}{\pi+2}\arctan x+1\right), & \xi\leqslant x\leqslant 1. \end{cases}$

3. $\frac{2}{\pi+2\sqrt{3}}\begin{cases} \left(1+\sqrt{3}\arctan\frac{x}{\sqrt{3}}\right)\left(\frac{\pi}{6}-\arctan\frac{\xi}{\sqrt{3}}\right), & 0\leqslant x\leqslant\xi, \\ \left(1+\sqrt{3}\arctan\frac{\xi}{\sqrt{3}}\right)\left(\frac{\pi}{6}-\arctan\frac{x}{\sqrt{3}}\right), & \xi\leqslant x\leqslant 1. \end{cases}$

4. $\begin{cases} \frac{1}{21}\left[\frac{1}{(1+x)^2}-(x+1)\right]\left[\frac{8}{(\xi+1)^2}-(\xi+1)\right], & 0\leqslant x\leqslant\xi, \\ \frac{1}{21}\left[\frac{1}{(\xi+1)^2}-(\xi+1)\right]\left[\frac{8}{(x+1)^2}-(x+1)\right], & \xi\leqslant x\leqslant 1. \end{cases}$

5. $\frac{1}{60}\begin{cases} \left[(x-2)^3-\frac{16}{x-2}\right]\left[\frac{1}{\xi-2}-(\xi-2)^3\right], & 0\leqslant x\leqslant\xi, \\ \left[(\xi-2)^3-\frac{16}{\xi-2}\right]\left[\frac{1}{x-2}-(x-2)^3\right], & \xi\leqslant x\leqslant 1. \end{cases}$

6. $\begin{cases} -\frac{1}{4\ln 3}\ln\frac{2+x}{2-x}\left(\ln\frac{2+\xi}{2-\xi}-\ln 3\right), & 0\leqslant x\leqslant\xi, \\ -\frac{1}{4\ln 3}\ln\frac{2+\xi}{2-\xi}\left(\ln\frac{2+x}{2-x}-\ln 3\right), & \xi\leqslant x\leqslant 1. \end{cases}$

7. $\begin{cases} \frac{x^2}{4}\left(\frac{1}{\xi^2}-\xi^2\right), & 0\leqslant x\leqslant\xi, \\ \frac{\xi^2}{4}\left(\frac{1}{x^2}-x^2\right), & \xi\leqslant x\leqslant 1. \end{cases}$ 8. $\begin{cases} x^2(\xi-\xi^2), & 0\leqslant x\leqslant\xi, \\ \xi^2(x-x^2), & \xi\leqslant x\leqslant 1. \end{cases}$

15.5. 1. $\begin{cases} -\cot\xi-\xi+\left(1+\frac{\pi}{4}\right), & 0\leqslant x\leqslant\xi, \\ -\cot x-x+\left(1+\frac{\pi}{4}\right), & \xi\leqslant x\leqslant\frac{\pi}{4}. \end{cases}$

2. $\begin{cases} \ln(\sqrt{2}\sin\xi), & 0\leqslant x\leqslant\xi, \\ \ln(\sqrt{2}\sin x), & \xi\leqslant x\leqslant\frac{\pi}{4}. \end{cases}$

15.6. 1. $\begin{cases} -\tan x, & 0\leqslant x\leqslant\xi, \\ -\tan\xi, & \xi\leqslant x\leqslant\frac{\pi}{2}. \end{cases}$

2. $\begin{cases} \cot\xi, & 0\leqslant x\leqslant\xi, \\ \cot x, & \xi\leqslant x\leqslant\frac{\pi}{2}. \end{cases}$ 3. $\begin{cases} \cot\xi+1, & 0\leqslant x\leqslant\xi, \\ \cot x+1, & \xi\leqslant x\leqslant\frac{\pi}{2}. \end{cases}$

15.7. 1. $\begin{cases} \frac{x^2}{5\xi^3}, & 0\leqslant x\leqslant\xi, \\ \frac{\xi^2}{5x^3}, & \xi\leqslant x\leqslant 1. \end{cases}$

2. $\begin{cases} \frac{x^2}{3}\left(\frac{1}{\xi}-\xi^2\right), & 0\leqslant x\leqslant \xi, \\ \frac{\xi^2}{3}\left(\frac{1}{x}-x^2\right), & \xi\leqslant x\leqslant 1. \end{cases}$ 3. $\begin{cases} \frac{x}{3}\left(2\xi+\frac{1}{\xi^2}\right), & 0\leqslant x\leqslant \xi, \\ \frac{\xi}{3}\left(2x+\frac{1}{x^2}\right), & \xi\leqslant x\leqslant 1. \end{cases}$

4. $\begin{cases} -\ln \xi, & 0\leqslant x\leqslant \xi, \\ -\ln x, & \xi\leqslant x\leqslant 1. \end{cases}$ 5. $\begin{cases} 1-\ln \xi, & 0\leqslant x\leqslant \xi, \\ 1-\ln x, & \xi\leqslant x\leqslant 1. \end{cases}$

6. $\begin{cases} \frac{x}{6}(\xi+2\xi^{-2}), & 0\leqslant x\leqslant \xi, \\ \frac{\xi}{6}(x+2x^{-2}), & \xi\leqslant x\leqslant 1. \end{cases}$ 7. $\begin{cases} \frac{1}{3}x\xi^{-2}, & 0\leqslant x\leqslant \xi, \\ \frac{1}{3}\xi x^{-2}, & \xi\leqslant x\leqslant 1. \end{cases}$

8. $\begin{cases} \frac{1}{1-2a}x^a(\xi^a-\xi^{1-a}), & 0\leqslant x\leqslant \xi, \\ \frac{1}{1-2a}\xi^a(x^a-x^{1-a}), & \xi\leqslant x\leqslant 1. \end{cases}$ 9. $\begin{cases} e^{x+\xi}\int_\xi^1 \frac{e^{-2t}}{t}\,dt, & 0\leqslant x\leqslant \xi, \\ e^{x+\xi}\int_x^1 \frac{e^{-2t}}{t}\,dt, & \xi\leqslant x\leqslant 1. \end{cases}$

15.8. $\begin{cases} \frac{1}{x\xi^2}, & 1\leqslant x\leqslant \xi, \\ \frac{1}{\xi x^2}, & \xi\leqslant x\leqslant 3. \end{cases}$

15.9. 1. $\begin{cases} \frac{1}{\Phi(0)-\Phi(1)}e^{(x^2+\xi^2)/2}(\Phi(x)-\Phi(0))(\Phi(\xi)-\Phi(1)), & 0\leqslant x\leqslant \xi, \\ \frac{1}{\Phi(0)-\Phi(1)}e^{(x^2+\xi^2)/2}(\Phi(\xi)-\Phi(0))(\Phi(x)-\Phi(1)), & \xi\leqslant x\leqslant 1, \end{cases}$

where $\Phi(x)=\int_{-\infty}^x e^{-\xi^2/2}\,d\xi$.

2. $\begin{cases} -\frac{1}{2+\int_0^1 e^{-t^2}\,dt}\left(\int_0^x e^{-t^2}\,dt+2\right)\int_1^\xi e^{-t^2}\,dt, & 0\leqslant x\leqslant \xi, \\ -\frac{1}{2+\int_0^1 e^{-t^2}\,dt}\left(\int_0^\xi e^{-t^2}\,dt+2\right)\int_1^x e^{-t^2}\,dt, & \xi\leqslant x\leqslant 1. \end{cases}$

3. $\begin{cases} Ky_1(x)\,y_2(\xi), & 0\leqslant x\leqslant \xi, \\ Ky_1(\xi)\,y_2(x), & \xi\leqslant x\leqslant 1, \end{cases}$

where $K=\left(e^{-1}+\int_0^1 e^{-t^2}\,dt\right)^{-1}$, $y_1(x)=e^{x^2/2}\int_0^x e^{-t^2}\,dt$,

$y_2(x)=e^{x^2/2}\left(e^{-1}+\int_x^1 e^{-t^2}\,dt\right)$.

15.10. $\begin{cases} \dfrac{x^2}{28\sqrt{2}}(\xi^2 - 8\sqrt{2}\xi^{-3/2}), & 0 \leqslant \xi \leqslant x, \\ \dfrac{\xi^2}{28\sqrt{2}}(x^2 - 8\sqrt{2}x^{-3/2}) & \xi \leqslant x \leqslant 2. \end{cases}$

15.11. $\begin{cases} -\ln(\xi+1), & -1 \leqslant x \leqslant \xi, \\ -\ln(x+1), & \xi \leqslant x \leqslant 0. \end{cases}$

15.12. $\begin{cases} \dfrac{x^n}{2n}(\xi^{-n} - \xi^n), & 0 \leqslant x \leqslant \xi, \\ \dfrac{\xi^n}{2n}(x^{-n} - x^n), & \xi \leqslant x \leqslant 1. \end{cases}$

15.13. $\begin{cases} x\left[1 + \dfrac{\xi}{2}\ln\dfrac{\xi-1}{\xi+1} + \dfrac{\xi}{2}(\ln 3 - 1)\right] & 1 \leqslant x \leqslant \xi, \\ \xi\left[1 + \dfrac{x}{2}\ln\dfrac{x-1}{x+1} + \dfrac{x}{2}(\ln 3 - 1)\right], & \xi \leqslant x \leqslant 2. \end{cases}$

15.14. 1. $y(x) = \lambda \int_0^1 G(x, \xi)\xi^2 y(\xi)\, d\xi,$

where $G(x, \xi) = \begin{cases} x - \ln(1 + e^x) + 1 + \ln 2, & 0 \leqslant x \leqslant \xi, \\ \xi - \ln(1 + e^\xi) + 1 + \ln 2, & \xi \leqslant x \leqslant 1. \end{cases}$

2. $y(x) = \lambda \int_0^1 G(x, \xi)\, y(\xi)\, d\xi,$

where $G(x, \xi) = \begin{cases} -\xi(1 + x\arctan x), & 0 \leqslant x \leqslant \xi, \\ -x(1 + \xi\arctan\xi), & \xi \leqslant x \leqslant 1. \end{cases}$

3. $y(x) = \lambda \int_0^1 G(x, \xi)\, \xi y(\xi)\, d\xi,$

where $G(x, \xi) = \begin{cases} -\ln\left(e^{-x} + \sqrt{1 + e^{-2x}}\right) + 1 + \ln\left(1 + \sqrt{2}\right), & 0 \leqslant x \leqslant \xi, \\ -\ln\left(e^{-\xi} + \sqrt{1 + e^{-2\xi}}\right) + 1 + \ln\left(1 + \sqrt{2}\right), & \xi \leqslant x \leqslant 1. \end{cases}$

4. $y(x) - \lambda \int_0^1 G(x, \xi)\, y(\xi)\, d\xi,$

where $G(x, \xi) = \begin{cases} \xi\left(\dfrac{x}{2}\ln\dfrac{1+x}{1-x} - 1\right), & 0 \leqslant x \leqslant \xi, \\ x\left(\dfrac{\xi}{2}\ln\dfrac{1+\xi}{1-\xi} - 1\right), & \xi \leqslant x \leqslant 1. \end{cases}$

5. $y(x) = \lambda \int_0^{\pi/2} G(x, \xi)\, \xi y(\xi)\, d\xi,$ where $G(x, \xi)$

$$= \begin{cases} \dfrac{1}{3}\tan^3 x + \tan x + \dfrac{1}{2}, & 0 \leqslant x \leqslant \xi, \\ \dfrac{1}{3}\tan^3 \xi + \tan\xi + \dfrac{1}{2}, & \xi \leqslant x \leqslant \dfrac{\pi}{2}. \end{cases}$$

6. $y(x)=(\lambda-1)\int_1^2 G(x,\xi)\cos 2\xi y(\xi)\,d\xi,$

where $G(x,\xi)=\frac{1}{15}\begin{cases}(x-x^{-2})(\xi+4\xi^{-2}), & 1\leqslant x\leqslant\xi,\\ (\xi-\xi^{-2})(x+4x^{-2}), & \xi\leqslant x\leqslant 2.\end{cases}$

7. $y(x)=(\lambda-a)\int_0^1 G(x,\xi)\,y(\xi)\,d\xi,$

where $G(x,\xi)$

$$=-\frac{1}{\sqrt{a}\sin\sqrt{a}}\begin{cases}\cos\sqrt{a}x\cdot\cos\sqrt{a}(\xi-1) & 0\leqslant x\leqslant\xi,\\ \cos\sqrt{a}\xi\cdot\cos\sqrt{a}(x-1), & \xi\leqslant x\leqslant 1,\end{cases}$$

$a>0$, $a\neq(\pi n)^2$, with n an integer.

15.15. $y(x)=\lambda\int_0^1 G(x,\xi)\,y(\xi)\,d\xi,$ where $G(x,\xi)$

$$=\begin{cases}2(-1+\sqrt{\xi}), & 0\leqslant x\leqslant\xi,\\ 2(-1+\sqrt{x}), & \xi\leqslant x\leqslant 1.\end{cases}$$

15.16. $y(x)=\lambda\int_1^2 G(x,\xi)\,\frac{y\xi}{\xi^2}\,d\xi,$

where $G(x,\xi)=\begin{cases}\frac{1}{2}(x^2-1), & 1\leqslant x\leqslant\xi,\\ \frac{1}{2}(\xi^2-1), & \xi\leqslant x\leqslant 2.\end{cases}$

15.17. 1. $y(x)=-\lambda\int_0^1 G(x,\xi)\,y(\xi)\,d\xi,$

where $G(x,\xi)=\begin{cases}\arctan x, & 0\leqslant x\leqslant\xi,\\ \arctan\xi, & \xi\leqslant x\leqslant 1.\end{cases}$

2. $y(x)=-\lambda\int_0^1 G(x,\xi)\,y(\xi)\,d\xi,$

where $G(x,\xi)=\begin{cases}(-e^{-x}+1)\,e^{-\xi}, & 0\leqslant x\leqslant\xi,\\ (-e^{-\xi}+1)\,e^{-x}, & \xi\leqslant x\leqslant 1.\end{cases}$

3. $y(x)=\lambda\int_0^1 G(x,\xi)\,y(\xi)\,d\xi+\int_0^1 G(x,\xi)\,f(\xi)\,d\xi,$

where $G(x,\xi)=\begin{cases}\frac{-\xi+1}{h+1}(x+h), & 0\leqslant x\leqslant\xi,\\ \frac{-x+1}{h+1}(\xi+h), & \xi\leqslant x\leqslant 1.\end{cases}$

4. $y(x) = -\lambda \int_0^1 G(x, \xi)\, \xi y(\xi)\, d\xi,$

where $G(x, \xi) = \begin{cases} \ln \xi, & 0 \leqslant x \leqslant \xi, \\ \ln x, & \xi \leqslant x \leqslant 1. \end{cases}$

15.18. 1. $y(x) = \int_1^e G(x, \xi) f(\xi)\, d\xi,$

where $G(x, \xi) = \begin{cases} (x + \ln x - 1)(\xi + \ln \xi), & 0 \leqslant x \leqslant \xi, \\ (\xi + \ln \xi - 1)(x + \ln x), & \xi \leqslant x \leqslant 1. \end{cases}$

2. $y(x) = \int_1^2 G(x, \xi) f(\xi)\, d\xi,$

where $G(x, \xi) = \begin{cases} \left(\frac{1}{x} - \frac{1}{x^2}\right) \frac{1}{\xi^2}, & 1 \leqslant x \leqslant \xi, \\ \left(\frac{1}{\xi} - \frac{1}{\xi^2}\right) \frac{1}{x^2}, & \xi \leqslant x \leqslant 1. \end{cases}$

3. $y(x) = \int_{-1}^0 G(x, \xi) f(\xi)\, d\xi,$ where $G(x, \xi)$

$= \begin{cases} \ln |x| - x, & -1 \leqslant x \leqslant \xi, \\ \ln |\xi| - \xi, & \xi \leqslant x \leqslant 0. \end{cases}$

4. $y(x) = \int_0^{\frac{\pi}{2}} G(x, \xi) f(\xi)\, d\xi,$ where $G(x, \xi)$

$= \begin{cases} \frac{1}{2}\left(\tan \frac{x}{2} + 1\right)\left(1 - \tan \frac{\xi}{2}\right), & 0 \leqslant x \leqslant \xi, \\ \frac{1}{2}\left(\tan \frac{\xi}{2} + 1\right)\left(1 - \tan \frac{x}{2}\right), & \xi \leqslant x \leqslant \frac{\pi}{2}. \end{cases}$

5. $y(x) = \int_1^2 G(x, \xi) f(\xi)\, d\xi,$ where $G(x, \xi) = \begin{cases} \frac{x^2}{3\xi}, & 1 \leqslant x \leqslant \xi, \\ \frac{\xi^2}{3x}, & \xi \leqslant x \leqslant 2. \end{cases}$

16 Fourier's Method for Laplace's and Poisson's Equations

We start with problems in the plane. The solution of boundary value problems in the case of simple regions (disks, annuli, rectangles, etc.) can be found by Fourier's method of variable separation. Let us discuss this method for a disk (circle), namely, we wish to find a function $u = u(r, \varphi)$ that satisfies Laplace's equation

$$\nabla^2 u = 0 \tag{16.1}$$

inside the circle and assumes specified values on the boundary, that is,

$$u|_{r=R} = f(\varphi). \tag{16.2}$$

Equation (16.1) in terms of polar coordinates (r, φ) has the form

$$\frac{1}{r}\frac{\partial}{\partial r}\left(r\frac{\partial u}{\partial r}\right)+\frac{1}{r^2}\frac{\partial^2 u}{\partial \varphi^2}=0. \tag{16.3}$$

We look for particular solutions of Eq. (16.3) in the form

$$u = Z(r)\,\Phi(\varphi). \tag{16.4}$$

Substitution of (16.4) into (16.3) yields

$$\Phi''(\varphi) + \lambda\Phi(\varphi) = 0, \tag{16.5}$$

$$r\frac{d}{dr}\left(r\frac{dZ}{dr}\right)-\lambda Z=0. \tag{16.6}$$

Since $u(r, \varphi + 2\pi) = u(r, \varphi)$, we can write $\Phi(\varphi + 2\pi) = \Phi(\varphi)$, and from Eq (16.5) we find that $\sqrt{\lambda} = n$ (with n an integer) and

$$\Phi_n(\varphi) = A_n \cos n\varphi + B_n \sin n\varphi.$$

Then if we assume that $Z(r) = r^\alpha$ and substitute this into Eq. (16.6), we find that $\alpha^2 = n^2$, or $\alpha = \pm n$ $(n > 0)$, and hence

$$Z_n(r) = ar^n + br^{-n}.$$

For $n = 0$ $(\lambda = 0)$, Eq. (16.6) yields $Z(r) = C_0 \ln r + C$.

To solve the interior Dirichlet problem, we must put $Z_n(r) = ar^n$ $(n = 1, 2, \ldots)$ and $Z_0(r) = C$, since r^{-n} tends to ∞ and $\ln r$ tends to $-\infty$ as $r \to +0$. We look for the solution of the interior Dirichlet problem in the form of a series,

$$u(r, \varphi)=C+\sum_{n=1}^{\infty}\frac{r^n}{R^n}(A_n \cos n\varphi+B_n \sin n\varphi), \tag{16.7}$$

where the coefficients A_n and B_n are determined by the boundary condition (16.2):

$$A_n=\frac{1}{\pi}\int_{-\pi}^{\pi} f(\psi)\cos n\psi\, d\psi,$$

$$C=\frac{1}{2\pi}\int_{-\pi}^{\pi} f(\psi)\, d\psi,$$

$$B_n=\frac{1}{\pi}\int_{-\pi}^{\pi} f(\psi)\sin n\psi\, d\psi.$$

Summing the series (16.7), we arrive at the solution of the interior Dirichlet problem inside the circle in the form of Poisson's integral formula:

$$u(r, \varphi)=\frac{1}{2\pi}\int_{-\pi}^{\pi} f(\psi)\frac{R^2-r^2}{r^2-2Rr\cos(\varphi-\psi)+R^2}\,d\psi.$$

The solution of the exterior Dirichlet problem is also sought in the form of a series,

$$u(r, \varphi)=C+\sum_{n=1}^{\infty}\frac{R^n}{r^n}(A_n\cos n\varphi+B_n\sin n\varphi).$$

Finally, the solution of Eq. (16.1) in the region $R_1<r<R_2$ under specified boundary conditions at the circumferences $r=R_1$ and $r=R_2$ is sought in the form of a series, too:

$$u(r, \varphi)=\sum_{n=1}^{\infty}\left(A_n r^n+\frac{C_n}{r^n}\right)\cos n\varphi$$
$$+\sum_{n=1}^{\infty}\left(B_n r^n+\frac{D_n}{r^n}\right)\sin n\varphi+a\ln r+b.$$

16.1. Find a function that is harmonic inside the unit circle and is such that $u|_{r=1}=f(\varphi)$, where

1. $f(\varphi)=\cos^2\varphi$. 2. $f(\varphi)=\sin^3\varphi$.
3. $f(\varphi)=\cos^4\varphi$. 4. $f(\varphi)=\sin^6\varphi+\cos^6\varphi$.

16.2. Find a function that is harmonic inside a circle of radius R, centered at the origin of coordinates, and is such that

1. $\left.\frac{\partial u}{\partial r}\right|_{r=R}=A\cos\varphi$.

2. $\left.\frac{\partial u}{\partial r}\right|_{r=R}=A\cos 2\varphi$.

3. $\left.\frac{\partial u}{\partial r}\right|_{r=R}=\sin^3\varphi$.

16.3. Find the steady-state temperature distribution $u(r, \varphi)$ inside an infinitely long cylinder of radius R if

(1) the temperature at its surface is kept constant,

$$u(r, \varphi)|_{r=R}=A\sin\varphi;$$

(2) the temperature of half of the surface ($0\leqslant\varphi\leqslant\pi$) is kept at $-T_0$ and the temperature of the other half ($-\pi\leqslant\varphi<0$) is kept at T_0.

16.4. Find a function that is harmonic in the annulus $1<r<2$ and is such that

$$u|_{r=1}=f_1(\varphi),\quad u|_{r=2}=f_2(\varphi),$$

where

1. $f_1(\varphi) = u_1 = \text{const}$, $f_2(\varphi) = u_2 = \text{const}$.
2. $f_1(\varphi) = 1 + \cos^2\varphi$, $f_2(\varphi) = \sin^2\varphi$.

16.5. Solve the equation $\nabla^2 u = A$ in the annulus $R_1 < r < R_2$ provided $u|_{r=R_1} = u_1$ and $u|_{r=R_2} = u_2$ (A, u_1, and u_2 are given numbers).

16.6. Solve Poisson's equation

$$\nabla^2 u = -Axy \quad (A \text{ is a constant})$$

in a circle of radius R centered at the origin of coordinates provided $u|_{r=R} = 0$.

16.7. Solve Laplace's equation $\nabla^2 u = 0$ in the rectangle $0 < x < a$, $0 < y < b$ if at the boundary of this figure the function $u(x, y)$ assumes the following values:

$$u|_{x=0} = A \sin \frac{\pi y}{b}, \quad u|_{x=a} = 0,$$

$$u|_{y=0} = B \sin \frac{\pi x}{a}, \quad u|_{y=b} = 0.$$

16.8. Find the distribution of the electrostatic field potential $u(x, y)$ inside the rectangle $[0 < x < a,\ 0 < y < b]$ if the potential along the side of the rectangle lying on the y axis is v_0 and the other three sides are grounded. It is assumed that there are no charges inside the rectangle.

16.9. Find the distribution of the electrostatic field potential $u(x, y)$ inside a rectangular box $-a < x < a$, $-b < y < b$ if the two opposite sides $x = a$, $x = -a$ have a potential v_0 and the other two sides $y = b$, $y = -b$ are grounded.

16.10. Find the steady-state temperature distribution $u(x, y)$ in a homogeneous rectangular plate $0 < x < a$, $0 < y < b$ if the sides $x = a$, $y = b$ are thermally isolated, the other two sides $x = 0$, $y = 0$ are kept at a zero temperature, and the plate emits heat with a constant flux density q.

The rest of this section is devoted to problems in space. Solution of Problems 16.11 and 16.12 via Fourier's method requires the use of Bessel functions (see the text between Problems 20.19 and 20.20).

16.11. Find the steady-state temperature distribution $u(r, z)$ inside a cylinder of base radius R and height h if

(1) the temperature of the lower base and the lateral surface is zero, while the temperature of the upper base depends solely on r (the distance from the cylinder's axis);

(2) the temperature of the lower base is zero, the lateral surface is thermally isolated, and the temperature of the upper base depends on r;

(3) the temperature of the lower base is zero, the lateral surface is freely cooled by air at zero temperature, and the temperature of the upper base is a function of r;

(4) the temperature of the upper and lower bases is zero: while the temperature of every point of the lateral surface depends only on the distance of the point from the lower base (i.e. on z);

(5) the bases are thermally isolated, while the temperature of the lateral surface is a given function of z.

16.12. Find the steady-state temperature distribution inside a solid in the form of a cylinder with base radius R and height h if

(1) a constant heat flux q is fed to the lower base $z = 0$, while the lateral surface $r = R$ and the upper base $z = h$ is kept at a zero temperature;

(2) a constant heat flux q is fed to the lower base $z = 0$, the upper base is kept at a zero temperature, and there is heat exchange across the lateral surface with a medium whose temperature is zero.

Let us apply Fourier's method to Laplace's equation in space for a ball of radius R for the case when solution u is independent of angle φ, that is, $u = u(r, \theta)$. Then

$$\nabla^2 u = \frac{1}{r^2}\frac{\partial}{\partial r}\left(r^2\frac{\partial u}{\partial r}\right) + \frac{1}{r^2\sin\theta}\frac{\partial}{\partial\theta}\left(\sin\theta\frac{\partial u}{\partial\theta}\right) = 0. \tag{16.8}$$

By putting

$$u = Z(r)\,W(\theta) \tag{16.9}$$

we find from (16.8) that

$$\frac{1}{Z}\frac{d}{dr}\left(r^2\frac{dZ}{dr}\right) = \lambda, \tag{16.10}$$

$$\frac{1}{W\sin\theta}\frac{d}{d\theta}\left(\sin\theta\frac{dW}{d\theta}\right) = -\lambda. \tag{16.11}$$

Introducing into Eqs. (16.10) and (16.11) a new variable ν instead of λ, with $\lambda = \nu(\nu + 1)$, we can write Eq. (16.10) in the following form

$$r^2\frac{d^2Z}{dr^2} + 2r\frac{dZ}{dr} - \nu(\nu+1)Z = 0. \tag{16.12}$$

Equation (16.12) has particular solutions $Z = r^\alpha$, with $\alpha_1 = \nu$ and $\alpha_2 = -(\nu + 1)$. Hence,

$$Z(r) = C_1 r^\nu + C_2 r^{-(\nu+1)}. \tag{16.13}$$

By introducing a new variable ξ via the formula $\xi = \cos\theta$ we can rewrite Eq. (16.11) thus:

$$\frac{d}{d\xi}\left[(1-\xi^2)\frac{dy}{d\xi}\right] + \nu(\nu+1)\,y = 0, \tag{16.14}$$

where $y = W(\arccos\xi)$. Equation (16.14) is known as *Legendre's differential equation*; it has solutions bounded on the segment $[-1, 1]$ if and only if $\nu = n$ (with n a nonnegative integer). Solutions of this equation at $\nu = n$ are known as *Legendre polynomials*

$$y = P_n(\xi) = \frac{1}{2^n n!} \frac{d^n (\xi^2 - 1)^n}{d\xi^n}.$$

Here are the first five Legendre polynomials:

$$P_0(\xi) = 1, \; P_1(\xi) = \xi,$$

$$P_2(\xi) = \frac{1}{2}(3\xi^2 - 1), \; P_3(\xi) = \frac{1}{2}(5\xi^3 - 3\xi),$$

$$P_4(\xi) = \frac{1}{8}(35\xi^4 - 30\xi^2 + 3).$$

Legendre polynomials form an orthogonal system in $L_2(-1, 1)$, that is,

$$\int_{-1}^{1} P_n(\xi) P_m(\xi)\, d\xi = 0 \quad (n \neq m),$$

and, in addition,

$$\| P_n \|^2 = \int_{-1}^{1} P_n^2(\xi)\, d\xi = \frac{2}{2n+1}.$$

Note that every function $f \in L_2(-1, 1)$ can be expanded in a Fourier series in Legendre polynomials:

$$f(\xi) = \sum_{n=0}^{\infty} \frac{2n+1}{2} (f, P_n) P_n(\xi),$$

and this series is convergent in $L_2(-1, 1)$.

From (16.13) and (16.14) we can find the particular solution of Eq. (16.8) in the form (16.9):

$$u_n(r, \theta) = [A_n r^n + B_n r^{-(n+1)}] P_n(\cos\theta),$$

where $P_n(\xi)$ are Legendre polynomials.

It is convenient to use the $u_n(r, \theta)$ to find the solutions to Laplace's equation when the boundary conditions are specified on a sphere (the interior and exterior problems) or on the boundary of a spherical layer $(R_1 < r < R_2)$.

When the boundary conditions are specified on the sphere $r = R$ and depend only on θ, the solution of the interior Dirichlet problem (and other interior problems for Laplace's equation) must be sought in the form

$$u(r, \theta) = \sum_{n=0}^{\infty} A_n r^n P_n(\cos\theta),$$

and that of the exterior problem in the form

$$u(r, \theta) = \sum_{n=0}^{\infty} B_n r^{-(n+1)} P_n(\cos\theta).$$

If the boundary conditions are specified on the boundary of the spherical layer $R_1 < r < R_2$ and depend only on θ, the solution must be sought in the form

$$u(r, \theta) = \sum_{n=0}^{\infty} [A_n r^n + B_n r^{-(n+1)}] P_n(\cos\theta).$$

The coefficients A_n and B_n can be determined from the boundary conditions.

16.13. Find a function u that is harmonic inside a ball of radius R centered at the origin of coordinates and is such that

$$u|_{r=R} = f(\theta),$$

where
1. $f(\theta) = \cos\theta$. 2. $f(\theta) = \cos^2\theta$.
3. $f(\theta) = \cos 2\theta$. 4. $f(\theta) = \sin^2\theta$.

16.14. Find a function u that is harmonic inside a ball of radius R and is such that

$$(u + u_r)|_{r=R} = 1 + \cos^2\theta.$$

16.15. Find a function that is harmonic outside a ball of radius R and is such that
1. $u_r|_{r=R} = \sin^2\theta$. 2. $(u - u_r)|_{r=R} = \sin^2\theta$. 3. $u_r|_{r=R} = A\cos\theta$.

16.16. Establish whether the interior Neumann problem for a ball of radius R has a solution if
1. $u_r|_{r=R} = A\cos\theta$. 2. $u_r|_{r=R} = \sin\theta$.
Find the solution.

16.17. Find a function that is harmonic inside the spherical layer $1 < r < 2$ and is such that

$$u|_{r=1} = f_1(\theta), \quad u|_{r=2} = f_2(\theta)$$

with the following data:

1. $f_1 = \cos^2\theta, \quad f_2 = \frac{1}{8}(\cos^2\theta + 1)$.
2. $f_1 = \cos^2\theta, \quad f_2 = 4\cos^2\theta - \frac{4}{3}$.
3. $f_1 = 1 - \cos 2\theta, \quad f_2 = 2\cos\theta$.
4. $f_1 = \frac{1}{2}\cos\theta, \quad f_2 = 1 + \cos 2\theta$.
5. $f_1 = 9\cos 2\theta, \quad f_2 = 3(1 - 7\cos^2\theta)$.

16.18. Find the steady-state temperature distribution inside a hemisphere of radius R if the temperature T_0 of the spherical surface is kept constant and if the base of the hemisphere is kept at a zero temperature.

16.19. Find the steady-state temperature distribution inside a homogeneous, isotropic ball of radius R if the ball's surface is kept at a temperature

$$u|_{r=R}=\begin{cases} u_1 \text{ at } 0\leqslant\theta<\pi/2, \\ u_2 \text{ at } \pi/2<\theta\leqslant\pi. \end{cases}$$

In spherical coordinates (r, φ, θ), Laplace's equation has the form

$$\frac{1}{r^2}\frac{\partial}{\partial r}\left(r^2\frac{\partial u}{\partial r}\right)+\frac{1}{r^2\sin\theta}\cdot\frac{\partial}{\partial\theta}\left(\sin\theta\frac{\partial u}{\partial\theta}\right)+\frac{1}{r^2\sin^2\theta}\frac{\partial^2 u}{\partial\varphi^2}=0. \qquad (16.15)$$

We will find the solution to Eq. (16.15) by Fourier's method. Assuming that $u(r, \theta, \varphi) = Z(r)\,Y(\theta, \varphi)$, we can write instead of Eq. (16.15) the following:

$$r^2Z''+2rZ'-\lambda Z=0, \qquad (16.16)$$

$$\frac{1}{\sin\theta}\cdot\frac{\partial}{\partial\theta}\left(\sin\theta\frac{\partial Y}{\partial\theta}\right)+\frac{1}{\sin^2\theta}\cdot\frac{\partial^2 Y}{\partial\varphi^2}+\lambda Y=0. \qquad (16.17)$$

If we require that the function $Y(\theta, \varphi)$ be bounded on the unit sphere and bear in mind that $Y(\theta, \varphi+2\pi) = Y(\theta, \varphi)$, we can look for the solution to Eq. (16.17) in the form $Y(\theta, \varphi) = W(\theta)\cdot\Phi(\varphi)$. We obtain

$$\Phi''+\mu\Phi=0, \quad \Phi(\varphi+2\pi)=\Phi(\varphi), \qquad (16.18)$$

whence $\mu = m^2$ (m is an integer) and

$$\Phi_m(\varphi) = C_m\cos m\varphi + D_m\sin m\varphi \qquad (16.19)$$

is the solution to problem (19.18).

The function $W(\theta)$ can be found by solving the equation

$$\frac{1}{\sin\theta}\cdot\frac{d}{d\theta}\left(\sin\theta\frac{dW}{d\theta}\right)+\left(\lambda-\frac{\mu}{\sin^2\theta}\right)W=0, \qquad (16.20)$$

and it must be bounded at $\theta = 0$ and $\theta = \pi$. Putting $\xi = \cos\theta$ in (16.20) and introducing the notation $W(\theta) = X(\cos\theta) = X(\xi)$, we can write Eq. (16.20) as follows:

$$\frac{d}{d\xi}\left[(1-\xi^2)\frac{dX}{d\xi}\right]+\left(\lambda-\frac{m^2}{1-\xi^2}\right)X=0. \qquad (16.21)$$

Equation (16.21) has solutions that are bounded on the segment $[-1, 1]$ only at $\lambda = n(n+1)$, with n an integer. The following functions are particular solutions of Eq. (16.21) at $\lambda = n(n+1)$;

$$P_n^{(m)}(\xi)=(1-\xi^2)^{m/2}\frac{d^mP_n(\xi)}{d\xi^n},$$

where $P_n(\xi)$ ($n = 0, 1, \ldots$) are Legendre polynomials.

Returning to the old variable θ, we arrive at the sought particular solutions of Eq. (16.20)

$$P_n^{(m)}(\cos\theta)=\sin^m\theta\frac{d^m}{d\cos\theta^m}[P_n(\cos\theta)], \qquad (16,22)$$

with $P_n^{(m)}(\cos\theta) = 0$ at $m > n$. The functions $P_n^{(m)}(\cos\theta)$ defined by (16.22) are known as associated Legendre polynomials.

Thus, the particular solutions of Eq. (16.17) bounded on a unit sphere have the form

$$Y_n(\theta, \varphi) = a_0 P_n(\cos\theta) + \sum_{k=1}^{n} (a_k \cos k\varphi + b_k \sin k\varphi) P_n^{(k)}(\cos\theta). \quad (16.23)$$

Since the general solution of Eq. (16.16) has the form

$$Z_n(r) = A_n r^n + \frac{B_n}{r^{n+1}},$$

the sought particular solutions to Eq. (16.15) are

$$u_n(r, \theta, \varphi) = Z_n(r) Y_n(\theta, \varphi) = \left(A_n r^n + \frac{B_n}{r^{n+1}}\right) Y_n(\theta, \varphi),$$

where $Y_n(\theta, \varphi)$ are given in (16.23).

Let us consider the interior Dirichlet problem for a sphere of radius R centered at the origin of coordinates, namely, we must find the solution to Eq. (16.15) under the condition that

$$u|_{r=R} = f(\theta, \varphi). \quad (16.24)$$

The solution of this problem (and other interior problems) must be sought in the form

$$u(r, \theta, \varphi) = \sum_{k=0}^{\infty} \left(\frac{r}{R}\right)^k Y_k(\theta, \varphi), \quad (16.25)$$

and in the case of the boundary value problem (16.15), (16.24) for $Y_k(\theta, \varphi)$ in (16.25) we must take those and only those functions that are present in the expansion of $f(\theta, \varphi)$ in a series in spherical functions $Y_k(\theta, \varphi)$:

$$f(\theta, \varphi) = \sum_{k=0}^{\infty} Y_k(\theta, \varphi).$$

The solution to problem (16.15), (16.24) at point $M_0(r_0, \theta_0, \varphi_0)$ can be represented via *Poisson's integral formula*

$$u(r_0, \theta_0, \varphi_0) = \frac{R}{4\pi} \int_0^{2\pi} \int_0^{\pi} f(\theta, \varphi) \frac{R^2 - r_0^2}{(R^2 - 2Rr_0 \cos\gamma + r_0^2)^{3/2}} \sin\theta \, d\theta \, d\varphi,$$

where $\cos\gamma = \cos\theta \cos\theta_0 + \sin\theta \sin\theta_0 \cos(\varphi - \varphi_0)$.

The solution of the exterior Dirichlet problem for a sphere of radius R (and other exterior problems) must be sought in the form

$$u(r, \theta, \varphi) = \sum_{k=0}^{\infty} \left(\frac{R}{r}\right)^{k+1} Y_k(\theta, \varphi).$$

Finally, a function that is harmonic in the spherical layer $R_1 < r < R_2$ and assumes specified values at the boundary of the layer must be sought in the form

$$u(r,\theta,\varphi)=\sum_{k=0}^{\infty}\left(\frac{r}{R_2}\right)^k Y_k(\theta,\varphi)+\sum_{k=0}^{\infty}\left(\frac{R_1}{r}\right)^{k+1}\tilde{Y}_k(\theta,\varphi),$$

where $\tilde{Y}_k(\theta,\varphi)$ are spherical functions of the type (16.23).

Here are some associated Legendre polynomials and spherical functions $Y_k(\theta,\varphi)$ in explicit form at $k=0, 1, 2, 3$:

$$P_1^{(1)}(\cos\theta)=\sin\theta;$$
$$P_2^{(1)}(\cos\theta)=3\sin\theta\cos\theta;$$
$$P_3^{(1)}(\cos\theta)=\sin\theta\,\frac{15\cos^2\theta-3}{2};$$
$$P_3^{(2)}(\cos\theta)=15\sin^2\theta\cos\theta;$$
$$P_3^{(3)}(\cos\theta)=15\sin^3\theta;$$
$$P_n^{(n)}(\cos\theta)=\frac{(2n)!}{2^n n!}\sin^n\theta;$$

$$\begin{aligned}
Y_0(\theta,\varphi)&=a_0,\\
Y_1(\theta,\varphi)&=a_1\cos\theta+(b_1\cos\varphi+c_1\sin\varphi)\sin\theta,\\
Y_2(\theta,\varphi)&=a_2(3\cos^2\theta-1)+(b_2\cos\varphi+c_2\sin\varphi)\\
&\quad\times\sin\theta\cos\theta+(d_2\cos 2\varphi+e_2\sin 2\varphi)\sin^2\theta,\\
Y_3(\theta,\varphi)&=a_3(5\cos^3\theta-3\cos\theta)+(b_3\cos\varphi\\
&\quad+c_3\sin\varphi)\sin\theta\,(15\cos^2\theta-3)\\
&\quad+(d_3\cos 2\varphi+e_3\sin 2\varphi)\sin^2\theta\cos\theta\\
&\quad+(f_3\cos 3\varphi+g_3\sin 3\varphi)\sin^3\theta.
\end{aligned}$$

16.20. Find a function that is harmonic inside the unit sphere and is such that

1. $u|_{r=1}=\cos\left(2\varphi+\frac{\pi}{3}\right)\sin^2\theta.$
2. $u|_{r=1}=(\sin\theta+\sin 2\theta)\sin\left(\varphi+\frac{\pi}{6}\right).$
3. $u|_{r=1}=\sin\theta(\sin\varphi+\sin\theta).$
4. $u_r|_{r=1}=\sin^{10}\theta\sin 10\varphi,\ u|_{r=0}=1.$

16.21. Find a function that is harmonic inside a sphere of radius R centered at the origin of coordinates and is such that

1. $u|_{r=R}=\sin\left(2\varphi+\frac{\pi}{6}\right)\sin^2\theta\cos\theta.$
2. $u|_{r=R}=\sin\left(3\varphi+\frac{\pi}{4}\right)\sin^3\theta.$
3. $u|_{r=R}=\sin^2\theta\cos\left(2\varphi-\frac{\pi}{4}\right)+\sin\theta\sin\varphi.$
4. $(u+u_r)|_{r=R}=\sin^2\theta\left[\sqrt{2}\cos\left(2\varphi+\frac{\pi}{4}\right)+2\cos^2\varphi\right].$
5. $(u+u_r)|_{r=R}=\sin\theta(\sin\varphi+\cos\varphi\cos\theta+\sin\theta).$

16.22. Find a function that is harmonic outside the unit sphere and is such that

1. $u_r|_{r=1} = \sin\left(\frac{\pi}{4} - \varphi\right)\sin\theta$.

2. $u|_{r=1} = \cos^2\theta\sin\theta\sin\left(\varphi + \frac{\pi}{3}\right)$.

16.23. Find a function that is harmonic outside a sphere of radius R centered at the origin of coordinates and is such that

1. $u|_{r=R} = \sin^3\theta\cos\theta\cos\left(3\varphi + \frac{\pi}{4}\right)$.

2. $u|_{r=R} = \sin 100\varphi \sin^{100}\theta$.

3. $(u - u_r)_{r=R} = \sin\theta\cos^2\frac{\theta}{2}\sin\left(\varphi + \frac{\pi}{6}\right)$.

16.24. Find a function that is harmonic inside the spherical layer $1 < r < 2$ and is such that $u|_{r=1} = f_1(\theta, \varphi)$ and $u|_{r=2} = f_2(\theta, \varphi)$, where

1. $f_1 = \sin\theta\sin\varphi$, $f_2 = 0$.
2. $f_1 = 3\sin 2\varphi\sin^2\theta$, $f_2 = 3\cos\theta$.
3. $f_1 = 7\sin\theta\cos\varphi$, $f_2 = 7\cos\theta$.
4. $f_1 = \sin^2\theta\,(3 - \sin 2\varphi)$, $f_2 = 4f_1$.
5. $f_1 = 12\sin\theta\cos^2\frac{\theta}{2}\cos\varphi$, $f_2 = 0$.
6. $f_1 = \sin 2\varphi\sin^2\theta$, $f_2 = \cos 2\varphi\sin^2\theta$.
7. $f_1 = \cos\varphi\sin 2\theta$, $f_2 = \sin\varphi\sin 2\theta$.
8. $f_1 = 31\sin 2\theta\sin\varphi$, $f_2 = 31\sin^2\theta\cos 2\varphi$.
9. $f_1 = \cos\theta$, $f_2 = \cos\varphi\,(12\sin\theta - 15\sin^3\theta)$.

16.25. Find a function that is harmonic inside the spherical layer $1 < r < 2$ and is such that

1. $(3u + u_r)|_{r=1} = 5\sin^2\theta\sin 2\varphi$, $u|_{r=2} = -\cos\theta$.

2. $u|_{r=1} = \sin\theta\sin\varphi\,(5 + 6\cos\theta)$, $u_r|_{r=2} = 12\sin 2\theta\sin\varphi$.

3. $u|_{r=1} = 1$, $u_r|_{r=2} = 15\cos\varphi\,(\cos^2\theta\sin\theta + \sin\varphi\sin^2\theta\cos\theta)$.

16.26. Find a function that is harmonic inside the spherical layer $1/2 < r < 1$ and is such that

1. $u|_{r=\frac{1}{2}} = 0$, $u|_{r=1} = 6\cos^2\varphi\sin^2\theta$.

2. $u|_{r=\frac{1}{2}} = 30\cos^2\varphi\sin^2\theta\cos\theta$, $u|_{r=1} = 0$.

Answers to Problems of Sec. 16

16.1. 1. $\frac{1}{2}(1 + r^2\cos 2\varphi)$. 2. $\frac{r}{4}(3\sin\varphi - r^2\sin 3\varphi)$.
3. $\frac{3}{8} + \frac{r^2}{2}\cos 2\varphi + \frac{r^4}{8}\cos 4\varphi$. 4. $\frac{5}{8} + \frac{3}{8}r^4\cos 4\varphi$.

16.2. 1. $Ar\cos\varphi + C$. 2. $\frac{A}{2R}r^2\cos 2\varphi + C$. 3. $\frac{1}{4}\left(3r\sin\varphi - \right.$ $\left. - \frac{r^2}{3R^2}\sin 3\varphi\right) + C$.

Here C is an arbitrary constant.

16.3. 1. $\frac{Ar}{R}\sin\varphi$. 2. $-\frac{4T_0}{\pi}\sum_{n=0}^{\infty}\left(\frac{r}{R}\right)^{2n+1}\frac{\sin(2n+1)\varphi}{2n+1}=$
$=\frac{2T_0}{\pi}\arctan\frac{R^2-r^2}{2rR\sin\varphi}-T_0$.

16.4. 1. $u_1+(u_2-u_1)\frac{\ln r}{\ln 2}$. 2. $\frac{3}{2}-\frac{\ln r}{\ln 2}+\left(\frac{2}{3r^2}-\frac{1}{6}r^2\right)\cos 2\varphi$.

16.5. $u_2+\frac{A}{4}(r^2-R_2^2)+\frac{u_1-u_2+\frac{A}{4}(R_2^2-R_1^2)}{\ln R_2-\ln R_1}\ln\frac{R_2}{r}$.

16.6. $\frac{Ar^2}{24}(R^2-r^2)\sin 2\varphi$. *Hint.* $u=v+w$, where $v=-\frac{Axy}{12}(x^2+$ $+y^2)=-\frac{Ar^4\sin 2\varphi}{24}$ is the particular solution of Poisson's equation and w is the solution of Laplace's equation that satisfies the condition $w|_{r=R}=\frac{A}{24}R^4\sin 2\varphi$.

16.7. $A\frac{\sinh\frac{\pi(a-x)}{b}}{\sinh\frac{\pi a}{b}}\sin\frac{\pi y}{b}+B\frac{\sinh\frac{\pi(b-y)}{a}}{\sinh\frac{\pi b}{a}}\sin\frac{\pi x}{a}$. *Hint.* Look for the solution in the form $u=v+w$, where v and w are harmonic functions such that $v|_{x=0}=A\sin(\pi y/b)$, $v|_{x=a}=v|_{y=0}=v|_{y=b}=0$, $w|_{x=0}=w|_{x=a}=w|_{y=b}=0$, $w|_{y=0}=B\sin(\pi x/a)$.

16.8. $\frac{4v_0}{\pi}\sum_{n=0}^{\infty}\frac{\sinh\frac{(2n+1)(a-x)\pi}{b}\sin\frac{(2n+1)\pi y}{b}}{(2n+1)\sinh\frac{(2n+1)\pi a}{b}}$.

16.9. $\frac{4v_0}{\pi}\sum_{n=0}^{\infty}(-1)^n\frac{\cosh\frac{(2n+1)\pi x}{2b}\cos\frac{(2n+1)\pi y}{2b}}{(2n+1)\cosh\frac{2n+1}{2b}\pi a}$.

16.10. $\frac{16qa^2}{k\pi^3}\sum_{n=0}^{\infty}\frac{1}{(2n+1)^3}\left[1-\frac{\cosh\frac{(2n+1)\pi(b-y)}{2a}}{\cosh\frac{(2n+1)\pi b}{2a}}\right]\sin\frac{(2n+1)\pi x}{2a}$,

with k the internal thermal conductivity. *Hint.* The problem amounts to solving the equation $\nabla^2 u=-q/k$ with the boundary conditions $u|_{x=0}=u|_{y=0}=0$, $u_x|_{x=a}=u_y|_{y=b}=0$.

16.11. 1. $\sum_{n=1}^{\infty}a_n\frac{\sinh\frac{\mu_n z}{R}}{\sinh\frac{\mu_n h}{R}}J_0\left(\mu_n\frac{r}{R}\right)$, where μ_n $(n=1, 2, \ldots)$ are

the positive roots of the equation $J_0(\mu)=0$ and $a_n=\frac{2}{R^2J_1^2(\mu_n)}\times$ $\times\int\limits_0^R ru_0(r)\,J_0\left(\frac{\mu_n r}{R}\right)dr$. 2. $\sum\limits_{n=1}^{\infty} a_n \frac{\sinh\frac{\mu_n z}{R}}{\sinh\frac{\mu_n h}{R}} J_0\left(\mu_n\frac{r}{R}\right)$,
where μ_n $(n=1, 2, \ldots)$ are the positive roots of the equation $J_1(\mu)=0$ and

$$a_n=\frac{2}{R^2J_0^2(\mu_n)}\int\limits_0^R ru_0(r)\,J_0\left(\mu_n\frac{r}{R}\right)dr.$$

Hint. The boundary conditions are $|u|_{r=0}|<\infty$, $u|_{z=0}=0$, $u_r|_{r=R}=0$, $u|_{z=h}=u_0(r)$.

3. $\sum\limits_{n=1}^{\infty} a_n \frac{\sinh\frac{\mu_n z}{R}}{\sinh\frac{\mu_n h}{R}} J_0\left(\frac{\mu_n r}{R}\right)$, where μ_n $(n=1, 2, \ldots)$ are the positive roots of the equation $\mu J_1(\mu)-h_1RJ_0(\mu)=0$, with $a_n=\frac{2}{R^2}\left(1+\frac{h_1^2R^2}{\mu_n^2}\right)^{-1}[J_0(\mu_n)]^{-2}\int\limits_0^R ru_0(r)\,J_0\left(\frac{\mu_n r}{R}\right)dr$. *Hint.* The boundary conditions are $|u|_{r=0}|<\infty$, $u_{z=0}=0$, $(u_r+h_1u)|_{r=R}=0$, and $u|_{z=h}=u_0(r)$. 4. $\frac{2}{h}\sum\limits_{n=1}^{\infty}\left[J_0\left(\frac{\pi iRn}{h}\right)\right]^{-1}\left(\int\limits_0^h f(\xi)\sin\frac{\pi n\xi}{h}\,d\xi\right)\times$ $\times\sin\frac{\pi nz}{h}J_0\left(\frac{\pi inr}{h}\right)$, where $J_0(ix)$ is the zeroth-order Bessel function of a pure imaginary argument. 5. $\frac{2}{h}\sum\limits_{n=1}^{\infty}\left[J_0\left(\frac{\pi niR}{h}\right)\right]^{-1}\int\limits_0^h f(\xi)\times$ $\times\cos\frac{\pi n\xi}{h}\,d\xi\cos\frac{\pi nz}{h}J_0\left(\frac{\pi nir}{h}\right)$. *Hint.* The boundary conditions are $u_z|_{z=0}=u_z|_{z=h}=0$ and $u|_{r=R}=f(z)$.

16.12. 1. $\sum\limits_{n=1}^{\infty} a_n \frac{\sinh\frac{\mu_n(h-z)}{R}}{\cosh\frac{\mu_n}{R}h} J_0\left(\frac{\mu_n}{R}r\right)$, where μ_n $(n=1, 2, \ldots)$ are the positive roots of the equation $J_0(\mu)=0$, $A_n=\frac{2aq}{k\mu_n^2J_1(\mu_n)}$ and k is the thermal conductivity. *Hint.* The problem amounts to solving the equation $\nabla^2u=0$ with the boundary conditions $-ku_z|_{z=0}=q$ and $u|_{r=R}=u|_{z=h}=0$. 2. $\sum\limits_{n=1}^{\infty} a_n \frac{\sinh\frac{\mu_n(h-z)}{R}}{\cosh\frac{\mu_n}{R}h} J_0\left(\frac{\mu_n}{a}r\right)$,
where μ_n $(n=1, 2, \ldots)$ are the positive roots of the equation $\mu J_1(\mu)-Rh_1J_0(\mu)=0$, with h_1 the heat exchange coefficient and $a_n=2h_1^2R^3qk^{-1}(R^2h_1^2+\mu_n^2)^{-1}[J_1(\mu_n)]^{-1}\mu_n^{-2}$. *Hint.* The boundary conditions are $-ku_z|_{z=0}=q$, $(u_r+h_1u)|_{r=R}=0$, $u|_{z=h}=0$.

16.13. 1. $\frac{1}{R}\cos\theta$. 2. $\frac{1}{3}\left(1-\frac{r^2}{R^2}\right)+\frac{r^2}{R^2}+\frac{r^2}{R^2}\cos^2\theta$.
3. $\frac{4}{3}\left(\frac{r}{R}\right)^2 P_2(\cos\theta)-\frac{1}{3}$. 4. $\frac{2}{3}-\frac{2}{3}\left(\frac{r}{R}\right)^2 P_2(\cos\theta)$.

16.14. $\frac{4}{3}+\frac{2r^2}{3R(R+2)}P_2(\cos\theta)$.

16.15. 1. $-\frac{2R^2}{3r}+\frac{R^4}{9r^3}(3\cos^2\theta-1)+C$, where C is an arbitrary constant. 2. $C+\left(\frac{2}{3}-C\right)\frac{R^2}{r(R+1)}-\frac{R^4}{(R+3)r^3}\left(\cos^2\theta-\frac{1}{3}\right)$, where C is an arbitrary constant. 3. $C-(A/2)(R^3/r^2)\cos\theta$, with C an arbitrary constant.

16.16. 1. The problem has a solution: $u=Ar\cos\theta$, with C an arbitrary constant. 2. The problem has no solution.

16.17. 1. $\frac{1}{3r}+\frac{3\cos^2\theta-1}{3r^3}$. 2. $\frac{1}{3}\left[\frac{2}{r}-1+r^2(3\cos^2\theta-1)\right]$.
3. $\frac{8}{3r}-\frac{4}{3}+\left(\frac{8}{7}r-\frac{8}{7r^2}\right)P_1(\cos\theta)+\left(\frac{4}{93}r^2-\frac{128}{93r^3}\right)P_2(\cos\theta)$.
4. $\frac{4}{3}\left(1-\frac{1}{r}\right)+\frac{1}{14}\left(\frac{8}{r^2}-r\right)P_1(\cos\theta)+\frac{32}{93}\left(r^2-\frac{1}{r^3}\right)P_2(\cos\theta)$.
5. $\frac{2}{r}-5+4\left(\frac{4}{r^3}-r^2\right)P_2(\cos\theta)$.

16.18. $T_0\sum_{n=0}^{\infty}(-1)^n\frac{1\cdot 3\cdot 5\ldots(2n-1)}{2\cdot 4\cdot 6\ldots(2n+2)}(4n+3)\left(\frac{r}{R}\right)^{2n+1}P_{2n+1}(\cos\theta)$,
$0\leqslant\theta\leqslant\frac{\pi}{2}$.

16.19. $\frac{u_1+u_2}{2}+\frac{u_1-u_2}{2}\sum_{n=0}^{\infty}(-1)^n\frac{3\cdot 5\cdot 7\ldots(2n-1)}{2\cdot 4\cdot 6\ldots(2n+2)}(4n+3)\times$
$\times P_{2n+1}(\cos\theta)\left(\frac{r}{R}\right)^{2n+1}$.

16.20. 1. $r^2\cos\left(2\varphi+\frac{\pi}{3}\right)\sin^2\theta$. 2. $\left(r\sin\theta+r^2\sin 2\theta\right)\sin\left(\varphi+\right.$ $\left.+\frac{\pi}{6}\right)$. 3. $\frac{2}{3}-\frac{r^2}{6}(3\cos 2\theta+1)+r\sin\theta\sin\varphi$.
4. $1+\frac{r^{10}}{10}\sin^{10}\theta\sin 10\varphi$.

16.21. 1. $\left(\frac{r}{R}\right)^3\sin\left(2\varphi+\frac{\pi}{6}\right)\sin^2\theta\cos\theta$. 2. $\left(\frac{r}{R}\right)^3\sin\left(3\varphi+\frac{\pi}{4}\right)\times$ $\times\sin^3\theta$. 3. $\left(\frac{r}{R}\right)^2\sin^2\theta\cos\left(2\varphi-\frac{\pi}{4}\right)+\frac{r}{R}\sin\theta\sin\varphi$. 4. $\frac{2}{3}+$ $+\left(\frac{r}{R}\right)^2\left[-\frac{R}{3(2+R)}(3\cos^2\theta-1)+\frac{R}{2+R}(2\cos 2\varphi-\sin 2\varphi)\sin^2\theta\right]$.
Hint. $(u_r+u)|_{r=R}=\frac{1}{3}P_2^{(2)}(\cos\theta)(2\cos 2\varphi-\sin 2\varphi)+\frac{2}{3}-$ $-\frac{2}{3}P_2(\cos\theta)$, $u=A+Br^2P_2(\cos\theta)+r^2(C\cos 2\varphi+D\sin 2\varphi)P_2^{(2)}\times$

$\times(\cos\theta)$. 5. $\frac{2}{3}+\frac{r}{R+1}\sin\varphi\sin\theta+\frac{r^2\sin\theta\cos\theta\cos\varphi}{R(R+2)}-\frac{r^2}{3}\times$ $\times\frac{3\cos^2\theta-1}{R(R+2)}$. *Hint.* $(u+u_r)|_{r=R}=\sin\varphi P_1^{(1)}(\cos\theta)+\frac{1}{3}\cos\varphi P_2^{(1)}\times$ $\times(\cos\theta)+\frac{2}{3}-\frac{2}{3}P_2(\cos\theta)$, $u=A+B\left(\frac{r}{R}\right)\sin\varphi\sin\theta+C\left(\frac{r}{R}\right)^2\times$ $\times\cos\varphi P_2^{(1)}(\cos\theta)+D\left(\frac{r}{R}\right)^2P_2(\cos\theta)$.

16.22. 1. $C-\frac{1}{2r^2}\sin\theta\sin\left(\frac{\pi}{4}-\varphi\right)$, where C is an arbitrary constant.

2. $\left[\frac{2}{15r^4}P_3^{(1)}(\cos\theta)+\frac{1}{5r^2}P_1^{(1)}(\cos\theta)\right]\sin\left(\varphi+\frac{\pi}{3}\right)$,
$u|_{r=1}=\left[\frac{2}{15}P_3^{(1)}(\cos\theta)+\frac{2}{3}P_1^{(1)}(\cos\theta)\right]\sin\left(\varphi+\frac{\pi}{3}\right)$.

16.23. 1. $\left(\frac{R}{r}\right)^5\sin^3\theta\cos\theta\cos\left(3\varphi+\frac{\pi}{4}\right)$. 2. $\left(\frac{R}{r}\right)^{101}\sin 100\varphi\sin^{100}\theta$. 3. $\left[\frac{R}{2+R}\left(\frac{R}{r}\right)^2P_1^{(1)}(\cos\theta)+\frac{6}{6(R+3)}\left(\frac{R}{r}\right)^3P_2^{(1)}(\cos\theta)\right]\sin\left(\varphi+\frac{\pi}{3}\right)$. *Hint.* $(u-u_r)|_{r=R}=\left[\frac{1}{2}P_1^{(1)}(\cos\theta)+\frac{1}{6}P_2^{(1)}(\cos\theta)\right]\sin\left(\varphi+\frac{\pi}{6}\right)$, $u=\left[A\left(\frac{R}{r}\right)^2P_1^{(1)}(\cos\theta)+B\left(\frac{R}{r}\right)^3P_2^{(1)}(\cos\theta)\right]\sin\left(\varphi+\frac{\pi}{6}\right)$.

16.24. 1. $\frac{1}{7}\left(-r+\frac{8}{r^2}\right)\sin\varphi\sin\theta$. 2. $\frac{12}{7}\left(r-\frac{1}{r^2}\right)\cos\theta+\left(\frac{96}{31r^3}-\right.$ $\left.-\frac{3r^2}{31}\right)\sin 2\varphi\sin^2\theta$. 3. $4\left(r-\frac{1}{r^2}\right)\cos\theta+\left(\frac{8}{r^2}-r\right)\sin\theta\cos\varphi$. 4. $\left(14-\frac{12}{r}\right)P_0(\cos\theta)+r^2(1-3\cos^2\theta-\sin^2\theta\cdot\sin 2\varphi)$. 5. $\frac{12}{7}\cos\varphi\times$ $\times\sin\theta\left(\frac{4}{r^2}-\frac{r}{2}\right)+\frac{12}{31}\left(\frac{8}{r^3}-\frac{r^2}{4}\right)\cos\varphi\sin 2\theta$. 6. $\left[\left(\frac{8}{31}\cos 2\varphi-\right.\right.$ $\left.\left.-\frac{1}{31}\sin 2\varphi\right)r^2+\frac{1}{r^3}\left(-\frac{8}{31}\cos 2\varphi+\frac{32}{31}\sin 2\varphi\right)\right]\sin^2\theta$.
7. $\left[r^2\left(-\frac{1}{31}\cos\varphi+\frac{8}{31}\sin\varphi\right)+\frac{1}{r^3}\left(\frac{32}{31}\cos\varphi-\frac{8}{31}\sin\varphi\right)\right]\sin 2\theta$.
8. $\left(\frac{32}{r^3}-r^2\right)\sin 2\theta\sin\varphi+\left(8r^2-\frac{8}{r^3}\right)\sin^2\theta\cos 2\varphi$. 9. $\frac{1}{7}\left(\frac{8}{r^2}-r\right)\times$ $\times\cos\theta+\frac{32}{127}\left(r^3-\frac{1}{r^4}\right)\frac{12\sin\theta-15\sin^3\theta}{2}\cos\varphi$.

16.25. 1. $\left(\frac{4}{r^2}-r\right)\cos\theta+\left(r^2-\frac{32}{r^3}\right)\sin^2\theta\sin 2\varphi$. 2. $\left(r+\frac{4}{r^2}\right)\sin\theta\times$ $\times\sin\varphi+3r^2\sin 2\theta\sin\varphi$. 3. $1+\frac{12}{5}\left(r-\frac{1}{r^2}\right)P_1^{(1)}(\cos\theta)\cos\varphi+$ $+\frac{16}{97}\left(r^3-\frac{1}{r^4}\right)\cos\varphi P_3^{(1)}(\cos\theta)+\frac{4}{97}\left(r^3-\frac{1}{r^4}\right)\sin 2\varphi P_3^{(2)}(\cos\theta)$. *Hint.* $u_r|_{r=2}=2P_3^{(1)}(\cos\theta)\cos\varphi+\frac{1}{2}P_3^{(2)}(\cos\theta)\sin 2\varphi+3P_1^{(1)}(\cos\theta)\cos\varphi$,

$u=\left(ar+\frac{b}{r^2}\right)\sin\theta\cos\varphi+C+\frac{d}{r}+\left(fr^3+\frac{h}{r^4}\right)P_3^{(1)}(\cos\theta)\cos\varphi+$ $+\left(lr^3+\frac{m}{r^4}\right)P_3^{(2)}(\cos\theta)\sin 2\varphi.$

16.26. 1. $4-\frac{2}{r}+\frac{2}{31}\left(\frac{1}{r^3}-32r^2\right)P_2(\cos\theta)+\frac{1}{31}\left(32r^2-\frac{1}{r^3}\right)\times$ $\times P_2^{(2)}(\cos\theta)\cos 2\varphi$. *Hint.* $u|_{r=1}=2-2P_2(\cos\theta)+P_2^{(2)}(\cos\theta)\cos 2\varphi.$
2. $\frac{12}{7}\left(\frac{1}{r^2}-r\right)\cos\theta+\frac{8}{127}\left(\frac{1}{r^4}-r^3\right)P_3^{(2)}(\cos\theta)\cos 2\varphi+\frac{48}{127}\left(r^3-\right.$ $\left.-\frac{1}{r^4}\right)P_3(\cos\theta)$. *Hint.* $u|_{r=\frac{1}{2}}=-6P_3(\cos\theta)+6P_1(\cos\theta)+$ $+P_3^{(2)}(\cos\theta)\cos 2\varphi,\quad u=\left(ar+\frac{b}{r^2}\right)P_1(\cos\theta)+\left(cr^3+\frac{d}{r^4}\right)\times$ $\times P_3^{(2)}(\cos\theta)\cos 2\varphi+\left(gr^3+\frac{h}{r^4}\right)P_3(\cos\theta).$

17 Green's Functions of the Dirichlet Problem

The *Green's function of the* (interior) *Dirichlet problem* for a region $G \in R^3$ is a function $\mathcal{G}(x, y)$, $x \in \bar{G}$, $y \in G$, with the following properties:

1. $\mathcal{G}(x, y)=\frac{1}{4\pi|x-y|}+g(x, y)$, where g is harmonic in G and continuous in $\bar{G}$ in x for every $y \in G$.

2. $\mathcal{G}(x, y)|_{x\in S}=0$ for every $y \in G$, where S is the boundary of G. For unbounded regions G we must ensure that $g(x, y)\to 0$ as $|x|\to\infty$.

If G is a bounded region and S is a sufficiently smooth surface, then $\mathcal{G}$ exists, is unique, has a regular normal derivative $\partial\mathcal{G}/\partial\mathbf{n}_x$ on S for every $y \in G$, and is symmetric, that is, $\mathcal{G}(x, y)=\mathcal{G}(y, x)$, $x \in G$, $y \in G$; $g(x, y)$ is continuous in the set of (x, y) in $\bar{G}\times G$.

If the solution of the interior Dirichlet problem for Poisson's equation $\nabla^2 u=-f(x)$, $u|_S=u_0(x)$, with $f \in C(\bar{G})$ and $u_0 \in C(S)$, has a regular normal derivative on S, then it is given by the following formula:

$$u(x)=-\int_S \frac{\partial\mathcal{G}(x, y)}{\partial\mathbf{n}_y}u_0(y)\,ds_y+\int_G \mathcal{G}(x, y)f(y)\,dy. \qquad (17.1)$$

For some regions the Green's function can be found via the *reflection method.*

17.1. Construct the Green's function for the following regions in R^3:
(1) The half-space $x_3>0$.
(2) The dihedral angle $x_2>0$, $x_3>0$.
(3) The octant $x_1>0$, $x_2>0$, $x_3>0$.

17.2. Construct the Green's function for the following regions in R^3:
(1) The ball $|x|<R$.
(2) The half-ball $|x|<R$, $x_3>0$.

(3) The quarter of the ball $|x| < R$, $x_2 > 0$, $x_3 > 0$.
(4) The one-eighth of the ball $|x| < R$, $x_1 > 0$, $x_2 > 0$, $x_3 > 0$.

17.3. Using the reflection method, construct the Green's function for the part of space enclosed between two parallel planes, $x_3 = 0$ and $x_3 = 1$.

Problems 17.4-17.10 deal with boundary value problems for Laplace's and Poisson's equations whose solutions can be found with the help of appropriate Green's functions taken from Problems 17.1-17.3 and formula (17.1).

17.4. Find the solution of the Dirichlet problem

$$\nabla^2 u = -f(x), \quad x_3 > 0; \quad u|_{x_3=0} = u_0(x),$$

with the following f and u_0:
1. f and u_0 are continuous and bounded.
2. $f = 0$, $u_0 = \cos x_1 \cos x_2$.
3. $f = e^{-x_3}$, $\sin x_1 \cos x_2$, $u_0 = 0$.
4. $f = 0$, $u_0 = \theta(x_2 - x_1)$.
5. $f = 0$, $u_0 = (1 + x_1^2 + x_2^2)^{-1/2}$.
6. $f = 2[x_1^2 + x_2^2 + (x_3 + 1)^2]^{-2}$, $u_0 = (1 + x_1^2 + x_2^2)^{-1}$.
7. $f = 0$, $u_0 = \begin{cases} -1, & x_1 < 0, \\ +1, & x_1 > 0. \end{cases}$

17.5. Find the solution to the Dirichlet problem

$$\nabla^2 u = 0, \quad x_2 > 0, \quad x_3 > 0, \quad u|_S = u_0(x),$$

with u_0 piecewise smooth and bounded.

17.6. Solve Problem 17.5 with the following data:
1. $u_0|_{x_2=0} = 0$, $u_0|_{x_3=0} = e^{-4x_1} \sin 5x_2$.
2. $u_0|_{x_2=0} = 0$, $u_0|_{x_3=0} = x_2(1 + x_1^2 + x_2^2)^{-3/2}$.
3. $u_0|_{x_2=0} = 0$, $u_0|_{x_3=0} = \theta(x_2 - |x_1|)$.

17.7. Find the solution of the following Dirichlet problem for the ball $|x| < R$:

$$\nabla^2 u = -f(x), \quad |x| < R, \quad u|_{|x|=R} = u_0(x).$$

17.8. Solve Problem 17.7 with the following data:
1. $f = a = \text{const}$, $u_0 = 0$. 2. $f = |x|^n$, $n = 0, 1, 2, \ldots$, $u_0 = a$.
3. $f = e^{|x|}$, $u_0 = 0$.

17.9. Solve the Dirichlet problem for Laplace's equation for the half-ball $|x| < R$, $x_3 > 0$.

17.10. Find the solution to Poisson's equation $\nabla^2 u = -f(|x|)$, $f \in C(a \leqslant |x| \leqslant b)$ in the spherical layer $a < |x| < b$ with the following conditions:

$$u|_{|x|=a} = 1, \quad u|_{|x|=b} = 0.$$

The *Green's function of the Dirichlet problem for a region* $G \subset R^2$ is the function

$$\mathcal{G}(z, \zeta) = \frac{1}{2\pi} \ln \frac{1}{|z-\zeta|} + g(z, \zeta),$$

with $z = x + iy \in \bar{G}$ and $\zeta = \xi + i\eta \in G$, and possesses all the properties of the Green's function in R^3 (see the beginning of Sec. 17). The solution to the Dirichlet problem $\nabla^2 u = -f(z)$, $z \in G$, $u|_S = u_0(z)$, in R^2 (provided it exists) is given by a formula similar to (17.1) in R^2. When G is a simply connected region with a sufficiently smooth boundary S and we know the function $w = w(z)$ that maps G conformally onto the unit circle $|w| < 1$, the Green's function is given by the formula

$$\mathcal{G}(z, \zeta) = \frac{1}{2\pi} \ln \frac{1}{|\omega(z, \zeta)|}, \quad \omega(z, \zeta) = \frac{w(z) - w(\zeta)}{1 - w(z)\, w(\zeta)}.$$

17.11. Find the Green's function for the following regions:
(1) The half-plane $\operatorname{Im} z > 0$.
(2) The quarter of the plane $0 < \arg z < \pi/2$.
(3) The circle $|z| < R$.
(4) The semicircle $|z| < R$, $\operatorname{Im} z > 0$.
(5) The quarter of the circle $|z| < 1$, $0 < \arg z < \pi/2$.
(6) The strip $0 < \operatorname{Im} z < \pi$.
(7) The half-strip $0 < \operatorname{Im} z < \pi$, $\operatorname{Re} z > 0$.

17.12. Find the solution to the Dirichlet problem

$$\nabla^2 u = 0, \quad y > 0; \quad u|_{y=0} = u_0(x),$$

where
1. $u_0(x)$ is piecewise continuous and bounded.
2. $u_0(x) = \theta(x - a)$.
3. $u_0(x) = \begin{cases} 1, & x \in [a, b], \\ 0, & x \bar{\in} [a, b]. \end{cases}$
4. $u_0(x) = \dfrac{1}{1+x^2}$. 5. $u_0(x) = \dfrac{x}{1+x^2}$.
6. $u_0(x) = \dfrac{x^2-1}{(1+x^2)^2}$. 7. $u_0(x) = \cos x$.

17.13. Find the solutions of the equation $\nabla^2 u = 0$ in the first quadrant $x > 0$, $y > 0$ with the following boundary conditions:
1. $u|_S = u_0(x, y)$ is a piecewise continuous, bounded function, with S consisting of the rays $\{x = 0, y \geqslant 0\}$ and $\{y = 0, x \geqslant 0\}$.
2. $u|_{x=0} = 0$, $u|_{y=0} = 1$. 3. $u|_{x=0} = a$, $u|_{y=0} = b$.
4. $u|_{x=0} = 0$, $u|_{y=0} = \theta(x-1)$.
5. $u|_{x=0} = 0$, $u|_{y=0} = \dfrac{x}{(1+x^2)^2}$. 6. $u|_{x=0} = \sin y$, $u|_{y=0} = \sin x$.

17.14. Find the solution to the Dirichlet problem for the equation $\nabla^2 u = 0$, with $0 < y < \pi$ with the following boundary conditions:

1. $u|_S = u_0(x)$ is a piecewise continuous, bounded function with S the boundary of the strip $0 < y < \pi$.
2. $u|_{y=0} = \theta(x)$, $u|_{y=\pi} = 0$.
3. $u|_{y=0} = \theta(x)$, $u|_{y=\pi} = \theta(x)$.
4. $u|_{y=0} = \theta(x)$, $u|_{y=\pi} = -\theta(x)$.
5. $u|_{y=0} = \theta(x)$, $u|_{y=\pi} = \theta(-x)$.
6. $u|_{y=0} = \cos x$, $u|_{y=\pi} = 0$.

17.15. Find the solution to Laplace's equation $\nabla^2 u = 0$ in the half-strip $0 < y < \pi$, $x > 0$ with the following boundary conditions:
1. $u|_{x=0} = 1$, $u|_{y=0} = 0$, $u|_{y=\pi} = 0$.
2. $u|_{x=0} = 0$, $u|_{y=0} = \sin x$, $u|_{y=\pi} = 0$.
3. $u|_{x=0} = 0$, $u|_{y=0} = \tanh x$, $u|_{y=\pi} = \tanh x$.
4. $u|_{x=0} = 0$, $u|_{y=0} = 0$, $u|_{y=\pi} = \tanh x$.

17.16. Find the solution to Poisson's equation $\nabla^2 u = -f(z)$ in the circle $|z| < R$ with the boundary condition $u|_{|z|=R} = u_0(z)$ with the following data.
1. f and u_0 are continuous functions.
2. $f = a$, $u_0 = b$. 3. $f = |z|^n$, $n = 1, 2, \ldots$, $u_0 = 0$.
4. $f = \sin |z|$, $u_0 = 0$. 5. $f = 0$, $u_0 = \cos \varphi$, where $\varphi = \arg z$; $0 \leqslant \varphi < 2\pi$.

17.17. Find the solution to Laplace's equation $\nabla^2 u = 0$ in the semicircle $|z| < 1$, $\operatorname{Im} z > 0$ with the boundary condition $u|_S = u_0(z)$, where S is the boundary of the semicircle, with the following data:
1. $u_0(z)$ is a piecewise continuous function.
2. $u_0|_{r=1} = \sin \varphi$, $u_0|_{\varphi=0} = 0$, $u_0|_{\varphi=\pi} = 0$, where $r = |z|$, $\varphi = \arg z$, $0 \leqslant \varphi < 2\pi$.
3. $u_0|_{r=1} = 0$, $u_0|_{\varphi=0} = 1$, $u_0|_{\varphi=\pi} = 1$.
4. $u_0|_{r=1} = \cos \frac{\varphi}{2}$, $u_0|_{\varphi=0} = \sqrt{r}$, $u_0|_{\varphi=\pi} = 0$.

17.18. Find the solution to the Dirichlet problem $\nabla^2 u = 0$, $\operatorname{Re} z > 0$, $|z - 5| > 3$; $u|_{\operatorname{Re} z=0} = 0$, $u|_{|z-5|=3} = 1$.

Answers to Problems of Sec. 17

In the answers to Problems 17.1-17.10 the following notations are used:

$$y_{mnk} = ((-1)^m y_1,\ (-1)^n y_2,\ (-1)^k y_3).$$

17.1. 1. $\frac{1}{4\pi} \sum_{k=0}^{1} \frac{(-1)^k}{|x - y_{00k}|}$. 2. $\frac{1}{4\pi} \sum_{n,\,k=0}^{1} \frac{(-1)^{n+k}}{|x - y_{0nk}|}$.

3. $\frac{1}{4\pi} \sum_{m,\,n,\,k=0}^{1} \frac{(-1)^{m+n+k}}{|x - y_{mnk}|}$.

17.2. 1. $\frac{1}{4\pi}\left(\frac{1}{|x-y|}-\frac{R}{|y|\,|x-y^*|}\right)$, where (as well as the other cases in this problem) $y^*_{mnk}=\frac{R^2}{|y|^2}y_{mnk}$, $|y_{mnk}|\,|y^*_{mnk}|=R^2$.

2. $\frac{1}{4\pi}\sum_{k=0}^{1}(-1)^k\left(\frac{1}{|x-y_{00k}|}-\frac{R}{|y|\,|x-y^*_{00k}|}\right)$.

3. $\frac{1}{4\pi}\sum_{n,\,k=0}^{1}(-1)^{n+k}\left(\frac{1}{|x-y_{0nk}|}-\frac{R}{|y|\,|x-y^*_{0nk}|}\right)$.

4. $\frac{1}{4\pi}\sum_{m,\,n,\,k=0}^{1}(-1)^{m+n+k}\left(\frac{1}{|x-y_{mnk}|}-\frac{R}{|y|\,|x-y^*_{mnk}|}\right)$.

17.3. $\frac{1}{4\pi}\sum_{n=-\infty}^{\infty}\left(\frac{1}{\sqrt{(x_1-y_1)^2+(x_2-y_2)^2+[x_3-(2n+y_3)]^2}}-\right.$

$\left.-\frac{1}{\sqrt{(x_1-y_1)^2+(x_2-y_2)^2+[x_3-(2n-y_3)]^2}}\right)$.

Hint (to Problem 17.4 and the problems that follow). For the case where f and u_0 are piecewise continuous, bounded functions and S is a piecewise smooth surface, the statement of the Dirichlet problem can be generalized in a way such that the solution can be determined via formula (17.1), too.

17.4. 1. $\frac{x_3}{2\pi}\int\limits_{y_3=0}\frac{u_0(y)}{|x-y|^3}\,dS_y+\int\limits_{y_3>0}f(y)\left(\frac{1}{|x-y|}-\frac{1}{|x-y_{001}|}\right)dy$.

2. $e^{-\sqrt{2}\,x_3}\cos x_1\cos x_2$. 3. $(e^{-\sqrt{2}\,x_3}-e^{-x_3})\sin x_1\cos x_2$.

4. $\frac{1}{2}+\frac{1}{\pi}\arctan\frac{x_2-x_1}{\sqrt{2}\,x_3}$. 5. $[x_1^2+x_2^2+(x_3+1)^2]^{-1/2}$.

6. $[x_1^2+x_2^2+(x_3+1)^2]^{-1}$. 7. $\frac{2}{\pi}\arctan\frac{x_1}{x_3}$.

17.5. $\frac{x_2}{2\pi}\int\limits_{y_2=0,\ y_3\geqslant 0}u_0(y)\left(\frac{1}{|x-y|^3}-\frac{1}{|x-y_{001}|^3}\right)dS_y+\frac{x_3}{2\pi}\times$

$\times\int\limits_{y_2\geqslant 0,\ y_3=0}u_0(y)\left(\frac{1}{|x-y|^3}-\frac{1}{|x-y_{010}|^3}\right)dS_y$.

17.6. 1. $e^{-4x_1-3x_3}\sin 5x_2$. 2. $x_2[x_1^2+x_2^2+(x_3+1)^2]^{-3/2}$.

3. $\frac{1}{\pi}\arctan\frac{x_2+x_1}{x_3\sqrt{2}}+\frac{1}{\pi}\arctan\frac{x_2-x_1}{x_3\sqrt{2}}$.

17.7. $\frac{1}{4\pi R}\int\limits_{|y|=R}\frac{R^2-|x|^2}{|x-y|^3}u_0(y)\,dS_y+\frac{1}{4\pi}\int\limits_{|y|\leqslant R}\left(\frac{1}{|x-y|}-\frac{R}{|y|\,|x-y^*|}\right)f(y)\,dy$, where $y^*=yR^2/|y|^2$ is the point symmetric to point y with respect to the sphere $|y|=R$.

17.8. 1. $\frac{a}{6}(R^2-|x^2|)$. 2. $a+\frac{R^{n+2}-|x|^{n+2}}{(n+2)(n+3)}$. 3. $e^R-e^{|x|}-\frac{2}{R}(e^R-1)+\frac{2}{|x|}(e^{|x|}-1)$.

17.9. $u(x)=\frac{x_3}{2\pi}\int\limits_{|y|\leqslant R,\ y_3=0}u_0(y_1,\ y_2)\left(\frac{1}{|x-y|^3}-\frac{R^3}{|y|^3\,|x-y^*|^3}\right)\times dy_1\,dy_2+\frac{R^2-|x|^2}{4\pi R}\int\limits_{|y|=R,\ y_3>0}u_0(y)\left(\frac{1}{|x-y|^3}-\frac{1}{|x-y^{**}|^3}\right)dS_y$, where

$|x|<R,\ x_3>0$; y^* and y^{**} are points symmetric to point y with respect to the sphere $|y|=R$ and the plane $y_3=0$, respectively.

17.10. $\frac{a(b-|x|)}{|x|(b-a)}-\frac{1}{|x|}\int\limits_0^{|x|}(|x|-\rho)\,\rho f(\rho)\,d\rho+\frac{b-|x|}{|x|(b-a)}\int\limits_0^a(a-\rho)\times\rho f(\rho)\,d\rho-\frac{a-|x|}{|x|(b-a)}\int\limits_0^b(b-\rho)\,\rho f(\rho)\,d\rho$.

17.11. 1. $\frac{1}{2\pi}\ln\frac{|z-\bar\zeta|}{|z-\zeta|}$, where $z=x+iy$ and $\zeta=\xi+i\eta$.

2. $\frac{1}{2\pi}\ln\frac{|z^2-\bar\zeta^2|}{|z^2-\zeta^2|}$. 3. $\frac{1}{2\pi}\ln\frac{|R^2-z\bar\zeta|}{R\,|z-\zeta|}$. 4. $\frac{1}{2\pi}\ln\frac{|z-\bar\zeta|\,|R^2-z\zeta|}{|z-\zeta|\,|R^2-z\bar\zeta|}$.

5. $\frac{1}{2\pi}\ln\frac{|z^2-\bar\zeta^2|\,|R^4-(z\zeta)^2|}{|z^2-\zeta^2|\,|R^4-(z\bar\zeta)^2|}$. 6. $\frac{1}{2\pi}\ln\frac{|e^z-e^{\bar\zeta}|}{|e^z-e^{\zeta}|}$.

7. $\frac{1}{2\pi}\ln\frac{|\cosh z-\cosh\bar\zeta|}{|\cosh z-\cosh\zeta|}$.

17.12. 1. $\frac{y}{\pi}\int\limits_{-\infty}^{\infty}\frac{u_0(\xi)\,d\xi}{(x-\xi)^2+y^2}$. 2. $\frac{1}{2}+\frac{1}{\pi}\arctan\frac{x-a}{y}$.

3. $\frac{1}{2}-\frac{1}{\pi}\arctan\frac{y^2+(x-a)(x-b)}{y(b-a)}$. 4. $\frac{y+1}{x^2+(y+1)^2}$.

5. $\frac{x}{x^2+(y+1)^2}$. 6. $\frac{x^2-(y+1)^2}{[x^2+(y+1)^2]^2}$. 7. $e^{-y}\cos x$.

17.13. 1. $\frac{y}{\pi}\int\limits_0^{\infty}u_0(\xi,\ 0)\left[\frac{1}{(x-\xi)^2+y^2}-\frac{1}{(x+\xi)^2+y^2}\right]d\xi+\frac{x}{\pi}\times\int\limits_0^{\infty}u_0(0,\ \eta)\left[\frac{1}{x^2+(y-\eta)^2}-\frac{1}{x^2+(y+\eta)^2}\right]d\eta$. 2. $\frac{2}{\pi}\arctan\frac{x}{y}$.

3. $\frac{2}{\pi}\left(a \arctan\frac{y}{x} + b \arctan\frac{x}{y}\right)$. 4. $\frac{1}{2} - \frac{1}{\pi}\arctan\frac{y^2-x^2+1}{2xy}$.
5. $\frac{x(y+1)}{[x^2+(y+1)^2]^2}$. 6. $e^{-y}\sin x + e^{-x}\sin y$.

17.14. 1. $\sum_{k=0}^{1} \frac{1}{\pi} e^x \sin y \int_{-\infty}^{\infty} \frac{u_0(\xi, k\pi)\, e^{-\xi}}{e^{2(x-\xi)} - 2e^{x-\xi}\cos(y-k\pi)+1}\, d\xi$. 2. $\frac{1}{2} - \frac{1}{\pi}\arctan\frac{e^{-x}-\cos y}{\sin y}$. 3. $\frac{1}{2} + \frac{1}{\pi}\arctan\frac{\sinh x}{\sin y}$. 4. $\frac{2}{\pi} \times \arctan\cot y - \frac{1}{\pi}\arctan\frac{\sin 2x}{e^{2x}-\cos 2y}$. 5. $\frac{1}{2} + \frac{1}{\pi}\arctan\frac{\tanh x}{\tan y}$.
6. $\frac{\cos x \sinh(\pi-y)}{\sinh \pi}$.

17.15. 1. $\frac{1}{2} + \frac{1}{\pi}\arctan\frac{\sin^2 y - \sinh^2 x}{2\sin y \sinh x}$. 2. $\frac{\sin x \sinh(\pi-y)}{\sinh \pi}$.
3. $\frac{\sinh x}{\cosh x + \sin y}$. 4. $\frac{x \sin 2y + y \sinh 2x - \pi \sin y \sinh x}{\pi(\cosh 2x + \cos 2y)}$.

17.16. 1. $\frac{1}{2\pi R} \int_{|\zeta|=R} u_0(\zeta) \frac{R^2-|z|^2}{|z-\zeta|^2}\, dS_\zeta + \frac{1}{2\pi} \int_{|\zeta|\leqslant R} f(\zeta) \ln \frac{|\zeta|\,|z-\zeta^*|}{R\,|z-\zeta|} \times d\xi\, d\eta$, where $z = x+iy$, $\zeta = \xi + i\eta$, and $\zeta^* = \zeta R^2/|\zeta|^2$. *Hint.* In Problems 17.16.2 and those below use the formula of Problem 17.16.1 and transfer to polar coordinates $z = re^{i\varphi}$, $\zeta = \rho e^{i\theta}$, $0 < \varphi$, $\theta < 2\pi$. 2. $\frac{a}{4}(R^2-r^2)+b$. 3. $\frac{R^{n+2}-r^{n+2}}{(n+2)^2}$. 4. $\sin r - \sin R + \int_r^R \frac{\sin\rho}{\rho}\, d\rho$. 5. $\frac{r}{R}\cos\varphi$.

17.17. 1. $\frac{1}{2\pi} \int_{|\xi|=1,\ \mathrm{Im}\,\xi\geqslant 0} u_0(\zeta) \left(\frac{|z|^2-1}{|z-\zeta|^2} - \frac{|z|^2-1}{|z-\bar{\zeta}|^2}\right) d\xi\, d\eta + \frac{y}{\pi} \times \int_{-1}^{1} u_0(\xi, 0)\left(\frac{1}{|z-\xi|^2} - \frac{1}{|\xi z - 1|^2}\right) d\xi$, where $z = x+iy$, $\zeta = \xi + i\eta$.
2. $r\sin\varphi$. 3. $\frac{2}{\pi}\arctan\frac{2r\sin\varphi}{r^2-1}$. 4. $\sqrt{r}\cos\frac{\varphi}{2}$.

17.18. $\frac{1}{\ln 3}\ln\frac{|z+4|}{|z-4|}$.

18 The Method of Potentials

Suppose ρ belongs to $\mathscr{D}'(R^n)$. The convolution $V_n = \frac{1}{|x|^{n-2}} * \rho$, $n \geqslant 3$, is known as the *Newtonian potential*, while $V_2 = \ln\frac{1}{|x|} * \rho$, is known as the *logarithmic potential*; ρ is the density (for the definition of a convolution see Sec. 8).

The potential V_n satisfies Poisson's equation

$$\nabla^2 V_n = -(n-2)\,\sigma_n \rho, \quad n \geqslant 3; \quad \nabla^2 V_2 = -2\pi\rho.$$

If ρ is a finite and absolutely integrable function in R^n, the respective Newtonian (logarithmic) potential is called the *volume (surface) potential.*

Suppose S is a bounded, piecewise smooth, and two-sided surface in R^n, $\mathbf{n}$ is a unit normal vector to S, and $\mu\delta_S$ and $-\frac{\partial}{\partial \mathbf{n}}(\nu\delta_S)$ are the simple and double layers on S with densities μ and ν (for the definitions of layers see Secs. 6 and 7). The convolutions

$$V_n^{(0)} = \frac{1}{|x|^{n-2}} * \mu\delta_S \text{ and } V_n^{(1)} = -\frac{1}{|x|^{n-2}} * \frac{\partial}{\partial \mathbf{n}}(\nu\delta_S), \; n \geqslant 3,$$

are known as the *surface potentials of the simple and double layers* with densities μ and ν, respectively. The convolutions

$$V_2^{(0)} = \ln\frac{1}{|x|} * \mu\delta_S \text{ and } V_2^{(1)} = -\ln\frac{1}{|x|} * \frac{\partial}{\partial \mathbf{n}}(\nu\delta_S), \; n=2,$$

are called, respectively, the *logarithmic potentials of the simple and double layers.*

If S is a Lyapunov surface and ν belongs to $C\ (S)$, then the limiting values in R^3 of the double layer potential from the outside and the inside of S are given by the following formulas:

$$V_{\pm}^{(1)}(x) = \pm 2\pi\nu(x) + V^{(1)}x = \pm 2\pi\nu(x) + \int\limits_S \nu(y)\frac{\cos\varphi_{xy}}{|x-y|^2}\,dS_y, \quad (18.1)$$

where φ_{xy} is the angle between vector $\mathbf{x}-\mathbf{y}$ and the normal unit vector $\mathbf{n}_y$ at point $y \in S$.

If $\mu \in C(S)$, the simple layer potential has regular normal derivatives $\left(\frac{\partial V^{(0)}}{\partial \mathbf{n}}\right)_+$ and $\left(\frac{\partial V^{(0)}}{\partial \mathbf{n}}\right)_-$ on S from the outside and the inside of S (see the definition at the beginning of Chap. V), while on S we have

$$\left(\frac{\partial V^{(0)}}{\partial \mathbf{n}}\right)_{\pm}(x) = \mp 2\pi\mu(x) + \frac{\partial V^{(0)}x}{\partial \mathbf{n}}$$
$$= \mp 2\pi\mu(x) + \int\limits_S \mu(y)\frac{\cos\psi_{xy}}{|x-y|^2}\,dS_y, \quad (18.2)$$

where ψ_{xy} is the angle between vector $\mathbf{y}-\mathbf{x}$ and the unit normal $\mathbf{n}_x$.

Similar formulas for $V_{\pm}^{(1)}(x)$ and $\left(\frac{\partial V^{(0)}}{\partial \mathbf{n}}\right)_{\pm}(x)$ are valid in R^2 if we substitute π for 2π and $|x-y|$ for $|x-y|^2$.

18.1. Suppose ρ is an absolutely integrable function that vanishes outside $\bar{G} \subset R^n$. Prove that (a) the surface potential is given by the formula

$$V_n(x) = \int\limits_G \frac{\rho(y)}{|x-y|^{n-2}}\,dy, \quad n \geqslant 3; \quad (18.3)$$

(b) V_n is harmonic outside $\bar{G}$; and

(c) $V_3(x) = \frac{1}{|x|} \int\limits_G \rho(y)\,dy + O\left(\frac{1}{|x|^2}\right)$, $|x| \to \infty$.

What is the physical meaning of these potentials?

18.2. Suppose ρ is an absolutely integrable function that vanishes outside $\bar{G} \subset R^2$. Prove that (a) the surface potential is given by the formula

$$V_2(x) = \int\limits_G \rho(y) \ln \frac{1}{|x-y|}\,dy; \tag{18.4}$$

(b) V_2 is harmonic outside $\bar{G}$; and

(c) $V_2(x) = \ln \frac{1}{|x|} \int\limits_G \rho(y)\,dy + O\left(\frac{1}{|x|}\right)$, $|x| \to \infty$.

What is the physical meaning of these potentials?

18.3. Suppose S is a bounded, piecewise smooth, and two-sided surface and both μ and ν belong to $C(S)$. Prove that (a) the simple and double layer potentials are given by the formulas

$$V_3^{(0)}(x) = \int\limits_S \frac{\mu(y)}{|x-y|}\,dS_y,$$

$$V_3^{(1)}(x) = \int\limits_S \nu(y) \frac{\partial}{\partial n_y} \frac{1}{|x-y|}\,dS_y = \int\limits_S \nu(y) \frac{\cos\varphi_{xy}}{|x-y|^2}\,dS_y, \tag{18.5}$$

where the angle φ_{xy} was defined at the beginning of Sec. 18.

(b) $V_3^{(0)}$ and $V_3^{(1)}$ are harmonic outside S; and

(c) $V_3^{(0)}(x) = \frac{1}{|x|} \int\limits_S \mu(y)\,dS + O\left(\frac{1}{|x|^2}\right)$ and $V_3^{(1)}(x) = O\left(\frac{1}{|x|^2}\right)$

as $|x| \to \infty$. What is the physical meaning of these potentials?

18.4. Suppose S is a bounded and piecewise smooth curve and both μ and ν belong to $C(S)$. Prove that (a) the logarithmic potentials of the simple and double layers are given by the formulas

$$V_2^{(0)}(x) = \int\limits_S \mu(y) \ln \frac{1}{|x-y|}\,dS_y,$$

$$V_2^{(1)}(x) = \int\limits_S \nu(y) \frac{\partial}{\partial n_y} \ln \frac{1}{|x-y|}\,dS_y = \int\limits_S \nu(y) \frac{\cos\varphi_{xy}}{|x-y|}\,dS_y, \tag{18.6}$$

where angle φ was defined earlier;

(b) $V_2^{(0)}$ and $V_2^{(1)}$ are harmonic outside S; and

(c) $V_2^{(0)}(x) = \ln \frac{1}{|x|} \int\limits_S \mu(\zeta)\,ds + O(|x|^{-1})$, $\quad V_2^{(1)}(x) = O(|x|^{-1})$,

$x \to \infty$.

What is the physical meaning of these potentials?

18.5. 1. Calculate the Newtonian potential V_3 with the density δ_{S_R}.
2. Calculate the logarithmic potential V_2 with the density δ_{S_R}.

18.6. Calculate the volume potential V_3 for a ball $|x|<R$ with the following densities:
1. $\rho=\rho(|x|)\in C$. 2. $\rho=\rho_0=\text{const}$.
3. $\rho=|x|$. 4. $\rho=|x|^2$. 5. $\rho=\sqrt{|x|}$.
6. $\rho=e^{-|x|}$. 7. $\rho=\frac{1}{1+|x|^2}$.
8. $\rho=\sin|x|$. 9. $\rho=\cos|x|$.
10. $\rho=\ln\left(1+\frac{|x|}{R}\right)$.

18.7. Calculate the volume potential V_3 for a spherical layer $R_1<|x|<R_2$ with the following mass distributions:
1. $\rho=\rho_0=\text{const}$. 2. $\rho=\rho(|x|)\in C$ $(R_1\leqslant|x|\leqslant R_2)$.

18.8. Suppose the mass of the ball $r<R$ is distributed with a density ρ. Find the volume potential V_3 at a point lying on the axis $\theta=0$ $(0\leqslant\theta\leqslant\pi)$ with the following densities:
1. ρ is proportional to the square of the distance from the plane $\theta=\pi/2$.
2. $\rho=\cos\theta$. 3. $\rho=\sin\varphi$.
4. $\rho=\rho(\varphi)$ is a 2π-periodic continuous function, $0\leqslant\varphi<2\pi$.

18.9. Suppose a mass is distributed in the cylinder $\{x_1^2+x_2^2<R^2, 0<x_3<H\}$ with a constant density ρ_0. Find the volume potential at the points of the x_3 axis that satisfy the inequality $x_3\geqslant H$.

18.10. Find the surface potential for the circle $r<R$ with the following densities:
1. $\rho=\rho(r)\in C([0,R])$. 2. $\rho=\rho_0=\text{const}$. 3. $\rho=r$. 4. $\rho=r^2$.
5. $\rho=e^{-r}$. 6. $\rho=\frac{1}{1+r^2}$. 7. $\rho=\sqrt{r}$. 8. $\rho=\sin r$.
9. $\rho=\cos r$. 10. $\rho=\sin\varphi$, $0\leqslant\varphi\leqslant 2\pi$. 11. $\rho=\cos\varphi$.
12. $\rho=\rho(\varphi)$ is a 2π-periodic continuous function.

18.11. Find the logarithmic surface potential for the annulus $R_1<r<R_2$ with the following densities:
1. $\rho=\rho_0=\text{const}$. 2. $\rho=\rho(r)\in C([R_1,R_2])$.

18.12. Suppose $f(|y|)$ is continuous at $|y|\leqslant R$ and vanishes at $|y|>R$, $y\in R^3$. Prove that (a) the volume potential $V_3(x)$ with a density equal to $f(|y|)$ depends only on $|x|$ and

$$V_3(x)=\frac{1}{|x|}\int\limits_{|y|<R} f(|y|)\,dy, \quad |x|>R;$$

(b) $V_3(x)$ vanishes at $|x|>R$ if and only if

$$\int f(|y|)\,dy=0; \tag{18.7}$$

(c) $\int V_3(x)\,dx = -\frac{2\pi}{3}\int f(|y|)\,|y|^2\,dy$, provided (18.7) holds. What is the physical meaning of the above equations?

18.13. Prove that if $f_1(x)$ and $f_2(|x|)$ are continuous at $|x| \leqslant R$, $x \in R^3$, vanish at $|x| > R$, and satisfy the equation

$$\nabla^2 f_1(x) = D^\alpha f_2(|x|),$$

then the potential $V_3(x)$ with the density $f_2(|x|)$ vanishes at $|x| > R$.

18.14. Prove the results similar to those of Problems 18.12 and 18.14 for the surface potential, namely,

1. $V_2(x) = \ln\frac{1}{|x|}\int\limits_{|y|<R} f(|y|)\,dy, \ |y| > R.$

2. $\int V_2(x)\,dx = -\frac{\pi}{2}\int f(|y|)\,|y|^2\,dy$

if

$$\int f(|y|)\,dy = 0.$$

18.15. Extend Problems 18.12-18.14 to the case where density f is a generalized function. The integral $\int f(x)\,dx$ for a finite $f \in \mathscr{D}'$ is the number (f, η), with $\eta \in \mathscr{D}$ and equal to unity in the neighbourhood supp f (the number does not depend on the choice of the auxiliary function η).

18.16. Find the potential of a simple layer distributed over the sphere $|x| = R$ with a constant density μ_0.

18.17. At a point lying on the axis $\theta = 0$ $(0 \leqslant \theta \leqslant \pi)$ find the potential of a simple layer distributed over a sphere $r = R$ with the following densities:

1. μ is proportional to the square of the distance from the plane $\theta = \pi/2$.

2. $\mu = \sin\frac{\theta}{2}$. 3. $\mu = e^\varphi$, $0 \leqslant \varphi \leqslant \pi$, and $\mu = e^{2\pi-\varphi}$, $\pi \leqslant \leqslant \varphi < 2\pi$.

18.18. A simple layer with a density μ is distributed over a disk of radius R. Find the potential at a point lying on the disk's axis for the following densities:

1. $\mu = \mu_0 =$ const. 2. $\mu = r$. 3. $\mu = r^2$.
4. $\mu = \mu(\varphi)$ is a 2π-periodic continuous function.

18.19. Find the potential of a simple layer distributed with a density μ over the cylinder $\{x_1^2 + x_2^2 = R^2,\ 0 \leqslant x_3 \leqslant H\}$ at a point lying on the x_3 axis for the following densities:

1. $\mu = \mu_0 =$ const.
2. $\mu = \mu(\varphi)$ is a 2π-periodic continuous function.

18.20. Find the potential of a double layer with a constant density ν_0 for the sphere $|x| = R$.

18.21. Dipoles with a dipole moment density ν are distributed over a sphere $r = R$ and are oriented along outward normals. Find the double layer potential at a point on the axis $\theta = 0$ $(0 \leqslant \theta \leqslant \pi)$ for the following densities:

1. $\nu = \cos\theta$. 2. $\nu = \sin\frac{\theta}{2}$.
3. $\nu = e^{\varphi}$, $0 \leqslant \varphi \leqslant \pi$, and $\nu = e^{2\pi-\varphi}$, $\pi \leqslant \varphi < 2\pi$.
4. $\nu = \nu(\varphi)$ is a 2π-periodic continuous function.
5. ν is equal to the square of the distance from the plane $\theta = \pi/2$.

18.22. Dipoles with a dipole moment density ν are distributed over a disk of radius R and are oriented along the normal directed to $-\infty$ on the x_3 axis. Find the double layer potential at a point that lies on the disk's axis with the following densities:

1. $\nu = \text{const}$. 2. $\nu = \nu(r) \in C([0, R])$.
3. $\nu = \nu(\varphi)$ is a 2π-periodic continuous function.
4. $\nu = r + \varphi$, $0 \leqslant \varphi \leqslant \pi$ and $\nu = r + 2\pi - \varphi$, $\pi \leqslant \varphi < 2\pi$.

18.23. Find the logarithmic potential of a simple layer for a circle $r = R$ with the following densities:

1. $\mu = \mu_0 = \text{const}$. 2. $\mu = \cos^2\varphi$; $R = 2$.

18.24. Find the logarithmic potential of a double layer for a circle $r = R$ with the following densities:

1. $\nu = \text{const}$. 2. $\nu = \sin\varphi$.

18.25. Find the logarithmic potential of a simple layer for the segment $-a \leqslant x \leqslant a$, $y = 0$ with the following densities:

1. $\mu = \text{const}$.
2. $\mu = -\mu_0$, $-a \leqslant x < 0$, and $\mu = \mu_0$, $0 < x \leqslant a$.
3. $\mu = x$.

18.26. Find the logarithmic potential of a double layer for the segment $-a \leqslant x \leqslant a$, $y = 0$ with the following densities:

1. $\nu = \text{const}$.
2. $\nu = -\nu_0$, $-a \leqslant x < 0$, and $\nu = \nu_0$, $0 < x \leqslant a$.
3. $\nu = x$. 4. $\nu = x^2$.

Suppose $\rho(x)$ is a finite generalized function. The convolutions $V = -4\pi\mathscr{E} * \rho$ and $\bar{V} = -4\pi\bar{\mathscr{E}} * \rho$, where

$$\mathscr{E} = -\frac{e^{ik|x|}}{4\pi |x|} \quad \text{and} \quad \bar{\mathscr{E}} = -\frac{e^{-ik|x|}}{4\pi |x|}$$

are fundamental solutions of Helmholtz's operator $\nabla^2 + k^2$ in R^3, are analogs of the Newtonian potential. Both satisfy Helmholtz's equation $\nabla^2 u + k^2 u = -4\pi\rho$.

The analogs of the simple and double layer potentials can be defined in a similar manner.

The same is true for the operator $\nabla^2 - k^2$. Here the analog of the Newtonian potential is $V_* = -4\pi\, \mathscr{E}_* * \rho$, where $\mathscr{E}_* = -\frac{e^{-k|x|}}{4\pi |x|}$ is the fundamental solution of the operator $\nabla^2 - k^2$ in R^3.

18.27. Suppose ρ is an absolutely integrable function of x that vanishes when $x \in G_1 = R^3 \setminus \bar{G}$. Prove that

1. V, $\bar{V}$, and V_* are given by the formulas

$$V(x) = \int_G \frac{e^{ik|x-y|}}{|x-y|} \rho(y)\, dy,$$

$$\bar{V}(x) = \int_G \frac{e^{-ik|x-y|}}{|x-y|} \rho(y)\, dy, \tag{18.8}$$

$$V_*(x) = \int_G \frac{e^{-k|x-y|}}{|x-y|} \rho(y)\, dy.$$

2. V, $\bar{V}$, and $V_* \subset C^1(R^3) \cap C^\infty(G_1)$ satisfy, in G_1, the homogeneous equations $\nabla^2 u + k^2 u = 0$ and $\nabla^2 u - k^2 u = 0$, respectively.

3. V and $\bar{V}$ satisfy Sommerfeld's radiation conditions

$$u(x) = O(|x|^{-1}), \quad \frac{\partial u(x)}{\partial |x|} \mp iku(x) = o(|x|^{-1}), \tag{18.9}$$

$$|x| \to \infty \text{ and } V_*(x) \to 0 \text{ as } |x| \to \infty.$$

18.28. For the operator $\nabla^2 + k^2$ calculate the potential V for a ball $|x| < R$ with the following densities:

1. $\rho = \rho(|x|) \in C(\bar{U}_R)$. 2. $\rho = \rho_0 = \text{const}$. 3. $\rho = e^{-|x|}$.

18.29. For the operator $\nabla^2 + k^2$ calculate the potential V for the spherical layer $R_1 < |x| < R_2$ with a constant density ρ_0.

18.30. 1. For the operator $\nabla^2 + k^2$ calculate the potential $V^{(0)}$ for a simple layer distributed over a sphere with a constant density μ_0.

2. For the operator $\nabla^2 + k^2$ calculate the potential $V^{(1)}$ for a double layer distributed over a sphere with a constant density ν_0.

18.31. For the operator $\nabla^2 - k^2$ calculate the potential V_* for a ball $r < R$ with the following densities: 1. $\rho = \rho(|x|) \in C(\bar{U}_R)$. 2. $\rho = \rho_0 = \text{const}$. 3. $\rho = e^{-|x|}$.

18.32. 1. For the operator $\nabla^2 - k^2$ calculate the potential $V_*^{(0)}$ for a simple layer distributed over a sphere with a constant density μ_0.

2. For the operator $\nabla^2 - k^2$ calculate the potential $V_*^{(1)}$ for a double layer distributed over a sphere with a constant density ν_0.

18.33. 1. Assuming the boundary S of a region $G \subset R^3$ to be a Lyapunov surface, prove that

$$V_3^{(1)}(x) = \int_S \frac{\cos \varphi_{xy}}{|x-y|^2}\, dS_y = \begin{cases} -4\pi, & x \in G, \\ -2\pi, & x \in S, \\ 0 & x \in R^3 \setminus \bar{G}, \end{cases} \tag{18.10}$$

where the angle φ_{xy} was defined at the beginning of Sec. 18.

2. Assuming the boundary S of a region $G \subset R^2$ to be a Lyapunov curve, prove that

$$V_2^{(1)}(x) = \int_S \frac{\cos \varphi_{xy}}{|x-y|} dS_y = \begin{cases} -2\pi, & x \in G, \\ -\pi, & x \in S, \\ 0, & x \in R^2 \setminus \bar{G}. \end{cases} \quad (18.10_1)$$

18.34. Prove the following propositions:

1. The substitution $u = v + V_3$, with

$$V_3(x) = \frac{1}{4\pi} \int_G \frac{f(y)}{|x-y|} dy, \quad x \in R^3,$$

reduces the interior boundary value problems for Poisson's equation $\nabla^2 u = -f$ to respective interior boundary value problems for Laplace's equation, provided $f \in C^1(\bar{G}) \cap C(\bar{G})$.

2. The same is true of exterior problems, with the additional provision that f is finite.

18.35. By employing the double layer potential solve the Dirichlet problem for Laplace's equation inside and outside a circle.

18.36. Find the steady-state temperature distribution inside and outside an infinitely long cylinder of radius R if the cylinder's surface is kept at the following temperature u_0:

1. $u_0 = \text{const}$. 2. $u_0 = \sin \varphi$. 3. $u_0 = \cos \varphi$.
4. $u_0 = C = \text{const}$ at $-\pi/2 < \varphi < \pi/2$, and $u_0 = 0$ at $\pi/2 \leqslant \varphi \leqslant 3\pi/2$.

18.37. Find the steady-state temperature distribution inside an infinitely long, round cylinder $0 \leqslant r < R$ under the conditions that the cylinder emits heat with a density $f(r, \varphi)$ and that the boundary $r = R$ is kept at a temperature $u_0^-(R, \varphi)$, for the following f and u_0^-:

1. $f = f_0 = \text{const}$, $u_0^- = 0$. 2. $f = r$, $u_0^- = 0$. 3. $f = r^2$, $u_0^- = a$. 4. $f = e^{-r}$, $u_0^- = \sin \varphi$. 5. $f = \sin r$, $u_0^- = \cos \varphi$. 6. $f = \sin \varphi$, $u_0^- = \sin\left(\varphi + \frac{\pi}{4}\right)$. 7. $f = \cos \varphi$, $u_0^- = \cos\left(\varphi - \frac{\pi}{4}\right)$.

18.38. By employing the simple layer potential solve the Neumann problem for Laplace's equation inside and outside a circle.

18.39. Find the density $U(r, \varphi, z)$ of a substance diffusing in a steady-state process inside and outside an infinitely long cylinder of radius R under the conditions that no sources of substance are involved, that the diffusion coefficient D is constant, and that a given diffusion flux u_1 is maintained at the boundary, for the following u_1:

1. $u_1 = \text{const}$. 2. $u_1 = \sin \varphi$. 3. $u_1 = \cos \varphi$.

18.40. Find the steady-state temperature distribution inside an infinitely long cylinder of radius R under the conditions that the cylinder emits heat with a density $f(r, \varphi)$ and that a given heat flux $u_1^-(R, \varphi)$ is maintained at the surface, for the following f and u_1^-:

1. $f=f_0=\text{const}$, $u_1^-=-\frac{f_0R}{2k}$.
2. $f=r$, $u_1^-=-\frac{R^3}{3}$; thermal conductivity $k=1$.
3. $f=\frac{1}{1+r^2}$, $u_1^-=\frac{\ln(1+R^2)}{R}$; $k=1$.
4. $f=\sin\varphi$, $u_1^-=\sin\varphi$; $k=1$.
5. $f=\cos\varphi$, $u_1^-=\cos\varphi$; $k=1$.

18.41. By employing the double and simple layer potentials find the steady-state temperature at the points in the half-plane $y>0$ if
1. The boundary $y=0$ is kept at a given temperature $u_0(x)$.
2. A given heat flux is maintained at $y=0$, that is,

$$\left.\frac{\partial u}{\partial \mathbf{n}}\right|_{y=0}=u_1(x).$$

No heat sources are present.

18.42. Find the distribution of the electrostatic field potential inside a dihedral angle under the conditions that the angle's boundary is charged to a potential $V_0=\text{const}$, for the following cases:
1. $x>0$, $y>0$, $-\infty<z<\infty$.
2. $0<\varphi<\varphi_0$, $\varphi_0<\frac{\pi}{2}$, $0\leqslant r<\infty$.

18.43. By employing the double layer potential solve the Dirichlet problem for Laplace's equation inside and outside the ball $|x|<R$.

18.44. Find the steady-state temperature distribution in the ball $r<R$ under the conditions that the ball emits heat with a density f and that the boundary $r=R$ is kept at a temperature u_0^-, for the following f and u_0^-:
1. $f=f_0=\text{const}$, $u_0^-=0$. 2. $f=r$, $u_0^-=a$.
3. $f=\sqrt{r}$, $u_0^-=\frac{2}{7}R^{5/2}$, $k=1$.

18.45. Prove that the solution to the interior Neumann problem for Laplace's equation for a ball $r<R$ is given by the formula

$$U(r,\theta,\varphi)=-R\int_0^r u(\rho,\theta,\varphi)\frac{d\rho}{\rho},$$

where u is given by Poisson's integral formula for a ball, i.e.

$$u(\rho,\theta,\varphi)=\frac{R}{4\pi}\int_0^{2\pi}\int_0^{\pi}u_0^-(R,\theta_1,\varphi_1)\times\frac{R^2-\rho^2}{(R^2+\rho^2-2R\rho\cos\gamma)^{3/2}}\sin\theta_1\,d\theta_1\,d\varphi_1,$$

with γ the angle between the radius vectors of the points (ρ,θ,φ) and (R,θ_1,φ_1) and $u_0^-=\left.\frac{\partial U}{\partial \mathbf{n}}\right|_{r=R}=u|_{\rho=R}$

Hint. Prove that if $u(\rho, \theta, \varphi)$, $u(0) = 0$, is a harmonic function in a region containing the origin of coordinates, then the function $U(r, \theta, \varphi) = -R \int_0^r u(\rho, \theta, \varphi) \frac{d\rho}{\rho}$ is harmonic, too. Then employ the solvability condition, or $\int_{r=R} u_0^- dS = 0$.

18.46. Prove that the solution to the exterior Neumann problem for Laplace's equation for a ball is given by the formula

$$U(r, \theta, \varphi) = R \int_\infty^r u(\rho, \theta, \varphi) \frac{d\rho}{\rho},$$

where $u(\rho, \theta, \varphi)$ is the solution to the exterior Dirichlet problem for a ball, that is,

$$u(\rho, \theta, \varphi) = \frac{R}{4\pi} \int_0^{2\pi} \int_0^{\pi} u_0^+ (R, \theta_1, \varphi_1) \frac{\rho^2 - R^2}{(\rho^2 + R^2 - 2\rho R \cos\gamma)^{3/2}} \sin\theta_1 \, d\theta_1 \, d\varphi_1;$$

$$u_0^+ = \frac{\partial U}{\partial n}\Big|_{r=R} = u|_{\rho=R}.$$

Hint. See the hint to Problem 18.45.

18.47. Solve the interior and exterior Neumann problems for Laplace's equation for $r < R$ at $u_0^- = u_0^+ = a = \text{const}$.

18.48. By employing surface potentials solve the Dirichlet and Neumann problems for Laplace's equation for the half-space $x_3 > 0$.

18.49. Find the density $u(x_1, x_2, x_3)$ of a substance diffusing in a steady-state process under the conditions that no sources are involved and that the diffusion coefficient D is constant, for the following regions G and boundary conditions $u|_S$:

1. $x_3 > 0$, $u|_{x_3=0} = u_0 = \text{const}$.
2. $x_3 > 0$, $u|_{x_3=0} = \begin{cases} -1, & x_1 < 0, \\ +1, & x_1 > 0. \end{cases}$
3. $x_2, x_3 > 0$, $-\infty < x_1 < \infty$, $u|_S = u_0 = \text{const}$.

Boundary value problems for Helmholtz's equation $\nabla^2 u + k^2 u = -f(x)$ and $\nabla^2 u - k^2 u = -f(x)$ in space are formulated in the same manner as those for Poisson's equation. The solutions to the exterior problems must satisfy Sommerfeld's radiation condition (Eq. (18.9)) at infinity for the equation $\nabla^2 u + k^2 u = -f$ and vanish for $\nabla^2 u - k^2 u = -f$.

18.50. Solve the Dirichlet problem for the equation $\nabla^2 u + k^2 u = 0$ inside and outside of the sphere $|x| = R$ with the boundary condition $u|_{|x|=R} = a$.

18.51. Solve the Neumann problem for the equation $\nabla^2 u + k^2 u = 0$ inside and outside of the sphere $|x| = R$ with the boundary condition $\left.\frac{\partial u}{\partial \mathbf{n}}\right|_{|x|=R} = a$.

18.52. Solve the boundary value problem

$$\nabla^2 u + k^2 u = -f(x), \quad u|_{|x|=R} = u_0^-(x)$$

inside the sphere $|x| = R$ for the following f and u_0^-:

1. $f = f_0 = \text{const}, \ u_0^- = 0; \ k = R = 1$.
2. $f = 1, \ u_0^- = \sqrt{2}\, e^{i(1-\pi/4)} \sin 1 - 1; \ k = R = 1$.

18.53. Solve the Dirichlet problem for the equation $\nabla^2 u - k^2 u = 0$ inside and outside of the sphere $|x| = R$ with the boundary condition $u|_{|x|=R} = a$.

18.54. Solve the Dirichlet problem for the equation $\nabla^2 u - k^2 u = 0$ inside and outside of the sphere $|x| = R$ with the boundary condition $u|_{|x|=R} = a \cos \theta, \ 0 \leqslant \theta \leqslant \pi$.

18.55. Solve the Neumann problem for the equation $\nabla^2 u - k^2 u = 0$ inside and outside of the sphere $|x| = R$ with the boundary condition $\left.\frac{\partial u}{\partial \mathbf{n}}\right|_{|x|=R} = a$.

18.56. Solve the boundary value problem

$$\nabla^2 u - k^2 u = -f(x), \quad u|_{|x|=R} = u_0^-(x)$$

inside the sphere $|x| = R$ for the following f and u_0^-:

1. $f = f_0 = \text{const}, \ u_0^- = 0, \ k = R = 1$.
2. $f = 1, \ u_0^- = 1 - 2e^- \sinh 1, \ k = R = 1$.

18.57. Find the steady-state concentration distribution in a non-equilibrium gas placed inside an infinitely long cylinder of radius R if a constant concentration u_0 is maintained at the cylinder's surface.

Answers to Problems of Sec. 18

18.3. *Solution.* In view of (8.7) and the definition of a simple layer (see Sec. 6) we have

$$\begin{aligned}(V_3^{(0)}, \varphi) &= \left(\frac{1}{|\xi|} \cdot \mu(y)\,\delta_S(y), \ \eta(y), \ \varphi(\xi + y)\right) \\ &= \left(\frac{1}{|\xi|}, \ (\mu(y)\,\delta_S(y), \ \eta(y)\,\varphi(\xi + y))\right) \\ &= \int_{R^3} \frac{1}{|\xi|} \left(\int_S \mu(y)\,\varphi(\xi + y)\, dS_y\right) d\xi \\ &= \int_{R^3} \left(\int_S \frac{\mu(y)}{|x - y|}\, dS_y\right) \varphi(x)\, dx.\end{aligned}$$

18.5.1. In view of (18.5) we find that $V_3 = 4\pi R, \ |x| \leqslant R; \ 4\pi R^2/|x|, \ |x| \geqslant R$. 2. $-2\pi \ln R, \ |x| \leqslant R; \ -2\pi \ln |x|, \ |x| \geqslant R$.

18.6. *Hint.* Employ formula (18.3) and introduce spherical coordinates. 1. $\frac{4\pi}{|x|}\int_0^R \rho(r)r^2\,dr$, $|x|\geqslant R$; $\frac{4\pi}{x}\int_0^{|x|}\rho(r)r^2\,dr+4\pi\int_{|x|}^R \rho(r)r\,dr$, $|x|\leqslant R$. 2. $\frac{4\pi R^3\rho_0}{3|x|}$, $|x|\geqslant R$, $2\pi R^2\rho_0-\frac{2}{3}\pi|x|^2\rho_0$, $|x|\leqslant R$. 3. $\frac{\pi R^4}{|x|}$, $|x|\geqslant R$; $\frac{\pi}{3}(4R^3-|x|^3)$, $|x|\leqslant R$. 4. $\frac{4\pi R^5}{5|x|}$, $|x|\geqslant R$; $\pi\left(R^4-\frac{|x|^4}{5}\right)$, $|x|\leqslant R$. 5. $\frac{8\pi R^{7/2}}{7|x|}$; $|x|\geqslant R$; $\frac{8\pi}{35}(7R^{5/2}-2|x|^{5/2})$, $|x|\leqslant R$. 6. $\frac{4\pi}{|x|}[2-e^{-R}(2+2R+R^2)]$, $|x|\geqslant R$; $4\pi\left[\frac{2}{|x|}(1-e^{-|x|})-e^{-R}(1+R)-e^{-|x|}\right]$, $|x|\leqslant R$. 7. $\frac{4\pi}{|x|}(R-\arctan R)$, $|x|\geqslant R$; $4\pi\left(1-\frac{\arctan|x|}{|x|}+\ln\sqrt{\frac{1+R^2}{1+|x|^2}}\right)$, $|x|\leqslant R$. 8. $\frac{4\pi}{|x|}\times[(2-R^2)\cos R-2(1-R\sin R)]$, $|x|\geqslant R$, $4\pi\left[\frac{2}{|x|}(\cos|x|-1)+\sin|x|+\sin R-R\cos R\right]$, $|x|\leqslant R$. 9. $\frac{4\pi}{|x|}[2R\cos R+(R^2-2)\sin R]$, $|x|\geqslant R$; $4\pi\left(\cos|x|-\frac{2\sin|x|}{|x|}+R\sin R+\cos R\right)$, $|x|\leqslant R$. 10. $\frac{2\pi R^3}{9|x|}(12\ln 2-5)$, $|x|\geqslant R$; $\frac{\pi}{3}\left[\left(\frac{2R^3}{|x|}+3R^2-|x|^2\right)\times\ln\left(1+\frac{|x|}{R}\right)^2+\frac{5}{3}|x|^2+2|x|(R-3)-R^2\right]$, $|x|\leqslant R$.

18.7. 1. $2\pi(R_2^2-R_1^2)\rho_0$, $|x|\leqslant R_1$; $2\pi R_2^2\rho_0-\frac{2}{3}\pi\rho_0\left(|x|^2+\frac{2R_1^3}{|x|}\right)$, $R_1\leqslant|x|\leqslant R_2$; $\frac{4\pi\rho_0}{3|x|}(R_2^3-R_1^3)$, $|x|\geqslant R_2$. 2. $4\pi\int_{R_1}^{R_2}\rho(r)r\,dr$, $|x|\leqslant R_1$, $\frac{4\pi}{x}\int_{R_1}^{|x|}\rho(r)r^2\,dr+4\pi\int_{|x|}^{R_2}\rho(r)r\,dr$, $R_1\leqslant|x|\leqslant R_2$; $\frac{4\pi}{|x|}\int_{R_1}^{R_2}\rho(r)r^2\,dr$, $|x|\geqslant R_2$.

18.8. 1. $\frac{4}{15}\pi R^4C\left(\frac{R}{r}+\frac{2}{7}\frac{R^3}{r^3}\right)$, $r\geqslant R$; $2\pi C\left(\frac{R^4}{6}+\frac{2}{15}R^2r^2-\frac{9}{70}r^4\right)$, $r\leqslant R$; C is the proportionality factor. 2. $\frac{\pi R^4}{3r^2}$, $r\geqslant R$; $\frac{4}{3}\pi Rr-\pi r^2$, $r\leqslant R$. 3. 0. 4. $\frac{2R^3}{3r}\int_0^{2\pi}\rho(\varphi)\,d\varphi$, $r\geqslant R$; $\left(R^2-\frac{r^2}{3}\right)\times\int_0^{2\pi}\rho(\varphi)\,d\varphi$, $r\leqslant R$.

18.9. $\pi[(H-x_3)\sqrt{R^2+(H-x_3)^2}+x_3\sqrt{R^2+x_3^2}+H^2-2Hx_3+R^2\ln(H-x_3+\sqrt{R^2+(H-x_3)^2})-R^2\ln(-x_3+\sqrt{R^2+x_3^2})]$.

18.10. 1. $\int_0^R\int_0^{2\pi}\rho(r_1)\ln\frac{1}{\sqrt{r^2+r_1^2-2rr_1\cos(\varphi_1-\varphi)}}\,r_1\,dr_1\,d\varphi_1$.

2. $-\pi R^2\rho_0\ln r$, $r\geqslant R$; $-\pi\rho_0\left(R^2\ln R-\frac{R^2-r^2}{2}\right)$, $r\leqslant R$. *Solution.* Suppose $r\geqslant R$. $V_2(r,\varphi)=\rho_0\int_0^R r_1\,dr_1\int_0^{2\pi}\left[\ln\frac{1}{r}+\right.$ $\left.+\ln\frac{1}{\sqrt{1+\left(\frac{r_1}{r}\right)^2-2\frac{r_1}{r}\cos(\varphi_1-\varphi)}}\right]d\varphi_1=-\pi\rho_0R^2\ln r$, since $\int_0^{2\pi}\ln[1+\lambda^2-2\lambda\cos(\varphi_1-\varphi)]\,d\varphi=\int_0^{2\pi}\left[\int_0^{\lambda}\frac{2\lambda-2\cos(\varphi_1-\varphi)}{1+\lambda^2-2\lambda\cos(\varphi_1-\varphi)}\,d\lambda\right]\times$ $\times\,d\varphi=\int_0^{2\pi}\left[\int_0^{\lambda}-\frac{2}{\lambda}\operatorname{Re}\frac{\lambda e^{i(\varphi_1-\varphi)}}{1-\lambda e^{i(\varphi_1-\varphi)}}\,d\lambda\right]d\varphi=-2\int_0^{2\pi}\left[\sum_{n=1}^{\infty}\frac{\lambda^n}{n}\cos n(\varphi_1-\right.$ $\left.-\varphi)\right]d\varphi_1=0$, where $\lambda=\frac{r_1}{r}<1$. 3. $-\frac{2}{3}\pi R^2\ln r$, $r\geqslant R$; $\frac{2\pi}{9}[R^3(1-3\ln R)-r^3]$, $r\leqslant R$. 4. $-\frac{\pi}{2}R^4\ln r$, $r\geqslant R$. $\frac{\pi}{8}[R^4(1-4\ln R)-r^4]$, $r\leqslant R$. 5. $-2\pi[1-(1+R)e^{-R}]\ln r$, $r\geqslant R$; $-2\pi\left[e^{-r}-e^{-R}+\ln r-(1+R)e^{-R}\ln R+\int_r^R\frac{e^{-r_1}}{r_1}dr_1\right]$, $r\leqslant R$.

6. $-2\pi\ln r\ln\sqrt{1+R^2}$, $r\geqslant R$; $-2\pi\left[\ln R\ln\sqrt{1+R^2}-\frac{1}{2}\times\right.$ $\left.\times\int_r^R\frac{\ln(1+r_1^2)}{r_1}dr_1\right]$, $r\leqslant R$. 7. $-\frac{4}{5}\pi R^{5/2}\ln r$, $r\geqslant R$; $-\frac{4}{5}\pi\times$ $\times\left[R^{5/2}\ln R+\frac{2}{5}(r^{5/2}-R^{5/2})\right]$, $r\leqslant R$. 8. $2\pi(R\cos R-\sin R)\ln r$, $r\geqslant R$; $2\pi\left(R\ln R\cos R-\ln R\sin R+\sin r-\sin R+\int_r^R\frac{\sin r_1}{r_1}dr_1\right)$, $r\leqslant R$. 9. $2\pi[\ln r(1-R\sin R-\cos R)]$, $r\geqslant R$; $2\pi\left[\ln r-\ln R\times\right.$ $\left.\times(R\sin R+\cos R)+\cos r-\cos R+\int_r^R\frac{\cos r_1}{r_1}dr_1\right]$, $r\leqslant R$.

10. $\frac{\pi R^3\sin\varphi}{3r}$, $r\geqslant R$; $\pi\left(rR-\frac{2r^2}{3}\right)\sin\varphi$, $r\leqslant R$. 11. $\frac{\pi R^3\cos\varphi}{3r}$, $r\geqslant R$; $\pi\left(rR-\frac{2r^2}{3}\cos\varphi\right)$, $r\leqslant R$. 12. $-\frac{R^2}{2}\ln r\int_0^{2\pi}\rho(\varphi)\,d\varphi$, $r\geqslant R$; $\left(\frac{R^2-r^2}{4}-\frac{R^2}{2}\ln R\right)\int_0^{2\pi}\rho(\varphi)\,d\varphi$, $r\leqslant R$.

18.11. *Hint.* See the solution to 18.10.2. 1. $\pi\rho_0(R_1^2-R_2^2)\ln r$, $r\geqslant R_2$; $\pi\rho_0\left(R_1^2\ln r-R_2^2\ln R_2+\frac{R_2^2-r^2}{2}\right)$, $R_1\leqslant r\leqslant R_2$; $\pi\rho_0\times\left(R_1^2\ln R_1-R_2^2\ln R_2+\frac{R_2^2-R_1^2}{2}\right)$, $r\leqslant R_1$. 2. $-2\pi\ln r\times\int\limits_{R_1}^{R_2}\rho(x)\,dx$, $r\geqslant R_2$; $-2\pi\left(\ln r\int\limits_{R_1}^{r}\rho(x)\,x\,dx+\int\limits_{r}^{R_2}\rho(x)\,x\ln x\,dx\right)$, $R_1\leqslant r\leqslant R_2$; $-2\pi\int\limits_{R_1}^{R_2}\rho(x)\,x\ln x\,dx$, $r\leqslant R_1$.

18.16. *Hint.* Employ (18.5). $4\pi\mu_0R^2/|x|$, $|x|\geqslant R$; $4\pi\mu_0R$, $|x|\leqslant R$.

18.17. 1. $\frac{4\pi R^2C}{3r}\left(1+\frac{2R^2}{5r^2}\right)$, $r\geqslant R$; $\frac{4}{3}\pi RC\left(1+\frac{2r^2}{5R^2}\right)$, $r\leqslant R$, C is the proportionality factor. 2. $\frac{\pi R}{r}\left(r+R-\frac{(r-R)^2}{2\sqrt{rR}}\times\ln\frac{\sqrt{r}+\sqrt{R}}{\sqrt{r}-\sqrt{R}}\right)$, $r\geqslant R$; $\frac{\pi R}{r}\left(r+R-\frac{(r-R)^2}{2\sqrt{rR}}\ln\frac{\sqrt{R}+\sqrt{r}}{\sqrt{R}-\sqrt{r}}\right)$, $r\leqslant R$. 3. $\frac{2R^2}{r}(e^{\pi}-1)$, $r\geqslant R$; $2R(e^{\pi}-1)$, $r\leqslant R$.

18.18. 1. $2\pi\mu_0(\sqrt{x_3^2+R^2}-x_3)$. 2. $\pi R\sqrt{x_3^2+R^2}-\pi x_3^2\ln\frac{R+\sqrt{x_3^2+R^2}}{|x_3|}$. 3. $\frac{4\pi}{3}\left[x_3^3+\left(\frac{R^2}{2}-x_3^2\right)\sqrt{x_3^2+R^2}\right]$. 4 $(\sqrt{x_3^2+R^2}-x_3)\int\limits_0^{2\pi}\mu(\varphi)\,d\varphi$.

18.19. 1. $2\pi R\mu_0\ln\frac{H-x_3+\sqrt{R^2+(H-x_3)^2}}{-x_3+\sqrt{R^2+x_3^2}}$. 2. $R[\ln(H-x_3+\sqrt{R^2+(H-x_3)^2})-\ln(-x_3+\sqrt{R^2+x_3^2})]\int\limits_0^{2\pi}\mu(\varphi)\,d\varphi$.

18.20. *Hint.* Employ (18.5). 0, $|x|>R$; $-4\pi\nu_0$, $|x|<R$; $-2\pi\nu_0$, $|x|=R$.

18.21. 1. $\frac{4\pi R^2}{3r^2}$, $r>R$; $-\frac{8\pi r}{3R}$, $r<R$, $-\frac{2\pi}{3}$, $r=R$. 2. $\frac{\pi}{2r}\left[R-3r+(R+3r)\left(\sqrt{\frac{r}{R}}-\sqrt{\frac{R}{r}}\right)\ln\frac{\sqrt{r}+\sqrt{R}}{\sqrt{r}-\sqrt{R}}\right]$, $r>R$; $\frac{\pi}{2r}\left[R-3r+(R+3r)\left(\sqrt{\frac{r}{R}}-\sqrt{\frac{R}{r}}\right)\ln\frac{\sqrt{R}+\sqrt{r}}{\sqrt{R}-\sqrt{r}}\right]$, $r<R$. 3. 0, $r>R$; $-4(e^{\pi}-1)$, $r<R$; $-2(e^{\pi}-1)$, $r=R$. 4. 0, $r>R$; $-2\int\limits_0^{2\pi}\nu(\varphi)\,d\varphi$, $r<R$; $-\int\limits_0^{2\pi}\nu(\varphi)\,d\varphi$, $r=R$. 5. $\frac{16\pi R^5}{15r^3}$, $r>R$; $-\frac{4\pi R^2}{3}\left(1+\frac{6r^2}{5R^2}\right)$, $r<R$; -2π, $r=R$.

18.22. 1. $2\pi\nu_0 x_3\left(\frac{1}{\sqrt{R^2+x_3^2}}-\frac{1}{|x_3|}\right)$, $x_3\neq 0$. 2. $-2\pi x_3\int\limits_0^R\frac{\nu(r)\,r\,dr}{(x_3^2+r^2)^{3/2}}$, $x_3\neq 0$. 3. $x_3\left(\frac{1}{\sqrt{R^2+x_3^2}}-\frac{1}{|x_3|}\right)\int\limits_0^{2\pi}\nu(\varphi)\,d\varphi$, $x_3\neq 0$.

4. $\pi x_3\left(\frac{\pi+2R}{\sqrt{R^2+x_3^2}}-\frac{\pi}{|x_3|}-2\ln\frac{R+\sqrt{R^2+x_3^2}}{|x_3|}\right)$, $x_3\neq 0$.

18.23. *Hint*. Employ (18.6). 1. $-2\pi R\mu_0\ln R^2$, $r\leqslant R$, $-2\pi R\mu_0\left(\ln R^2+\ln\frac{r}{R}\right)$, $r\geqslant R$. 2. $-2\pi\ln 2+\frac{\pi}{8}r^2\cos 2\varphi$, $r\leqslant 2$, $-2\pi\ln r+\frac{2\pi}{r^2}\cos 2\varphi$, $r\geqslant 2$.

18.24. 1. 0, $r>R$; $-\pi\nu_0$, $r=R$; $-2\pi\nu_0$, $r<R$, 2. $V_2^{(1)}(r,\varphi)=$

$$=\begin{cases}\frac{r^2-R^2}{2Rr}\left[-\pi\sin\varphi+\frac{R^2+r^2}{r^2-R^2}\arctan\left(\frac{R+r}{r-R}\,2\cot\varphi\right)\right], & r>R;\\ \frac{r^2-R^2}{2Rr}\left[-\pi\sin\varphi+\frac{R^2+r^2}{R^2-r^2}\arctan\left(\frac{R+r}{R-r}\,2\cot\varphi\right)\right], & r<R;\\ 0, & r=R.\end{cases}$$

18.25. 1. $\mu_0\left[2a-y\arctan\frac{2ay}{x^2+y^2-a^2}-\frac{(a+x)}{2}\ln((a+x^2)+y^2)-\right.$ $\left.-\frac{(a-x)}{2}\ln((a-x)^2+y^2)\right]$. 2. $\mu_0\left[\frac{a+x}{2}\ln((a+x)^2+y^2)-\right.$ $\left.-\frac{a-x}{2}\ln((a-x)^2+y^2)-x\ln(x^2+y^2)+\frac{1}{y}\arctan\frac{2x(a^2-x^2)}{y(x^2+y^2-a^2)}\right]$.

3. $\frac{a^2-x^2+y^2}{4}\ln\frac{(a+x)^2+y^2}{(a-x)^2+y^2}-xy\arctan\frac{2ay}{x^2+y^2-a^2}$.

18.26. 1. $-\nu_0\left[\arctan\frac{a-x}{y}+\arctan\frac{a+x}{y}\right]$, $y\neq 0$; 0 at $y=0$. $\lim V_2^{(1)}=\mp\nu_0\pi$, $y\to\pm 0$, $-a<x<a$. 2. $-\nu_0\left[2\arctan\frac{x}{y}-\right.$ $\left.-\arctan\frac{a+x}{y}+\arctan\frac{a-x}{y}\right]$, $y\neq 0$; 0, $y=0$; $\lim V_2^{(1)}=\mp\nu_0\pi$, $y\to\pm 0$, $0<x<a$; $\lim V_2^{(1)}=\pm\nu_0\pi$, $y\to\pm 0$, $-a<x<0$.
3. $-x\left[\arctan\frac{a-x}{y}+\arctan\frac{a+x}{y}\right]+\frac{y}{2}\ln\frac{(a+x)^2+y^2}{(a-x)^2+y^2}$, $y\neq 0$, 0 at $y=0$; $\lim V_2^{(1)}(x,y)=\mp x\pi$, $y\to\pm 0$, $-a<x<a$.
4. $(y^2-x^2)\left(\arctan\frac{a-x}{y}+\arctan\frac{a+x}{y}\right)+xy\ln\frac{(a+x)^2+y^2}{(a-x)^2+y^2}$, $y\neq 0$; 0 at $y=0$; $\lim V_2^{(1)}(x,y)=\mp x^2\pi$, $y\to\pm 0$, $-a<x<a$.

18.28. 1. $\frac{4\pi}{k|x|}e^{ik|x|}\int\limits_0^R r\rho(r)\sin kr\,dr$, $|x|\geqslant R$.

$\frac{4\pi}{k|x|}\left(e^{ik|x|}\int\limits_0^{|x|}r\rho(r)\sin kr\,dr+\sin k|x|\int\limits_{|x|}^R r\rho(r)e^{ikr}\,dr\right)$, $|x|\leqslant R$.

2. $\frac{4\pi\rho_0}{k^2|x|}e^{ik|x|}\left(-R\cos kR+\frac{\sin kR}{k}\right)$, $|x|\geqslant R$; $\frac{4\pi\rho_0}{k^2|x|}\times$
$\times\left[\sin k|x|\left(-iR+\frac{1}{k}\right)e^{ikR}-|x|\right]$, $|x|\leqslant R$. 3. $\frac{4\pi\rho_0}{k|x|}e^{ik|x}\times$
$\times\left\{-\frac{Re^{-R}}{k^2+1}(\sin k+k\cos k)+\frac{1}{(1+k^2)^2}[2k(1-e^{-R}\cos k)-(1-k^2)\times\right.$
$\left.\times e^{-R}\sin k]\right\}$, $|x|\geqslant R$; $-\frac{2\pi\rho_0 i}{|x|}[e^{-1}\cos(1-|x|)-2e^{-1}\sin(1-|x|)+$
$+ie^{i|x|}-\sqrt{5}e^{i(|x|+1+i+\arctan 2)}]$, $|x|\leqslant 1$, $k=R=1$.

18.29. $\frac{4\pi\rho_0}{k^2|x|}e^{ik|x|}\left(R_1\cos kR_1-R_2\cos kR_2+\frac{\sin kR_2-\sin kR_1}{k}\right)$,
$|x|\geqslant R_2$; $\frac{4\pi\rho_0}{k^2|x|}\sin k|x|\left[-iR_2e^{ikR_2}+iR_1e^{ikR_1}+\frac{1}{k}(e^{ikR_2}-e^{ikR_1})\right]$,
$|x|\leqslant R_1$; $\frac{4\pi\rho_0}{k^2|x|}\left[e^{ik|x|}\left(R_1\cos kR_1-|x|\cos k|x|-\frac{\sin kR_1}{k}+\right.\right.$
$\left.\left.+i|x|\sin k|x|\right)+e^{ikR_2}\left(\frac{\sin k|x|}{k}-iR_2\sin k|x|\right)\right]$, $R_1\leqslant|x|\leqslant R_2$.

18.30. 1. $\frac{4\pi R\mu_0}{k|x|}e^{ik|x|}\sin kR$, $|x|\geqslant R$; $\frac{4\pi R\mu_0}{k|x|}e^{ikR}\sin k|x|$, $x\leqslant R$.
2. $\frac{4\pi\nu_0}{R}e^{ik|x|}\left(R\cos kR-\frac{1}{k}\sin kR\right)$, $|x|>R$;
$\frac{4\pi\nu_0}{R}e^{ikR}\left(iR\sin k|x|-\frac{1}{k}\sin k|x|\right)$, $|x|<R$;
$\frac{4\pi\nu_0}{R}e^{ikR}\left(\frac{iR}{2}\sin kR+\frac{R}{2}\cos kR-\frac{1}{k}\sin kR\right)$, $|x|=R$.

18.31. 1. $\frac{4\pi e^{-k|x|}}{k|x|}\int_0^R r\rho(r)\sinh kr\,dr$, $|x|\geqslant R$;

$\frac{4\pi}{k|x|}\left(e^{-k|x|}\int_0^{|x|}r\rho(r)\sinh kr\,dr+\sinh k|x|\int_{|x|}^R r\rho(r)e^{-kr}\,dr\right)$, $|x|\leqslant R$.
2. $\frac{4\pi\rho_0}{k^2|x|}e^{-k|x|}\left(R\cosh kR-\frac{1}{k}\sinh kR\right)$, $|x|\geqslant R$;
$\frac{4\pi\rho_0}{k^2|x|}\left[|x|-\left(R+\frac{1}{k}\right)e^{-kR}\sinh k|x|\right]$, $|x|\leqslant R$.
3. $\frac{4\pi}{k|x|}\left[\frac{R}{k^2-1}e^{-(R+k|x|)}(k\cosh kR+\sinh kR)+\frac{\sinh R}{(k+1)^2}e^{-k(R+|x|)}\right]$,
$|x|\geqslant R$, $k\neq-1$.

18.32. 1. $\frac{4\pi R\mu_0}{k|x|}e^{-k|x|}\sinh kR$, $|x|\geqslant R$; $\frac{4\pi R\mu_0}{k|x|}e^{-kR}\sinh k|x|$, $|x|\leqslant R$.
2. $\frac{4\pi\nu_0}{|x|}e^{-k|x|}\left(R\cosh kR-\frac{1}{k}\sinh kR\right)$, $|x|>R$;
$\frac{4\pi\nu_0}{2R}e^{-kR}\left[R\cosh kR-\left(R+\frac{2}{k}\right)\sinh kR\right]$, $|x|=R$;
$-\frac{4\pi\nu_0}{|x|}e^{-kR}\left(R+\frac{1}{k}\right)\sinh k|x|$, $|x|<R$.

18.35. *Hint.* Employ (18.1), (18.10$_1$), and (8.4).

$$\frac{1}{2\pi R}\int\limits_{|y|=R} u_0^-(y)\frac{R^2-|x|^2}{|x-y|^2}\,dS_y,\quad |x|<R;$$

$$\frac{1}{2\pi R}\int\limits_{|y|=R} u_0^+(y)\frac{|x^2|-R^2}{|x-y|^2}\,dS_y,\quad |x|>R.$$

18.36. *Hint.* Employ the result of Problem 18.35.

1. u_0, $r\leqslant R$; u_0, $r\geqslant R$. 2. $\frac{r}{R}\sin\varphi$, $r\leqslant R$; $\frac{R}{r}\sin\varphi$, $r\geqslant R$. 3. $\frac{r}{R}\cos\varphi$, $r\leqslant R$; $\frac{R}{r}\cos\varphi$, $r\geqslant R$. 4. $\frac{c}{2}\left(1+\frac{2}{\pi}\arctan\frac{Rr\cos\varphi}{R^2-r^2}\right)$, $r\leqslant R$; $\frac{c}{2}\left(1+\frac{2}{\pi}\arctan\frac{Rr\cos\varphi}{r^2-R^2}\right)$, $r\geqslant R$.

18.37. 1. *Solution.* The problem $\nabla^2 u(x)=-f_0/k$, $|x|<R$; $u|_{|x|=R}=u_0^-=0$, where $x=(x_1, x_2)$ and k is the thermal conductivity is reduced, via the substitution $u=v+V_2$, where $V_2(x)=\frac{1}{2\pi k}\int\limits_{|y|\leqslant R} f_0\ln\frac{1}{\sqrt{(x_1-y_1)^2+(x_2-y_2)^2}}\,dy_1\,dy_2$, to the problem $\nabla^2 v(x)=0$, $|x|<R$; $v|_{|x|=R}=(u-V_2)_{|x|=R}$. In view of the result of Problem 18.11.2 we have $V_2(r,\varphi)=\frac{f_0}{2k}\left(\frac{R^2-r^2}{2}-R^2\ln R\right)$, where (r,φ) are the polar coordinates of point x. Then the formula in the answer to Problem 18.35 implies that $V(r,\varphi)=\frac{f_0}{2k}R^2\ln R$. Thus, $u(r,\varphi)=v+V_2=\frac{f_0}{4k}(R^2-r^2)$. 2. $\frac{R^3-r^3}{9k}$. 3. $a+\frac{R^4-r^4}{16k}$. 4. $\frac{r}{R}\sin\varphi+\frac{1}{k}\left(e^{-R}-e^{-r}+\ln R-\ln r-\int\limits_r^R\frac{e^{-\rho}}{\rho}\,d\rho\right)$. 5. $\frac{r}{R}\cos\varphi+\frac{1}{k}\left(\sin r-\sin R+\int\limits_r^R\frac{\sin\rho}{\rho}\,d\rho\right)$. 6. $\frac{r}{R}\sin\left(\varphi+\frac{\pi}{4}\right)+\left(\frac{rR}{2k}-\frac{r^2}{3k}\right)\sin\varphi$. 7. $\frac{r}{R}\cos\left(\varphi-\frac{\pi}{4}\right)+\left(\frac{rR}{2k}-\frac{r^2}{3k}\right)\cos\varphi$.

18.38. *Hint.* Look for the solution in the form of a simple layer potential (see (18.6)). Then employ (18.2) and the solvability condition $\int\limits_{r=R} u_1^-(y)\,dS_y=0$. $\frac{1}{\pi}\int\limits_{|y|=R} u_1^-(y)\ln\frac{1}{|x-y|}\,dS_y+\text{const}$, $|x|\leqslant R$; $x=(x_1, x_2)$; $\frac{1}{\pi}\int\limits_{|y|=R} u_1^+(y)\ln|x-y|\,dS_y+\text{const}$, $|x|\geqslant R$.

18.39. *Hint.* Employ the formulas in the answer to Problem 18.38.

1. The problem has no solution since $\int\limits_{r=R} u_1\,dS\neq 0$. 2. $r\sin\varphi+$

$+\text{const},\ r<R;\ -\frac{R^2}{r}\sin\varphi+\text{const},\quad r>R.\quad$ 3. $r\cos\varphi+\text{const},$ $r<R;\ -\frac{R^2}{r}\cos\varphi+\text{const},\ r>R.$

18.40. *Hint.* The problem $\nabla^2 u=-f/k,\ r\leqslant R,\ \left.\frac{\partial u}{\partial \mathbf{n}}\right|_{r=R}=u_1^-$ is reduced, via the substitution $u=v+V_2$ (see the solution to Problem 18.37), to the boundary value problem $\nabla^2 v=0,\ r<R,\ \left.\frac{\partial v}{\partial \mathbf{n}}\right|_{r=R}=$ $=\left.\frac{\partial(u-V_2)}{\partial \mathbf{n}}\right|_{r=R}.$

1. $\frac{f_0}{2k}\left(\frac{R^2-r^2}{2}-R^2\ln R\right)+\text{const}.$

2. $\frac{1}{9}(R^3-r^3-3R^2\ln R)+\text{const}.$

3. $\ln R\ln\sqrt{1+R^2}-\frac{1}{2}\int\limits_r^R\frac{\ln(1+\rho^2)}{\rho}\,d\rho+\text{const}.$ 4. $\left(r+\frac{2}{3}rR-\frac{r^2}{3}\right)\sin\varphi+\text{const}.$ 5. $\left(r+\frac{2}{3}rR-\frac{r^2}{3}\right)\cos\varphi+\text{const}.$

18.41. 1. $\frac{y}{\pi}\int\limits_{-\infty}^{\infty}\frac{u_0(\xi)\,d\xi}{(x-\xi)^2+y^2}.\qquad$ 2. $\frac{1}{\pi}\int\limits_{-\infty}^{\infty}u_1(\xi)\ln\frac{1}{\sqrt{(x-\xi)^2+y^2}}\,d\xi.$

18.42. 1. $\frac{2v_0}{\pi}\left[\arctan\frac{x}{y}+\arctan\frac{y}{x}\right].\qquad$ 2. $\frac{y}{x}=\tan\varphi_0$: $\frac{v_0}{2\pi}\left(\pi+\frac{\sin\varphi_0}{x}+\arctan\frac{x}{y}\right)$; $\frac{y}{x}<\tan\varphi_0$: $\frac{v_0}{\pi}\arctan\frac{(y^2-x^2)\sin\varphi_0+2xy\cos\varphi_0}{(y^2-x^2)\cos\varphi_0-2xy\sin\varphi_0}=$ $=\frac{v_0}{\pi}F(x,\ y,\ \varphi_0)$; $\frac{y}{x}>\tan\varphi_0$: $\frac{v_0}{\pi}(\pi+F(x,\ y,\ \varphi_0)).$

18.43. $\frac{1}{4\pi R}\int\limits_{|y|=R}\frac{R^2-|x|^2}{|x-y|^3}u_0^-(y)\,dS_y,\quad |x|<R;$

$\frac{1}{4\pi R}\int\limits_{|y|=R}\frac{|x|^2-R^2}{|x-y|^3}u_0^+(y)\,dS_y,\quad |x|>R.$

18.44. See the hints in the answer to Problem 18.37 and the results of Problem 18.6.

1. $\frac{f_0}{6k}(R^2-r^2)$. 2. $a+\frac{R^3-r^3}{12k}$. 3. 0.

18.47. *Hint.* Employ the results of Problems 18.45 and 18.46. $-R^2a/r,\ r>R$; in the region $r<R$ the problem has no solution.

18.48. $\frac{x_3}{2\pi}\int\limits_{y_3=0}\frac{u_0(y)}{|x-y|^3}\,dS_y;\ \frac{1}{2\pi}\int\limits_{y_3=0}\frac{u_1(y)}{|x-y|}\,dS_y.$

18.49. 1. u_0. 2. $\frac{2}{\pi}\arctan\frac{x_1}{x_3}$. 3. $\frac{u_0}{\pi}\left(\frac{\pi}{2}+\arctan\frac{x_2}{x_3}+\arctan\frac{x_3}{x_2}\right).$

18.50. $\frac{aR}{|x|}\frac{\sin k|x|}{\sin kR}$, $|x|\leqslant R$; $\frac{aR}{|x|}\frac{e^{ik|x|}}{e^{ikR}}$, $|x|\geqslant R$. *Hint.* Look for the solution in the form of a double layer potential,

$$u(x)=V^{(1)}(x)=\int\limits_{r=R}\nu(y)\frac{\partial}{\partial \mathbf{n}_y}\frac{e^{ik|x-y|}}{|x-y|}dS_y. \qquad (18.11)$$

The density can be found from the following integral equations:

$$u|_{r=R}=V^{(1)}_{\mp}(x)=\mp 2\pi\nu(x)+\int\limits_{r=R}\nu(y)\frac{\partial}{\partial \mathbf{n}_y}\frac{e^{ik|x-y|}}{|x-y|}dS_y=a,$$
$$x\in\{r=R\}.$$

The answer is as follows: $\nu(x)=\frac{akR}{4\pi(kR+i)\sin kR}$ for the interior problem and $\nu(x)=\frac{ae^{-ikR}}{4\pi\left(\cos kR-\frac{1}{kR}\sin kR\right)}$ for the exterior problem.

18.51. *Hint.* Look for the solution in the form of a simple layer potential.

$$\frac{aR^2}{|x|}\frac{\sin k|x|}{(kR\cos kR-\sin kR)},\quad |x|\leqslant R;$$
$$\frac{aR^2}{|x|}\frac{e^{ik|x|}}{(ikR-1)},\quad |x|\geqslant R.$$

18.52. See the hints in the solution to Problem 18.37 and the result of Problem 18.28.2.

1. $\frac{f_0}{|x|}\left(\frac{\sin|x|}{\sin 1}-|x|\right)$.

2. $\sqrt{2}e^{i\left(1-\frac{\pi}{4}\right)}\frac{\sin|x|}{|x|}-1$.

18.53. See the hint in the answer to Problem 18.50.

$$\frac{aR}{|x|}\frac{\sinh k|x|}{\sinh kR},\quad |x|\leqslant R;$$
$$\frac{aR}{|x|}\frac{e^{-k|x|}}{e^{-kR}},\quad |x|\geqslant R.$$

18.54. $a\left(\frac{R}{|x|}\right)^2\frac{k|x|\cosh k|x|-\sinh k|x|}{kR\cosh kR-\sinh kR}\cos\theta,\quad |x|\leqslant R;$

$a\left(\frac{R}{|x|}\right)^3\frac{k|x|+1}{kR+1}\frac{\cosh k|x|-\sinh k|x|}{\cosh kR-\sinh kR}\cos\theta,\quad x\geqslant R.$

18.55. $\frac{aR^2}{|x|}\frac{\sinh k|x|}{kR\cosh kR-\sinh kR}$, $|x|\leqslant R$; $-\frac{aR^2}{|x|}\frac{e^{k(R-|x|)}}{1+kR}$, $|x|\geqslant R$.

18.56. 1. $f_0\left(1-\frac{\sinh|x|}{|x|\sinh 1}\right)$. 2. $1-2e^{-1}\frac{\sinh|x|}{|x|}$.

18.57. $u(x,y)=u_0\frac{J_0(kr)}{J_0(kR)}$. *Hint.* u is the solution to the problem $\nabla^2u-k^2u=0$, $r<R$, $u|_{r=R}=u_0$.

19 Variational Methods

Suppose Poisson's equation

$$\nabla^2 u = -f \tag{19.1}$$

is specified in a bounded region $Q \subset R^n$ and one of the following boundary conditions on the smooth boundary Γ of G:

$$u|_\Gamma = g, \tag{19.2}$$

$$\left.\frac{\partial u}{\partial n}\right|_\Gamma = g, \tag{19.3}$$

$$\left.\left(\frac{\partial u}{\partial n} + \sigma u\right)\right|_\Gamma = g, \tag{19.4}$$

with the function $f(x)$ specified in Q and the functions $g(x)$ and $\sigma(x)$ on Γ; here $\sigma \in C(\Gamma)$. A function $u(x) \in C^2(Q) \cap C(\bar{Q})$ is said to be the *classical solution to the boundary value problem* (19.1), (19.2) if it satisfies Eq. (19.1) in Q and the boundary condition (19.2) on Γ. A function $u(x) \in C^2(Q) \cap C^1(\bar{Q})$ is said to be the *classical solution to the boundary value problem* (19.1), (19.3) (*or* (19.4)) if it satisfies Eq. (19.1) in Q and the boundary condition (19.3) (or (19.4)) on Γ. While looking for the generalized solutions of these boundary value problems, we will assume that $f \in L_2(Q)$ and $g(x) \in L_2(\Gamma)$, where in the case involving the boundary condition (19.2) the function $g(x)$ is, in addition, a trace on Γ of a function belonging to $H^1(Q)$ (in particular, $g \in C^1(\Gamma)$). A function $u \in H^1(Q)$ is said to be the *generalized solution to the boundary value problem* (19.1), (19.2) if its trace on Γ is g and if

$$\int_Q (\operatorname{grad} u \cdot \operatorname{grad} v)\, dx = \int_Q fv\, dx \tag{19.5}$$

for all $v \in \mathring{H}^1(Q)$. A function $u \in H^1(Q)$ is said to be the *generalized solution to the boundary value problem* (19.1), (19.3) (*or* (19.4)) if

$$\int_Q (\operatorname{grad} u \cdot \operatorname{grad} v)\, dx + \int_\Gamma \sigma uv\, dS = \int_Q fv\, dx + \int_\Gamma gv\, dS \tag{19.6}$$

for all $v \in H^1(Q)$. 1248

If the functions f, g, and σ are sufficiently smooth (e.g. continuously differentiable), the generalized solutions are the classical solutions of the respective boundary value problems.

The following theorem is important in investigating the generalized solutions of boundary value problems:

F. Riesz theorem *Suppose a bounded linear function $l(u)$ is specified over a Hilbert space H. Then there exists a unique element $h \in H$ such that $l(u) = (h, u)$ (here (h, u) denotes the scalar product of elements h and u in H).*

19.1. Suppose $u(x)$ is the classical solution of the boundary value problem (19.1), (19.2). Show that if $u \in C^1(Q)$ and $f \in L_2(Q)$, then $u(x)$ is the generalized solution to (19.1), (19.2).

19.2. Let $u(x)$ be the classical solution of the boundary value problem (19.1), (19.3) (or (19.4)). Show that $u(x)$ is also the generalized solution of the same boundary problem.

19.3. Show that if $u(x)$ is the generalized solution to the boundary value problem (19.1), (19.2) and belongs to $C^2(Q) \cap C(\bar{Q})$, then $u(x)$ is the classical solution to this problem.

19.4. Show that if $u(x)$ is the generalized solution to the boundary value problem (19.1), (19.3) (or (19.4)) and belongs to $C^2(Q) \cap C^1(\bar{Q})$, then $u(x)$ is the classical solution to this problem.

19.5. Employing Steklov's inequality (see Problem 4.104), prove the existence and uniqueness of the generalized solution to problem (19.1), (19.2) at $g = 0$.

19.6. Show that if g is the trace on Γ of a function belonging to $H^1(Q)$ (for instance, $g \in C^1(\Gamma)$), then the generalized solution to problem (19.1), (19.2) exists and is unique.

19.7. Suppose in a region G the following elliptic equation is given:

$$L(u) = -\text{div}\,(p \text{ grad } u) + q(x)\,u = f(x), \tag{19.7}$$

where $p \in C^1(\bar{Q})$, $\min p(x) = p_0 > 0$, $q \in C(\bar{Q})$, and $f \in L_2(Q)$. A function $u(x)$ that belongs to the space $H^1(Q)$ is called the generalized solution to problem (19.7), (19.2) if for all $v(x) \in \mathring{H}^1(Q)$ it satisfies the integral identity

$$\int_Q (p \text{ grad } u \text{ grad } v + quv)\, dx = \int_Q fv\, dx$$

and if its trace on Γ is g. Prove that the classical solution to problem (19.7), (19.2) that belongs to $H^1(Q)$ is also the generalized solution.

19.8. Prove the existence and uniqueness of the generalized solution to the boundary value problem (19.7), (19.2) for $q \geqslant 0$. *Hint.* Employ the result of Problem 4.106.

19.9. Suppose that in a region Q the following elliptic equation is given:

$$L(u) = -\sum_{i,\,j=1}^{n} \frac{\partial}{\partial x_j}\left(p_{ij}(x)\frac{\partial u}{\partial x_i}\right) + q(x)\,u = f(x), \tag{19.8}$$

where the real-valued functions p_{ij} belong to $C^1(\bar{Q})$, $p_{ij}(x) = p_{ji}(x)$, $(i, j = 1, \ldots, n)$, and that for all $x \in \bar{Q}$ and all real numbers $(\xi_1, \ldots, \xi_n)$ the following inequality holds:

$$\sum_{i,\,j=1}^{n} p_{ij}(x)\,\xi_i\xi_j \geqslant \gamma_0 |\xi|^2,$$

with γ_0 a positive constant; in Eq. (19.8) $q \in C(\bar{Q})$ and $f \in L_2(Q)$. A function $u(x)$ that belongs to the space $H^1(Q)$ is said to be a generalized solution to the boundary value problem (19.8), (19.2) if for all $v(x) \in \overset{\circ}{H}{}^1(Q)$ it satisfies the integral identity

$$\int_Q \left(\sum p_{ij}(x)\, u_{x_i} v_{x_j} + quv\right) dx = \int_Q fv\, dx$$

and if its trace on Γ is g. Prove that the classical solution to (19.8), (19.2) belonging to $H^1(Q)$ is also the generalized solution.

19.10. Prove the existence and uniqueness of the generalized solution to the boundary value problem (19.8), (19.2) for $q \geqslant 0$. *Hint.* Employ the result of Problem 4.112.

19.11. A function $u(x)$ that belongs to the space $H^1(Q)$ is said to be a generalized solution to the boundary value problem (19.7), (19.3) (or (19.4)) if for all $v(x) \in H^1(Q)$ it satisfies the integral identity

$$\int_Q (p \operatorname{grad} u \operatorname{grad} v + quv)\, dx + \int_\Gamma p\sigma uv\, ds = \int_Q fv\, dx.$$

Prove that the classical solution to (19.7), (19.3) (or (19.4)) is also the generalized solution.

19.12. Prove the existence of a unique generalized solution to the boundary value problem (19.7), (19.3) (or (19.4)), assuming that $\sigma(x) \geqslant 0$ on Γ, $q(x) \geqslant 0$ in Q, and either $\sigma(x) \not\equiv 0$ or $q(x) \not\equiv 0$. *Hint.* Employ the result of Problem 4.117.

19.13. Let $\tilde{L}_2(Q)$ and $\tilde{H}^1(Q)$ be subspaces of $L_2(Q)$ and $H^1(Q)$ consisting of those functions belonging to $L_2(Q)$ and $H^1(Q)$, respectively, for which $\int_Q f\, dx = 0$. Prove that at $g(x) \equiv 0$, $q(x) \equiv 0$, and $f \in \tilde{L}_2(Q)$ there is a unique solution to the boundary value problem (19.7), (19.3) that belongs to $\tilde{H}^1(Q)$. *Hint.* Employ the result of Problem 4.121.

Suppose $p \in C(\bar{Q})$, $q \in C(\bar{Q})$, $\sigma \in C(\Gamma)$, $\min p(x) = p_0 > 0$, $\sigma(x) \geqslant 0$, $q(x) \geqslant 0$, and either $q(x) \not\equiv 0$ or $\sigma(x) \not\equiv 0$. Then (see Problems 4.105 and 4.113) in $\overset{\circ}{H}{}^1(Q)$ and $H^1(Q)$ we can introduce scalar products (equivalent to the common one):

$$(f, g)_{\overset{\circ}{H}{}^1} = \int_Q [p(x)(\operatorname{grad} f \cdot \operatorname{grad} g) + q(x)\, fg]\, dx, \qquad (19.9)$$

$$(f, g)_{H^1} = \int_Q [p(x)(\operatorname{grad} f \cdot \operatorname{grad} g) + qfg]\, dx + \int_\Gamma p\sigma fg\, dS. \qquad (19.10)$$

A function $u \in \overset{\circ}{H}{}^1(Q)$ on which the functional

$$E(v) = \|v\|^2_{\overset{\circ}{H}{}^1} - 2(f, v)_{L_2},$$

considered for $v \in \overset{\circ}{H}{}^1(Q)$, attains its minimal value is a generalized solution to the boundary value problem (19.7), (19.2) at $g \equiv 0$ if the norm is generated by the scalar product (19.9).

A function $u \in H^1(Q)$ on which the functional

$$E(v) = \|v\|^2_{H^1} - 2(f, v)_{L_2},$$

considered for $v \in H^1(Q)$, attains its minimal value is a generalized solution to the boundary value problem (19.7), (19.4) at $g(x) = 0$ if the norm $\|v\|_{H^1}$ is generalized by the scalar product (19.10).

19.14. Let us consider, for $f \in L_2(Q)$, the functional

$$E_1(v) = \int_Q (\operatorname{grad} v)^2 \, dx - 2 \int_Q fv \, dx$$

on the set of functions $v \in H^1(Q)$ for which $v|_\Gamma = g$, where the function $g(x)$ is the trace on Γ of a function from $H^1(Q)$. Show that the function $u(x)$ for which the functional $E(v)$ attains its minimal value is the generalized solution to the boundary value problem (19.1), (19.2).

19.15. Consider, for $f \in L_2(Q)$, $p \in C(\overline{Q})$, $q \in C(\overline{Q})$, $\min p(x) = p_0 > 0$, and $q(x) \geqslant 0$, the functional

$$E_1(v) = \int_Q p\, |\operatorname{grad} v|^2 \, dx + \int_Q q(x)\, v^2 \, dx - 2 \int_Q fv \, dx$$

on the set of functions $v \in H^1(Q)$ for which $v|_\Gamma = g$, where the function $g(x)$ is the trace on Γ of a function from $H^1(Q)$. Show that the function $u(x)$ for which the functional attains its minimum is the generalized solution to the boundary value problem (19.7), (19.2).

19.16. Suppose p_{ij} $(i, j = 1, \ldots, n)$, q, and f are the functions introduced in Problem 19.9. Consider the functional

$$E_2(v) = \int_Q \Big[\sum_{i,j=1}^{n} p_{ij} v_{x_i} v_{x_j} \Big] dx + \int_Q qv^2 \, dx - 2 \int_Q fv \, dx$$

on the set of functions $v \in H^1(Q)$ for which $v|_\Gamma = g$, where the function $g(x)$ is the trace on Γ of a function from $H^1(Q)$. Show that the function $u(x)$ for which the functional attains its minimum is the generalized solution to the boundary value problem (19.8), (19.2).

19.17. Consider, for $f \in L_2(Q)$, $g(x) \in L_2(\Gamma)$, $\sigma \in C(\Gamma)$, $\sigma \geqslant 0$ on Γ, and $\sigma(x) \not\equiv 0$, the functional

$$\tilde{E}_1(v) = \int_Q |\operatorname{grad} v|^2 \, dx + \int_\Gamma \sigma v^2 \, dS - 2 \int_Q fv \, dx - 2 \int_\Gamma gv \, dS, \quad v \in H^1(Q).$$

Show that the function $u(x)$ for which the functional $\widetilde{E}_1(v)$ attains its minimum is the generalized solution to the boundary value problem (19.1), (19.4).

19.18. Let $f \in L_2(Q)$, $g(x) \in L_2(\Gamma)$, $p \in C(\overline{Q})$, $q \in C(\overline{Q})$, $\sigma \in C(\Gamma)$, $\min p(x) = p_0 > 0$, $q(x) \geqslant 0$, $\sigma(x) \geqslant 0$, and either $q(x) \not\equiv 0$ or $\sigma(x) \not\equiv 0$. Consider on $H^1(Q)$ the functional

$$E_2(v) = \int_Q p|\text{grad } v|^2\, dx + \int_Q qv^2\, dx + \int_\Gamma \sigma pv^2\, dS - 2\int_Q fv\, dx - 2\int_\Gamma pgv\, dS.$$

Show that the function $u(x)$ for which the functional attains its minimum is the generalized solution to the boundary value problem (19.7), (19.3) (or (19.4)). *Hint.* See Problem 4.117.

19.19. Consider, for $f \in L_2(Q)$, $p \in C(\overline{Q})$, and $\min p(x) = p_0 > 0$, the functional

$$E_1(v) = \int_Q (p\,|\text{grad } v|^2 + qv^2)\, dx - 2\int_Q fv\, dx$$

on the subspace $\widetilde{H}^1(Q)$ (for the definitions of sets $\widetilde{L}_2(Q)$ and $\widetilde{H}^1(Q)$ see Problem 19.13; also see Problems 4.118-4.120) of the space $H^1(Q)$. Show that the function $u \in \widetilde{H}^1(Q)$ for which this functional attains its minimum is the generalized solution to the boundary value problem (19.7), (19.3).

19.20. Find the function v_0 for which the functional $\int_0^1 (v'^2 + v^2)\, dx + 2\int_0^1 v\, dx$ attains its minimum in the class $\overset{\circ}{H}{}^1(0, 1)$.

19.21. Prove that $\int_0^1 (v'^2 + 2xv)\, dx + v^2(0) + v^2(1) \geqslant -\frac{41}{270}$ for all $v \in C^1([0, 1])$. Is there such a function for which this inequality turns into an equality?

19.22. Prove that $\int_0^1 v\, dx \leqslant \frac{5}{24} + \frac{v^2(0)}{4} + \frac{1}{4}\int_0^1 v'^2\, dx$ for all functions $v \in C^1[0, 1]$, $v(1) = 0$. Find the function belonging to this class that turns the inequality into an equality.

19.23. Find $\inf\limits_{v \in \overset{\circ}{H}_1(Q)} \left\{\int_Q [(\text{grad } v)^2 + 2\sin x_1 \sin x_2 v]\, dx\right\}$, where $Q = \{0 \leqslant x_1 \leqslant \pi, 0 \leqslant x_2 \leqslant \pi\}$.

19.24. Find $\inf\limits_{v\in \overset{\circ}{H}_1(|x|<1)} \left\{ \int\limits_{|x|<1} [(\operatorname{grad} v)^2+2|x|^2 v]\,dx \right\}$, where $x=(x_1, x_2)$.

19.25. Find $\inf\limits_{v\in H^1(|x|<1)} \int\limits_{|x|<1} |\operatorname{grad} v|^2\,dx$, where $x=(x_1, x_2)$, $x_1 = |x|\cos\varphi$, $x_2=|x|\sin\varphi$, $v|_{|x|=1}=\varphi(\pi-\varphi)(2\pi-\varphi)$.

19.26. Find $\inf \int\limits_{|x|<1} |\operatorname{grad} v|^2\,dx$ on the set of functions $v\in H^1(|x|<1)$, $x=(x_1, x_2)$, $x_1=|x|\cos\varphi$, $x_2=|x|\sin\varphi$, that satisfy the condition $v|_{|x|=1}=\varphi^2$, $-\pi<\varphi\leqslant\pi$.

19.27. May a function $\psi(\varphi)$ that is given on the circle $|x|=1$, $x_1=\cos\varphi$, $x_2=\sin\varphi$, be the boundary value of a function belonging to $H^1(|x|<1)$ if

(a) $\psi(\varphi)=\operatorname{sign}\varphi,\quad -\pi<\varphi\leqslant\pi$;

(b) $\psi(\varphi)=\sum\limits_{0}^{\infty} 2^{-n}\cos 2^{2n}\varphi$;

(c) $\psi(\varphi)=\sum\limits_{1}^{\infty} \frac{\cos n^4\varphi}{n^5}$.

19.28. Let Q be the square $\{0<x_1<1,\ 0<x_2<1\}$. Prove that

$$\int\limits_Q f^2\,dx \leqslant \frac{1}{2\pi^2}\int\limits_Q |\operatorname{grad} f|^2\,dx$$

for every f that belongs to $\overset{\circ}{H}{}^1(Q)$ and that the constant in the inequality is exact.

19.29. Let Q be the cube $\{0<x_1<1,\ 0<x_2<1,\ 0<x_3<1\}$. Prove that

$$\|f\|^2_{L_2}\leqslant\frac{1}{3\pi^2}\|\operatorname{grad} f\|^2_{L_2}$$

for every function $f\in \overset{\circ}{H}{}^1(Q)$.

19.30. Let Q be the annulus $\{1<|x|<2\}$. Find

$$\inf_{\substack{f\in H^1(Q)=0\\ f|_{|x|=1}}} \left\{ \int\limits_{1<|x|<2} [(\operatorname{grad} f)^2+4f]\,dx + \int\limits_{|x|=2} f^2\,dS \right\},\quad x=(x_1, x_2).$$

19.31. Let Q be the square $\{0<x_1<1,\ 0<x_2<1\}$. Find the function for which the functional

$$\inf_{u\in H^1}\left\{ \int\limits_Q [(\operatorname{grad} u)^2+4\sin x_1 \sin x_2 u]\,dx + 2\int\limits_0^{\pi} \sin x_1 u(x_1, \pi)\,dx_1 \right\}$$

attains its minimum in the class of functions $u\in H^1(Q)$, $u|_{x_2=0} = u|_{x_1=0}=u|_{x_1=\pi}=0$.

19.32. Let Q be the circle $\{|x| < 1\}$, $x = (x_1, x_2)$. Prove that for every function $u \in H^1(Q)$ the following inequality holds:

$$\| u \|^2_{L_2} \leqslant \frac{1}{\mu_{0,1}} \| \operatorname{grad} u \|^2_{L_2},$$

where $\mu_{0,1}$ is the smallest positive root of the function $J_0(\mu)$.

19.33. Prove that for all functions $u \in C^1 (0 < x_1 < 1, 0 < x_2 < 1)$ that satisfy the boundary conditions

$$u|_{x_1=0} = u|_{x_2=0} = 0, \quad u|_{x=1} = x_2, \quad u|_{x_2=1} = x_1,$$

the following inequality holds:

$$\int_0^1 \int_0^1 (\operatorname{grad} u)^2 \, dx_1 \, dx_2 \geqslant \frac{2}{3}.$$

Is there a function among the above-mentioned functions that turns this inequality into an equality?

19.34. Prove that for all functions $u \in \mathring{C}^1 (|x| < 1)$, $x = (x_1, x_2)$, the following inequality holds:

$$2 \int_{|x|<1} x_1 x_2 u(x) \, dx \leqslant \frac{\pi}{1152} + \int_{|x|<1} (\operatorname{grad} u)^2 \, dx.$$

19.35. Prove that for all functions $u \in \mathring{C}^1 (|x| < 1)$, $x = (x_1, x_2, x_3)$, the following inequality holds:

$$\int_{|x|<1} [(\operatorname{grad} u)^2 + u] \, dx \geqslant -\frac{\pi}{45}.$$

Is there a function among the above-mentioned functions that turns this inequality into an equality?

19.36. Show that for all functions $v \in C^1 (|x| \leqslant 1)$, $x_1 = |x| \cos \varphi$, $x_2 = |x| \sin \varphi$, that satisfy the condition $v|_{|x|=1} = \sin \varphi$, the following inequality holds:

$$\int_{|x|<1} [2|x|^2 v + (\operatorname{grad} v)^2] \, dx \geqslant \frac{63}{64} \pi.$$

Is there a function among the above-mentioned functions that turns this inequality into an equality?

19.37. Prove that for all functions $u \in C^1 (0 < x_1 < 1, 0 < x_2 < 1, 0 < x_3 < 1)$, $x = (x_1, x_2, x_3)$ that satisfy the boundary conditions

$$u|_{x_1=0} = x_2 x_3, \quad u|_{x_2=0} = x_1 x_3, \quad u|_{x_3=0} = x_1 x_2,$$
$$u|_{x_1=1} = x_2 + x_3 + x_2 x_3, \quad u|_{x_2=1} = x_1 + x_3 + x_1 x_3,$$
$$u|_{x_3=1} = x_1 + x_2 + x_1 x_2$$

the following inequality holds:

$$\int_0^1\int_0^1\int_0^1 |\operatorname{grad} u|^2\, dx_1\, dx_2\, dx_3 \geqslant \frac{7}{2}.$$

Is there a function among the above-mentioned functions that turns this inequality into an equality?

19.38. Show that for all functions $v \in C^1$ $(|x| \leqslant 1)$, $x = (x_1, x_2, x_3)$ that satisfy the condition $v|_{|x|=1} = \cos\theta$ the following inequality holds:

$$\int_{|x|<1} [2v + (\operatorname{grad} v)^2]\, dx \geqslant \frac{14\pi}{15}.$$

Is there a function among the above-mentioned functions that turns this inequality into an equality?

19.39. Let Q be the square $\{0 < x_1 < 1,\ 0 < x_2 < 1\}$. Prove that for every function $v \in \mathring{H}^1(Q)$ that satisfies the condition

$$\int_Q \sin \pi x_1 \sin \pi x_2 v(x)\, dx = 0$$

the following inequality holds:

$$\|v\|_{L_2}^2 \leqslant \frac{1}{5\pi^2} \|\operatorname{grad} v\|_{L_2}^2.$$

19.40. Let Q be the cube $\{0 < x_1 < 1,\ 0 < x_2 < 1,\ 0 < x_3 < 1\}$. Prove that for every function $v \in \mathring{H}^1(Q)$ that satisfies the condition

$$\int_Q \sin \pi x_1 \sin \pi x_2 \sin \pi x_3 v(x)\, dx = 0$$

the following inequality holds:

$$\|v\|_{L_2}^2 \leqslant \frac{1}{6\pi^2} \|\operatorname{grad} v\|_{L_2}^2.$$

19.41. Let Q be the cube $\{0 < x_1 < \pi,\ 0 < x_2 < \pi,\ 0 < x_3 < \pi\}$. Among the functions $u \in H^1(Q)$ that take on the boundary values

$$u|_{x_1=0} = u|_{x_2=0} = u|_{x_3=0} = u|_{x_1=\pi} = u|_{x_2=\pi} = 0$$

find the one for which the functional

$$E(u) = \int_Q (\operatorname{grad} u)^2\, dx + \int_0^\pi\int_0^\pi \sin x_1 \sin x_2 u(x_1, x_2, \pi)\, dx_1\, dx_2$$

attains its minimum.

19.42. Suppose Q is the spherical layer $\{1 < |x| < 2\}$, $x = (x_1, x_2, x_3)$. Among the functions $u \in H^1(Q)$ that take on the boundary value $u|_{|x|=2}=0$ find the one for which the functional

$$E(u)=\int_Q [(\operatorname{grad} u)^2+2u]\,dx+\int_{|x|=1} u^2\,dS$$

attains its minimum.

Answers to Problems of Sec. 19

19.20. $-1+\frac{2\sqrt{e}}{e+1}\cosh\left(x-\frac{1}{2}\right)$. **19.21.** The function is $\frac{x^3}{6}-\frac{2}{9}(x+1)$. **19.22.** $-x^2+\frac{x+1}{2}$. **19.23.** $\frac{-\pi^2}{8}$. **19.24.** $-\frac{\pi}{64}$.

19.25. $144\pi\sum_1^\infty k^{-5}$. **19.26.** $16\pi\sum_1^\infty k^{-3}$. **19.27.** (a) No, (b) no, (c) yes. **19.30.** $\frac{\pi}{2(1+\ln 4)}(51-94\ln 2)$. **19.31.** $-\sin x_1\sin x_2-2\frac{\sin x_1\cosh x_2}{\sinh\pi}$. **19.33.** Yes. **19.35.** Yes. **19.36.** The function is $\rho\sin\varphi+\frac{\rho^4-1}{16}$. **19.37.** The function is $x_1x_2+x_1x_3+x_2x_3$.

19.38. The function is $r\cos\theta+\frac{r^2-1}{6}$.

19.41. $-\frac{1}{\sqrt{2}\cosh(\sqrt{2}\pi)}\sin x_1\sin x_2\sinh(\sqrt{2}x_3)$.

19.42. $\frac{|x|^2}{6}+\frac{5}{9|x|}-\frac{17}{18}$.

Chapter VI

Mixed Problems

20 Fourier's Methods

We start with hyperbolic equations. What follows is a brief survey of Fourier's method of variable separation. Let us take a string whose ends are fixed. The problem of string vibrations in this case is given by the equation

$$\frac{\partial^2 u}{\partial t^2} = a^2 \frac{\partial^2 u}{\partial x^2} \tag{20.1}$$

with the initial conditions

$$u|_{t=0} = u_0(x), \quad \left.\frac{\partial u}{\partial t}\right|_{t=0} = u_1(x) \tag{20.2}$$

and boundary conditions

$$u|_{x=0} = 0, \quad u|_{x=l} = 0. \tag{20.3}$$

We will first look for particular solutions of Eq. (20.1) that are not identically equal to zero and satisfy conditions (20.3) in the form

$$u(x, t) = X(x)\,T(t). \tag{20.4}$$

Substituting (20.4) into (20.1) yields

$$T''(t) + a^2\lambda T(t) = 0, \tag{20.5}$$

$$X''(x) + \lambda X(x) = 0, \tag{20.6}$$

with $\lambda = \text{const}$, where to obtain nontrivial solutions (i.e. solutions that are not identically equal to zero) of the form (20.4) we must find the nontrivial solutions that satisfy the following conditions

$$X(0) = 0, \quad X(l) = 0. \tag{20.7}$$

We have thus arrived at the Sturm-Liouville problem (20.6), (20.7) (see the theory at the beginning of Chap. V).

The eigenvalues of this problem are the numbers

$$\lambda_k = \left(\frac{\pi k}{l}\right)^2 \ (k = 1, 2, \ldots)$$

(and only these numbers), and the corresponding (normalized) eigenfunctions are

$$X_k(x) = \sqrt{\frac{2}{l}} \sin \frac{\pi k x}{l}.$$

At $\lambda = \lambda_k$ the general solution of Eq. (20.5) is

$$T_k(t) = a_k \cos \frac{k\pi at}{l} + b_k \sin \frac{k\pi at}{l},$$

whence the function

$$u_k(x, t) = X_k(x) T_k(t) = \left(a_k \cos \frac{k\pi at}{l} + b_k \sin \frac{k\pi at}{l}\right) \sin \frac{k\pi x}{l}$$

satisfies Eq. (20.1) and the boundary conditions (20.3) for all values of a_k and b_k.

We look for the solution to Eq. (20.1) that satisfies conditions (20.2)-(20.3) in the form of a series:

$$u(x, t) = \sum_{k=1}^{\infty} \left(a_k \cos \frac{k\pi at}{l} + b_k \sin \frac{k\pi at}{l}\right) \sin \frac{k\pi x}{l}. \tag{20.8}$$

If this series is uniformly convergent and it can be differentiated twice, then its sum will satisfy Eq. (20.1) and the boundary conditions (20.3).

If we define the constants a_k and b_k in such a way that the sum (20.8) satisfies, in addition, the initial conditions (20.2), we arrive at the following series:

$$u_0(x) = \sum_{k=1}^{\infty} a_k \sin \frac{k\pi x}{l}, \tag{20.9}$$

$$u_1(x) = \sum_{k=1}^{\infty} \frac{k\pi a}{l} b_k \sin \frac{k\pi x}{l}; \tag{20.10}$$

these two formulas give the expansions of $u_0(x)$ and $u_1(x)$ in Fourier sine series in the interval $(0, l)$. The expansion coefficients are given by the well-known formulas

$$a_k = \frac{2}{l} \int_0^l u_0(x) \sin \frac{k\pi x}{l} dx.$$

$$b_k = \frac{2}{k\pi a} \int_0^l u_1(x) \sin \frac{k\pi x}{l} dx.$$

In Problems 20.1 and 20.2 use the Fourier method to find the string vibrations under the assumption that no external forces are present.

20.1. Solve the problem of the vibrations of a string $0 < x < l$ whose ends are fixed, provided the initial velocities of the points of the string are zero and the initial deflection u_0 of these points has the following form:

(1) a sinusoid $u_0(x) = A \sin(\pi nx/l)$, with n an integer;

(2) a parabola whose symmetry axis is the straight line $x = l/2$ and whose vertex is the point $M(l/2, h)$;

(3) a broken line OAB, with $O(0, 0)$, $A(c, h)$, $B(l, 0)$, $0 < c < l$. Consider the case with $c = l/2$.

20.2. Solve the problem of the vibrations of a string $0 < x < l$ whose ends are fixed, provided at time zero the string is at rest ($u_0 = 0$) and the initial velocity u_1 is

1. $u_1(x) = v_0 = \text{const}, \ x \in [0, l]$.

2. $u_1(x) = \begin{cases} v_0 \text{ if } x \in [\alpha, \beta], \\ 0 \text{ if } x \,\overline{\in}\, [\alpha, \beta], \end{cases}$ where $0 \leqslant \alpha < \beta \leqslant l$.

3. $u_1(x) = \begin{cases} A \cos \dfrac{\pi (x - x_0)}{2\alpha} & \text{if } x \in [x_0 - \alpha, \ x_0 + \alpha], \\ 0 & \text{if } x \,\overline{\in}\, [x_0 - \alpha, \ x_0 + \alpha], \end{cases}$

where $0 \leqslant x_0 - \alpha < x_0 + \alpha \leqslant l$.

Equation (20.1) describes free longitudinal vibrations of a rod. In Problems 20.3 and 20.4 find the longitudinal vibrations of a rod by employing Fourier's method.

20.3. Solve the problem of the longitudinal vibrations of a homogeneous rod under arbitrary initial conditions in each of the following cases:

(1) One end of the rod ($x = 0$) is rigidly fixed and the other ($x = l$) is free.

(2) Both ends are free.

(3) One end of the rod ($x = l$) is elastically fixed and the other ($x = 0$) is free.

20.4. Find the longitudinal vibrations of a rod whose one end ($x = 0$) is rigidly fixed and the other ($x = l$) is under a force P (at time $t = 0$ the force ceases to exist).

20.5. Find the current $i(x, t)$ in a conductor of length l in which there is an alternating current. Assume that the initial current in the conductor (at $t = 0$) is zero and the initial voltage is given by the formula $v|_{t=0} = E_0 \sin(\pi x/2l)$. The left end of the conductor ($x = 0$) is isolated, while the right end ($x = l$) is grounded.

The problem of finding the forced vibrations of a homogeneous string $0 < x < l$ that is rigidly fixed at its ends and with the external force having a density p can be reduced to solving the equation

$$\frac{\partial^2 u}{\partial t^2} = a^2 \frac{\partial^2 u}{\partial x^2} + g(x, t) \qquad (20.11)$$

($g = p/\rho$, where ρ is the linear density of the string) with the boundary conditions (20.3) and the initial conditions (20.2).

We look for the solution of the problem (20.11), (20.2), (20.3) in the form of a sum,

$$u = v + w,$$

where v is a solution of the nonhomogeneous equation (20.11) that satisfies the boundary conditions (20.3) and the initial conditions

$$v|_{t=0} = 0, \quad \left. \frac{\partial v}{\partial t} \right|_{t=0} = 0,$$

while w is a solution of the homogeneous equation (20.1) that satisfies the boundary conditions (20.3) and the initial conditions (20.2).

Solution v represents the forced vibrations of the string (the vibrations are due to the external force in the absence of initial perturbations), while w represents the free vibrations of the string (which are due to initial perturbations).

We look for v in the form of a series:

$$v(x, t) = \sum_{k=1}^{\infty} T_k(t) \sin \frac{k\pi x}{l}, \tag{20.12}$$

which is an expansion in the eigenfunctions of (20.6), (20.7).

Substituting (20.12) into (20.11) yields

$$\sum_{k=1}^{\infty} \left[T_k''(t) + \left(\frac{k\pi a}{l}\right)^2 T_k(t) \right] \sin \frac{k\pi x}{l} = g(x, t). \tag{20.13}$$

Expanding $g(x, t)$ in the interval $(0, l)$ in a Fourier sine series,

$$g(x, t) = \sum_{k=1}^{\infty} g_k(t) \sin \frac{k\pi x}{l}, \tag{20.14}$$

and comparing (20.13) with (20.14), we arrive at the differential equation

$$T_k''(t) + \left(\frac{k\pi a}{k}\right)^2 T_k(t) = g_k(t), \tag{20.15}$$

where $g_k(t) = \frac{2}{l} \int_0^l g(\xi, t) \sin \frac{k\pi \xi}{l} \, d\xi$ $(k = 1, 2, \ldots)$. Solving Eqs. (20.15) with the zero initial conditions

$$T_k(0) = 0, \quad T_k'(0) = 0 \quad (k = 1, 2, \ldots), \tag{20.16}$$

we find $T_k(t)$, and then determine v via (20.12). Note that the solutions $T_k(t)$ of Eqs. (20.15) combined with (20.16) can be represented in the form

$$T_k(t) = \frac{2}{k\pi a} \int_0^t \left[\int_0^l g(\xi, \tau) \sin \frac{k\pi a}{l} (t - \tau) \sin \frac{k\pi \xi}{l} \, d\xi \right] d\tau. \tag{20.17}$$

The solution of the problem (20.11), (20.2), (20.3) can be written as

$$u(x, t) = \sum_{k=1}^{\infty} T_k(t) \sin \frac{k\pi x}{l} + \sum_{k=1}^{\infty} \left(a_k \cos \frac{k\pi a t}{l} + b_k \sin \frac{k\pi a t}{l} \right) \sin \frac{k\pi x}{l},$$

where the $T_k(t)$ are determined via (20.17) and the coefficients a_k and b_k via the formulas

$$a_k = \frac{2}{l} \int_0^l u_0(x) \sin \frac{k\pi x}{l}\, dx,$$

$$b_k = \frac{2}{k\pi a} \int_0^l u_1(x) \sin \frac{k\pi x}{l}\, dx.$$

20.6. Employ Fourier's method to solve the following mixed problems:

1. $u_{tt} = u_{xx} + 2b$ $(b = \text{const},\quad 0 < x < l)$, $u|_{x=0} = 0$, $u|_{x=l} = 0$; $u|_{t=0} = u_t|_{t=0} = 0$.
2. $u_{tt} = u_{xx} + \cos t$ $(0 < x < \pi)$, $u|_{x=0} = u|_{x=\pi} = 0$, $u|_{t=0} = u_t|_{t=0} = 0$.

20.7. Solve the problem of the vibrations of a homogeneous string $0 < x < l$ whose ends $x = 0$ and $x = l$ are fixed, with an external continuously distributed force of a density $p(x, t) = A\rho \sin \omega t$, $\omega \neq k\pi a/l$ $(k = 1, 2, \ldots)$, acting on it. The initial conditions are zero.

20.8. Solve the problem of the longitudinal vibrations under the force of gravity of a rod hung by one of its ends $(x = 0)$, while the other end $(x = l)$ is free.

The problem of forced vibrations of a string of finite length under an external force in the case where the ends of the string move according to a definite law can be reduced to solving Eq. (20.11) with the boundary conditions

$$u|_{x=0} = \mu_1(t), \quad u|_{x=l} = \mu_2(t) \tag{20.18}$$

and the initial conditions (20.2). We look for the solution of problem (20.11), (20.2), (20.18) in the form

$$u = v + w,$$

where $w = \mu_1(t) + (x/l)(\mu_2(t) - (\mu_1(t))$ is a function that satisfies the given boundary conditions (20.18).

Then the function $v(x, t)$ satisfies the zero boundary conditions $v|_{x=0} = v|_{x=l} = 0$, the equation $v_{tt} - a^2 v_{xx} = g_1$, with $g_1(x, t) = g(x, t) - (w_{tt} - a^2 w_{xx})$, and the following initial conditions:

$$v|_{t=0} = u_0(x) - w|_{t=0},$$
$$v_t|_{t=0} = u_1(x) - w_t|_{t=0}. \tag{20.19}$$

We have thus arrived at a problem of the type (20.11), (20.2), (20.3) for v.

Remark. It is sometimes possible to find a function v that satisfies the nonhomogeneous equation (20.11) and the given boundary conditions (20.18). Then, looking for the solution of problem (20.11),

(20.2), (20.18) in the form $u = v + w$, we can find that w satisfies the homogeneous equation (20.1) and zero boundary and initial conditions (20.19).

20.9. Solve the following mixed problems:

1. $u_{xx} = u_{tt}$, $0 < x < l$; $u|_{x=0} = 0$, $u|_{x=l} = t$; $u|_{t=0} = u_t|_{t=0} = 0$.

2. $u_{xx} = u_{tt}$, $0 < x < 1$; $u|_{x=0} = t+1$, $u|_{x=1} = t^3 + 2$; $u|_{t=0} = x+1$, $\left.\frac{\partial u}{\partial t}\right|_{t=0} = 0$.

20.10. Solve the problem of forced transverse vibrations of a string with one end ($x = 0$) fixed and the other ($x = l$) under a force that results in a deflection $A \sin \omega t$, with $\omega \neq k\pi a/l$ ($k = 1, 2, \ldots$). At time $t = 0$ the deflections and velocities are zero.

20.11. Suppose a rod of length l with one end ($x = 0$) rigidly fixed is in the state of rest. At time $t = 0$ a force $Q = \text{const}$ is applied to the other end ($x = l$) and acts along the rod. Find the deflection $u(x, t)$ of the rod.

20.12. Solve the problem of the longitudinal vibrations of a homogeneous cylindrical rod one end of which is fixed and the other is under a force $Q = A \sin \omega t$ directed along the rod's axis ($\omega \neq a\pi(2k+1)/2l$, $k = 0, 1, 2, \ldots$).

20.13. Solve the problem of the free vibrations of a homogeneous string of length l fixed at both ends and vibrating in a medium whose resistance to the vibrations is proportional to the first power of velocity. The initial conditions are assumed to be zero.

20.14. Solve the following mixed problems:

1. $u_{tt} = u_{xx} - 4u$ $(0 < x < 1)$; $u|_{x=0} = u|_{x=1} = 0$; $u|_{t=0} = x^2 - x$, $u_t|_{t=0} = 0$.

2. $u_{tt} + 2u_t = u_{xx} - u$ $(0 < x < \pi)$; $u|_{x=0} = u|_{x=\pi} = 0$; $u|_{t=0} = \pi x - x^2$, $u_t|_{t=0} = 0$.

3. $u_{tt} + 2u_t = u_{xx} - u$ $(0 < x < \pi)$; $u_x|_{x=0} = 0$, $u|_{x=\pi} = 0$, $u|_{t=0} = 0$, $u_t|_{t=0} = x$.

4. $u_{tt} + u_t = u_{xx}$ $(0 < x < 1)$; $u|_{x=0} = t$, $u|_{x=1} = 0$; $u|_{t=0} = 0$, $u_t|_{t=0} = 1 - x$.

5. $u_{tt} = u_{xx} + u$ $(0 < x < 2)$; $u|_{x=0} = 2t$, $u|_{x=2} = 0$; $u|_{t=0} = u_t|_{t=0} = 0$.

6. $u_{tt} = u_{xx} + u$ $(0 < x < l)$; $u|_{x=0} = 0$, $u|_{x=l} = t$; $u|_{t=0} = 0$, $u_t|_{t=0} = \frac{x}{l}$.

20.15. Solve the following mixed problems:

1. $u_{tt} = u_{xx} + x$ $(0 < x < \pi)$; $u|_{x=0} = u|_{x=\pi} = 0$; $u|_{t=0} = \sin 2x$, $u_t|_{t=0} = 0$.

2. $u_{tt} + u_t = u_{xx} + 1$ $(0 < x < 1)$; $u|_{x=0} = u|_{x=1} = 0$; $u|_{t=0} = u_t|_{t=0} = 0$.

20.16. Solve the following mixed problems:

1. $u_{tt}-u_{xx}+2u_t=4x+8e^t\cos x\ (0<x<\pi/2)$; $u_x|_{x=0}=2t$, $u\big|_{x=\frac{\pi}{2}}=\pi t$; $u|_{t=0}=\cos x$, $u_t|_{t=0}=2x$.

2. $u_{tt}-u_{xx}-2u_t=4t(\sin x-x)\ (0<x<\pi/2)$, $u|_{x=0}=3$, $u_x|_{x=\pi/2}=$ $=t^2+t$, $u|_{t=0}=3$, $u_t|_{t=0}=x+\sin x$.

3. $u_{tt}-3u_t=u_{xx}+u-x(4+t)+\cos\frac{3x}{2}\ (0<x<\pi)$; $u_x|_{x=0}=$ $=t+1$, $u|_{x=\pi}=\pi(t+1)$; $u|_{t=0}=u_t|_{t=0}=x$.

4. $u_{tt}-7u_t=u_{xx}+2u_x-2t-7x-e^{-x}\sin 3x\ (0<x<\pi)$; $u|_{x=0}=0$, $u|_{x=\pi}=\pi t$; $u|_{t=0}=0$, $u_t|_{t=0}=x$.

5. $u_{tt}+2u_t=u_{xx}+8u+2x(1-4t)+\cos 3x\ (0<x<\pi/2)$; $u_x|_{x=0}=t$, $u\big|_{x=\frac{\pi}{2}}=\frac{\pi t}{2}$; $u|_{t=0}=0$, $u_t|_{t=0}=x$.

6. $u_{tt}=u_{xx}+4u+2\sin^2 x\ (0<x<\pi)$; $u_x|_{x=0}=u_x|_{x=\pi}=0$; $u|_{t=0}=u_t|_{t=0}=0$.

7. $u_{tt}=u_{xx}+10u+2\sin 2x\cos x\ (0<x<\pi/2)$; $u|_{x=0}=$ $=u_x|_{x=\pi/2}=0$; $u|_{t=0}=u_t|_{t=0}=0$.

8. $u_{tt}-3u_t=u_{xx}+2u_x-3x-2t\ (0<x<\pi)$; $u|_{x=0}=0$, $u|_{x=\pi}=$ $=\pi t$; $u|_{t=0}=e^{-x}\sin x$, $u_t|_{t=0}=x$.

In Problems 20.17-20.20 the reader is advised to employ Fourier's method in studying the vibrations of a membrane. The problem of the vibrations of a homogeneous membrane can be reduced to solving the equation $u_{tt}=a^2\nabla^2 u+f$ with certain initial and boundary conditions (see Example 3 in Sec. 1).

In particular, the problem of the free vibrations of a rectangular membrane $(0<x<p,\ 0<y<q)$ fixed at its edge can be reduced to solving the wave equation

$$\frac{\partial^2 u}{\partial t^2}=a^2\left(\frac{\partial^2 u}{\partial x^2}+\frac{\partial^2 u}{\partial y^2}\right)$$

with the boundary conditions

$$u|_{x=0}=u|_{x=p}=u|_{y=0}=u|_{y=q}=0$$

and the initial conditions

$$u|_{t=0}=u_0(x,\ y),\ \frac{\partial u}{\partial t}\bigg|_{t=0}=u_1(x,\ y).$$

20.17. Solve the problem of the free vibrations of a square membrane $(0<x<p,\ 0<y<p)$ fixed along its edge if $u|_{t=0}=$ $=A\sin\frac{\pi x}{p}\sin\frac{\pi y}{p}$, $\frac{\partial u}{\partial t}\Big|_{t=0}=0$.

20.18. Solve the following mixed problem:

$u_{tt}=\nabla^2 u\quad (0<x<\pi,\ 0<y<\pi)$,

$u|_{x=0}=u|_{x=\pi}=u|_{y=0}=u|_{y=\pi}=0$,

$u|_{t=0}=3\sin x\sin 2y,\quad u_t|_{t=0}=5\sin 3x\sin 4y$.

20.19. Solve the problem of the free vibrations of a rectangular membrane $(0 < x < p,\ 0 < y < q)$ fixed along its edge if $u|_{t=0} = Axy(x-p)(y-q)$, $\left.\frac{\partial u}{\partial t}\right|_{t=0} = 0$.

The problem of the free vibrations of a round membrane of radius R fixed at its edge can be reduced to the equation

$$\frac{1}{a^2}\frac{\partial^2 u}{\partial t^2} = \frac{\partial^2 u}{\partial r^2} + \frac{1}{r}\frac{\partial u}{\partial r} + \frac{1}{r^2}\frac{\partial^2 u}{\partial \varphi^2} \tag{20.20}$$

with the boundary condition

$$u|_{r=R} = 0 \tag{20.21}$$

and the initial conditions

$$u|_{t=0} = u_0(r, \varphi), \quad \left.\frac{\partial u}{\partial t}\right|_{t=0} = u_1(r, \varphi). \tag{20.22}$$

To apply Fourier's method we put

$$u(r, \varphi, t) = T(t)\, v(r, \varphi). \tag{20.23}$$

Substituting (20.23) into (20.20), we arrive at an equation for $T(t)$,

$$T''(t) + a^2\lambda^2 T(t) = 0, \tag{20.24}$$

and the following boundary value problem for $v(r, \varphi)$:

$$\frac{\partial^2 v}{\partial r^2} + \frac{1}{r}\frac{\partial v}{\partial r} + \frac{1}{r^2}\frac{\partial^2 v}{\partial \varphi^2} + \lambda^2 v = 0. \tag{20.25}$$

The physics of the problem requires that $v(r, \varphi)$ be a 2π-periodic function of φ, i.e.

$$v(r, \varphi) = v(r, \varphi + 2\pi), \tag{20.26}$$

and finite at the center of the circle, i.e.

$$|v|_{r=0}| < \infty. \tag{20.27}$$

In addition, condition (20.21) implies

$$v|_{r=R} = 0. \tag{20.28}$$

To apply Fourier's method to problem (20.25)-(20.28) we put

$$v(r, \varphi) = \Phi(\varphi)\, Z(r), \tag{20.29}$$

and from (20.25) we find that

$$\Phi''(\varphi) + \nu^2\Phi(\varphi) = 0, \tag{20.30}$$

$$Z''(r) + \frac{1}{r}Z'(r) + \left(\lambda^2 - \frac{\nu^2}{r^2}\right)Z(r) = 0, \tag{20.31}$$

where, in view of (20.27) and (20.28),

$$Z(R) = 0, \tag{20.32}$$

$$|Z(0)| < \infty. \tag{20.33}$$

From (20.30) and (20.26) we find ($\nu = n$ is an integer) that

$$\Phi_n(\varphi) = A_n \cos n\varphi + B_n \sin n\varphi. \tag{20.34}$$

Substituting x for λr in Eq. (20.31) ($Z(r) \equiv y(x)$), we arrive at Bessel's differential equation

$$x^2 y'' + xy' + (x^2 - \nu^2) y = 0,$$

whose general solution has the following form:

$$y_\nu(x) = C_1 J_\nu(x) + C_2 Y_\nu(x),$$

where $J_\nu(x)$ and $Y_\nu(x)$ are Bessel functions of order ν of the first and second kinds, respectively. Below we give the properties of Bessel functions.

(1) The roots of the equation

$$J_\nu(\mu) = 0 \tag{20.35}$$

at $\nu > -1$ are real and simple (except, perhaps, the root $\mu = 0$); they are symmetric with respect to point $\mu = 0$ (all lie on the μ axis) and have no finite limit points.

(2) $$\int_0^R x J_\nu\left(\frac{\mu_i(x)}{R}\right) J_\nu\left(\frac{\mu_j x}{R}\right) dx$$

$$= \begin{cases} 0, & i \neq j, \\ \frac{R^2}{2}[J_\nu'(\mu_i)]^2 = \frac{R^2}{2} J_{\nu+1}^2(\mu_i), & i = j, \end{cases} \tag{20.36}$$

where μ_i and μ_j are distinct positive roots of Eq. (20.35).

(3) Under certain conditions (see Vladimirov [2], p. 319), a function $f(x)$ can be expanded in a regularly convergent Fourier series in the system of functions $J_\nu(\mu_k x/R)$ ($k = 1, 2, \ldots$), where μ_1, μ_2, ... are the positive roots of Eq. (20.35).

Let us now go back to Eq. (20.31). Its general solution for $\nu = n$ has the form $Z_n(r) = C_n J_n(\lambda r) + D_n Y_n(\lambda r)$. Since in the neighborhood of point $x = 0$ the function $J_n(x)$ is bounded and the function $Y_n(x)$ is not, in view of (20.33) we have $D_n = 0$, that is,

$$Z_n(r) = C_n J_n(\lambda r). \tag{20.37}$$

Condition (20.32) yields $J_n(\lambda R) = 0$. If we put

$$\lambda R = \mu, \tag{20.38}$$

we arrive at Eq. (20.35). Suppose $\mu_1^{(n)}$, $\mu_2^{(n)}$, ..., are the positive roots of Eq. (20.35), that is,

$$J_n(\mu_m^{(n)}) = 0, \quad (m = 1, 2, \ldots). \tag{20.39}$$

Then (20.37)-(20.39) imply that the

$$Z_{nm}(r) = J_n(\mu_m^{(n)} r/R) \tag{20.40}$$

are solutions to problem (20.31)-(20.33).

In view of (20.23), (20.24), (20.29), (20.34), (20.38), and (20.40), the functions

$$u_{nm}(r, \varphi, t) = \left[\left(A_{nm}\cos\frac{a\mu_m^{(n)}t}{R} + B_{nm}\sin\frac{a\mu_m^{(n)}t}{R}\right)\cos n\varphi\right.$$

$$\left. + \left(C_{nm}\cos\frac{a\mu_m^{(n)}(t)}{R} + D_{nm}\sin\frac{a\mu_m^{(n)}(t)}{R}\right)\sin n\varphi\right] J_n\left(\frac{\mu_m^{(n)}r}{R}\right) \quad (20.41)$$

are particular solutions of Eq. (20.20) that satisfy the boundary condition (20.21).

We look for the solution to problem (20.20)-(20.22) in the form of the formal series

$$u(r, \varphi, t) = \sum_{n=0}^{\infty}\sum_{m=1}^{\infty} u_{nm}(r, \varphi, t).$$

where the u_{nm} are given by (20.41).

The problem is reduced to expanding certain functions in a series in the system of functions

$$J_n(\mu_m^{(n)}r/R) \quad (m = 1, 2, \ldots).$$

In view of (20.36), the coefficients a_m in the expansion

$$g(r) = \sum_{m=1}^{\infty} a_m J_n(\mu_m^{(n)}r/R)$$

are given by the following formulas:

$$a_m = \frac{2}{R^2 J_{n+1}^2(\mu_m^{(n)})}\int_0^R rg(r)\, J_n\left(\frac{\mu_m^{(n)}r}{R}\right) dr.$$

20.20. Solve the problem of the free vibrations of a homogeneous round membrane of radius R fixed at its edge in each of the following cases:

1. The initial deflection is $u|_{t=0} = AJ_0(\mu_k r/R)$, where μ_k is a positive root of the equation $J_0(\mu) = 0$; the initial velocity is zero.

2. The initial deflection and initial velocity depend only on r:

$$u|_{t=0} = f(r), \quad u_t|_{t=0} = F(r).$$

3. The initial deflection has the form of a paraboloid of revolution, and the initial velocity is zero.

20.21. Find the solution of the mixed problem

$$u_{tt} = u_{xx} + \frac{1}{x}u_x + f(t)\, J_0(\mu_k x),$$

where μ_k is a positive root of the equation $J_0(\mu) = 0$, $0 < x < 1$,

$$u|_{x=1} = u|_{t=0} = u_t|_{t=0} = 0, \quad |u|_{x=0}| < \infty$$

with

1. $f(t) = t^2 + 1$. 2. $f(t) = \sin t + \cos t$.

20.22. Find the solution of the mixed problem

$$u_{tt}=u_{xx}+\frac{1}{x}u_x,\quad 0<x<1,$$
$$|u|_{x=0}|<\infty,\quad u|_{x=1}=g(t),\quad u|_{t=0}=u_0(x),$$
$$u_t|_{t=0}=u_1(x)$$

with

1. $g(t)=\sin^2 t,\quad u_0(x)=\frac{1}{2}\left[1-\frac{J_0(2x)}{J_0(2)}\right],\quad u_1(x)=0,$
2. $g(t)=\cos 2t,\quad u_0(x)=\frac{J_0(2x)}{J_0(2)},\quad u_1(x)=0.$
3. $g(t)=t-1,\ u_0(x)=J_0(\mu_1 x)-1$, where μ_1 is a positive root of the equation $J_0(\mu)=0$, $u_1(x)=1$.

20.23. Find the solution of the mixed problem

$$u_{tt}+f(t)=u_{xx}+\frac{1}{x}u_x,\quad 0<x<1,$$
$$|u|_{x=0}|<\infty,\quad u|_{x=1}=g(t),\quad u|_{t=0}=u_0(x),$$
$$u_t|_{t=0}=u_1(x)$$

with

1. $f(t)=\cos t,\quad u_0(x)=1-\frac{J_0(x)}{J_0(1)},\ u_1(x)=0.$
2. $f(t)=\sin 3t,\ g(t)=u_0(x)=1,\ u_1(x)=\frac{1}{3}\left[1-\frac{J_0(3x)}{J_0(3)}\right].$
3. $f(t)=-2\cos 2t,\ g(t)=u_1(x)=0,\quad u_0(x)=\frac{1}{2}\left[\frac{J_0(2x)}{J_0(2)}-1\right]+$ $+J_0(\mu_1 x)$, where μ_1 is a positive root of the equation $J_0(\mu)=0$.

20.24. Solve the mixed problem

$$u_{xx}+\frac{1}{x}u_x=u_{tt}+u,\ 0<x<1,$$
$$|u|_{x=0}|<\infty,\ u|_{x=1}=\cos 2t+\sin 3t,$$
$$u|_{t=0}=\frac{J_0(x\sqrt{3})}{J_0(\sqrt{3})},\quad u_t|_{t=0}=\frac{3J_0(2x\sqrt{2})}{J_0(2\sqrt{2})}.$$

20.25. Solve the problem of the vibrations of a homogeneous round membrane of radius R fixed at its edge, assuming that the vibrations are generated by a uniformly distributed pressure $p=p_0\sin\omega t$ applied to one side of the membrane. The medium in which the membrane moves is assumed to exert no resistance, and $\omega\neq a\mu_n/R$, where μ_n $(n=1, 2, \ldots)$ are the positive roots of the equation $J_0(\mu)=0$ (no resonance is present).

20.26. Solve the mixed problem

$$_{tt}=u_{xx}+\frac{1}{x}u_x-\frac{u}{x^2},\quad 0<x<1,$$
$$|u|_{x=0}|<\infty,\quad u|_{x=1}=0,\quad u|_{t=0}=u_0(x),\quad u_t|_{t=0}=u_1(x)$$

with

1. $u_0(x) = J_1(\mu_k x) + J_1(\mu_m x)$, $u_1(x) = 0$.
2. $u_0(x) = J_1(\mu_k x)$, $u_1(x) = J_1(\mu_m x)$. Here μ_k and μ_m are two distinct positive roots of the equation $J_1(\mu) = 0$.

20.27. Solve the mixed problem

$$u_{tt} = u_{xx} + \frac{1}{x} u_x - \frac{u}{x^2} + e^t J_1(\mu_k x),$$

where μ_k is a positive root of the equation $J_1(\mu) = 0$, $0 < x < 1$,

$$|u|_{x=0}| < \infty, \quad u|_{x=1} = u|_{t=0} = u_t|_{t=0} = 0.$$

20.28. Solve the mixed problem

$$u_{tt} = u_{xx} + \frac{1}{x} u_x - \frac{u}{x^2}, \quad 0 < x < 1,$$

$$|u|_{x=0} < \infty, \quad u|_{x=1} = \sin 2t \cos t, \quad u|_{t=0} = 0,$$

$$u_t|_{t=0} = \frac{J_1(x)}{2J_1(1)} + \frac{3}{2} \frac{J_1(3x)}{J_1(3)}.$$

20.29. Solve the mixed problem

$$u_{tt} = u_{xx} + \frac{1}{x} u_x - \frac{4u}{x^2}, \quad 0 < x < 1,$$

$$|u|_{x=0}| < \infty, \quad u|_{x=1} = 0, \quad u|_{t=0} = u_0(x), \quad u_t|_{t=0} = u_1(x)$$

with

1. $u_0(x) = u_1(x) = J_2(\mu_k x)$.
2. $u_0(x) = \frac{1}{2} J_2(\mu_k x)$, $u_1(x) = \frac{3}{2} J_2(\mu_k x)$.

Here μ_k is a positive root of the equation $J_2(\mu) = 0$.

20.30. Solve the mixed problem

$$u_{tt} = u_{xx} + \frac{1}{x} u_x - \frac{4u}{x^2} + f(t) J_2(\mu_1 x), \quad 0 < x < 1,$$

where μ_1 is a positive root of the equation $J_2(\mu) = 0$,

$$|u|_{x=0} < \infty, \quad u|_{x=1} = u|_{t=0} = u_t|_{t=0} = 0$$

with

1. $f(t) = t$. 2. $f(t) = \cos t$.

20.31. Solve the mixed problem

$$u_{tt} = u_{xx} + \frac{1}{x} u_x - \frac{9u}{x^2}, \quad 0 < x < 1,$$

$$|u|_{x=0} < \infty, \quad u|_{x=1} = 0, \quad u_t|_{t=0} = J_3(\mu_1 x),$$

where μ_1 is a positive root of the equation $J_3(\mu) = 0$,

$$u|_{t=0} = u_0(x)$$

with

1. $u_0(x) = 0$ 2. $u_0(x) = J_3(\mu_1 x)$

20.32. Solve the mixed problem

$$u_{tt}=u_{xx}+\frac{1}{x}u_x-\frac{9u}{x^2}+f(t)J_3(\mu_k x),\quad 0<x<1,$$

$$|u|_{x=0}|<\infty,\quad u|_{x=1}=u|_{t=0}=u_t|_{t=0}=0,$$

where μ_k is a positive root of the equation $J_3(\mu)=0$, with
1. $f(t)=e^{-t}$. 2. $f(t)=t-t^2$.

20.33. Solve the mixed problem

$$(xu_x)_x=u_{tt},\quad 0<x<\frac{1}{4},$$

$$|u|_{x=0}|<\infty,\ u|_{x=\frac{1}{4}}=0,\ u|_{t=0}=J_0(2\mu_1\sqrt{x})$$

where μ_1 is a positive root of the equation $J_0(\mu)=0$, $u_t|_{t=0}=0$.

20.34. A heavy homogeneous thread of length l is hung by one of its ends ($x=l$). It is then taken out of the equilibrium state and is released without an initial velocity. Study the vibrations that the thread performs under the force of gravity; it is assumed that the medium does not exert any resistance.

20.35. A heavy homogeneous thread of length l is fixed by its upper end ($x=l$) to a vertical axis and rotates about this axis with a constant angular velocity ω. Find the deflection $u(x, t)$ of the thread from the equilibrium position.

20.36. Solve the mixed problem

$$u_{tt}=(xu_x)_x,\quad 0<x<1,$$

$$|u|_{x=0}<\infty,\quad u_x|_{x=1}=0,\quad u|_{t=0}=0,\ u_t|_{t=0}=J_0(\mu_k\sqrt{x}),$$

where μ_k is a positive root of the equation $J_1(\mu)=0$.

20.37. Solve the mixed problem

$$u_{tt}=xu_{xx}+u_x+f(t)J_0(\mu_1\sqrt{x}),\quad 0<x<1,$$

$$|u|_{x=0}|<\infty,\quad u|_{x=1}=u|_{t=0}=u_t|_{t=0}=0,$$

where μ_1 is a positive root of the equation $J_1(\mu)=0$ with
1. $f(t)=t$. 2. $f(t)=\sin t$.

20.38. Solve the mixed problem

$$u_{tt}=xu_{xx}+u_x-\frac{u}{x},\quad 0<x<1,$$

$$|u|_{x=0}|<\infty,\ u|_{x=1}=0,\ u|_{t=0}=0,$$

$$u_t|_{t=0}=J_2(\mu_1\sqrt{x}),$$

where μ_k is a positive root of the equation $J_2(\mu)=0$.

20.39. Solve the mixed problem

$$u_{tt}=xu_{xx}+u_x-\frac{9u}{4x},\quad 0<x<1,$$

$$|u|_{x=0}| < \infty, \quad u|_{x=1} = 0, \quad u|_{t=0} = 0, \quad u_t|_{t=0} = J_3(\mu_1 \sqrt{x}),$$

where μ_1 is a positive root of the equation $J_3(\mu) = 0$.

Let us now turn to equations of the parabolic type.

(a) The problem of the propagation of heat in a thin homogeneous rod $0 < x < l$ whose lateral surface is thermally isolated and whose ends $x = 0$ and $x = l$ are kept at zero temperature can be reduced to solving the heat conduction equation

$$u_t = a^2 u_{xx} \tag{20.42}$$

with the boundary conditions

$$u|_{x=0} = 0, \quad u|_{x=l} = 0 \tag{20.43}$$

and the initial condition

$$u|_{t=0} = u_0(x). \tag{20.44}$$

Since we wish to apply Fourier's method, we look for particular solutions of Eq. (20.42) in the form

$$u(x, t) = X(x) T(t). \tag{20.45}$$

Substituting (20.45) into Eq. (20.42) yields two equations:

$$T''(t) + a^2 \lambda T(t) = 0, \tag{20.46}$$

$$X''(x) + \lambda X(x) = 0. \tag{20.47}$$

To find the nontrivial solutions of Eq. (20.42) of the type (20.45) that satisfy the boundary conditions (20.43), we must find the nontrivial solutions of Eq. (20.47) that satisfy (20.43). For values of λ equal to (see the theory at the beginning of Sec. 20)

$$\lambda_n = \left(\frac{n\pi}{l}\right)^2 \quad (n = 1, 2, \ldots),$$

and only such values problem (20.47), (20.43) has the following nontrivial solutions:

$$X_n(x) = \sqrt{\frac{2}{l}} \sin \frac{\pi n x}{l}.$$

The values $\lambda = \lambda_n$ correspond to the following solutions of Eq. (20.46):

$$T_n(t) = a_n e^{-(\pi n a/l)^2 t}.$$

Then the function

$$u_n(x, t) = X_n(x) T_n(t) = a_n e^{-(\pi n a/l)^2 t} \sin \frac{\pi n x}{l}$$

satisfies Eq. (20.42) and the boundary conditions (20.43) for all values of a_n.

We look for the solution of Eq. (20.42) that satisfies (20.44) in the form of the formal series

$$u(x, t) = \sum_{n=1}^{\infty} u_n(x, t) = \sum_{n=1}^{\infty} a_n e^{-(\pi n a/l)^2 t} \sin \frac{\pi n x}{l}. \tag{20.48}$$

Combining (20.44) with (20.48), we find that

$$u_0(x)=\sum_{n=1}^{\infty} a_n \sin\frac{\pi n x}{l},$$

where $a_n=\frac{2}{l}\int_0^l u_0(x)\sin\frac{\pi n x}{l}\,dx$

(b) The problem of the temperature of a homogeneous rod of length l whose lateral surface is thermally isolated and through whose ends there is convective heat exchange with media that have constant temperatures (u_1 and u_2, respectively) can be reduced to solving Eq. (20.42) with the initial condition (20.44) and boundary conditions

$$\begin{aligned} u_x|_{x=0}-h_1[u|_{x=0}-u_1]&=0,\\ u_x|_{x=l}+h_2[u|_{x=l}-u_2]&=0, \end{aligned} \tag{20.49}$$

where h_1 and h_2 are positive.

If $h_1=h_2=0$, conditions (20.49) take the form

$$u_x|_{x=0}=u_x|_{x=l}=0, \tag{20.50}$$

which mean that the ends of the rod are thermally isolated.

We look for the solution of problem (20.42), (20.44), (20.49) in the form

$$u(x,t)=v(x)+w(x,t),$$

where $v(x)$ is the solution of Eq. (20.42) ($v''(x)=0$) that satisfies the boundary conditions (20.49). The equation $v''(x)=0$ has the general solution

$$v(x)=C_1x+C_2. \tag{20.51}$$

From (20.49) we find that

$$C_1=\frac{h_1(u_2-u_1)}{h_1+h_2+h_1h_2l},\qquad C_2=u_1+\frac{C_1}{h_1}. \tag{20.52}$$

The function $w(x,t)$ satisfies Eq. (20.42), the initial condition

$$w|_{t=0}=u|_{t=0}-v|_{t=0}=u_0(x)-v(x)=\tilde{u}_0(x), \tag{20.53}$$

where $v(x)$ is given by (20.51) and (20.52), and the following homogeneous boundary conditions:

$$(w_x-h_1w)_{x=0}=(w_x+h_2w)_{x=l}=0. \tag{20.54}$$

Solving the problem (20.42), (20.53), (20.54) via Fourier's method, we find that

$$w_n(x,t)=A_ne^{-a^2\lambda_n^2t}X_n(x),$$

where $\lambda_n^2=\frac{\mu_n^2}{l^2}$, μ_n $(n=1,2,\ldots)$ are the positive roots of the equation

$$\cot\mu=\frac{1}{l(h_1+h_2)}\left(\mu-\frac{h_1h_2l^2}{\mu}\right),$$

and

$$X_n(x) = \frac{\mu_n}{l} \cos \frac{\mu_n}{l} x + h_1 \sin \frac{\mu_n}{l} x.$$

Then $w(x, t) = \sum_{n=1}^{\infty} A_n e^{-a^2\lambda_n^2 t} X_n(x)$, where the expansion coefficients A_n can be found from the initial condition (20.53) using the orthogonality of the $X_n(x)$ in $[0, l]$:

$$A_n = \frac{1}{\|\Phi_n\|^2} \int_0^l \tilde{u}_0(x) \left(\frac{\mu_n}{l} \cos \frac{\mu_n}{l} x + h_1 \sin \frac{\mu_n}{l} x \right) dx,$$

$$\|\Phi_n\|^2 = \int_0^l \left(\frac{\mu_n}{l} \cos \frac{\mu_n}{l} x + h_1 \sin \frac{\mu_n}{l} x \right)^2 dx.$$

20.40. Given a thin homogeneous rod $0 < x < l$ whose lateral surface is thermally isolated, find the temperature distribution $u(x, t)$ in the rod under the following conditions:

1. The ends $x = 0$ and $x = l$ of the rod are kept at zero temperature, while the initial temperature of the rod is $u|_{t=0} = u_0(x)$. Consider the cases when (a) $u_0(x) = A = \text{const}$, and (b) $u_0(x) = Ax(l - x)$, with A a constant.
2. The end $x = 0$ is kept at zero temperature, while at the other end, $x = l$ there is heat exchange with the surrounding medium kept at zero temperature; the initial temperature of the rod is $u|_{t=0} = u_0(x)$.
3. Both ends of the rod ($x = 0$ and $x = l$) are involved in heat exchange with the surrounding medium, and the initial temperature of the rod is $u|_{t=0} = u_0(x)$.
4. Both ends ($x = 0$ and $x = l$) are thermally isolated, while the initial temperature is $u|_{t=0} = u_0 = \text{const}$.
5. Both ends of the rod are thermally isolated, while the initial temperature distribution in the rod is

$$u|_{t=0} = \begin{cases} u_0 = \text{const} & \text{if } 0 < x < l/2, \\ 0 & \text{if } l/2 < x < l. \end{cases}$$

Study the behavior of $u(x, t)$ as $t \to \infty$.

6. Both ends are thermally isolated, while

$$u|_{t=0} = \begin{cases} (2u_0/l)\, x & \text{if } 0 < x < l/2, \\ (2u_0/l)(l - x) & \text{if } l/2 \leqslant x < l, \end{cases}$$

where $u_0 = \text{const}$. Study the behavior of $u(x, t)$ as $t \to \infty$.

20.41. Solve the following mixed problems:

1. $u_t = u_{xx}$, $0 < x < 1$, $u_x|_{x=0} = 0$, $u|_{x=1} = 0$, $u|_{t=0} = x^2 - 1$.
2. $u_{xx} = u_t + u$, $0 < x < l$, $u|_{x=0} = u|_{x=l} = 0$, $u|_{t=0} = 1$.
3. $u_t = u_{xx} - 4u$, $0 < x < \pi$, $u|_{x=0} = u|_{x=\pi} = 0$, $u|_{t=0} = x^2 - \pi x$.

20.42. Given a thin homogeneous rod $0 < x < l$ whose lateral surface is thermally isolated, find the temperature distribution $u(x, t)$ in the rod under the following conditions:

1. Both ends of the rod are kept at constant temperatures, $u|_{x=0} = u_1$ and $u|_{x=l} = u_2$, while the initial temperature of the rod is $u|_{t=0} = u_0 = \text{const}$. Study the behavior of $u(x, t)$ as $t \to \infty$.

2. Both ends are kept at the same temperature $u|_{x=0} = u|_{x=l} = u_1$, while the initial temperature of the rod is

$$u|_{t=0} = u_0(x) = Ax(l-x),$$

where $A = \text{const}$. Study the behavior of $u(x, t)$ as $t \to \infty$.

3. The left end is thermally isolated, the right end is kept at a constant temperature $u|_{x=l} = u_2$, and the initial temperature distribution in the rod is $u|_{t=0} = (A/l)x$, with $A = \text{const}$.

4. The left end is kept at a constant temperature $u|_{x=0} = u_1$, at the right end there is a constant influx of heat, and the initial temperature distribution in the rod is $u|_{t=0} = u_0(x)$.

20.43. Given a thin homogeneous rod of length l with a lateral surface through which there is heat emission into the surrounding medium kept at zero temperature and with the left end kept at a constant temperature $u|_{x=0} = u_1$, determine the temperature distribution $u(x, t)$ in the rod under the following conditions:

1. The right end ($x = l$) is kept at a constant temperature $u|_{x=l} = u_2$, and the initial temperature distribution in the rod is $u|_{t=0} = u_0(x)$.

2. At the right end there is heat exchange with the surrounding medium (which is kept at zero temperature), while the initial temperature of the rod is zero.

In problems concerning heat propagation in a rod whose ends are kept at temperatures that are generally functions of t, the boundary conditions have the form

$$u|_{x=0} = \alpha_1(t), \quad u|_{x=l} = \alpha_2(t). \tag{20.49a}$$

In this case the solution to problem (20.42), (20.44), (20.49a) should be sought in the form $u = v + w$, where w is given by the formula $w = \alpha_1(t) + (x/l)(\alpha_2(t) - \alpha_1(t))$.

20.44. Find the temperature distribution in a rod $0 \leqslant x \leqslant l$ with a thermally isolated lateral surface if the right end of the rod ($x = l$) is kept at zero temperature, while the temperature at the left end is $u|_{x=0} = At$, with $A = \text{const}$. The initial temperature of the rod is zero.

20.45. Solve the following mixed problems:

1. $u_t = u_{xx}$, $0 < x < l$, $u_x|_{x=0} = 1$, $u|_{x=l} = 0$, $u|_{t=0} = 0$.

2. $u_t = u_{xx} + u + 2\sin 2x \sin x$, $0 < x < \frac{\pi}{2}$, $u_x|_{x=0} = u|_{x=\frac{\pi}{2}} = u|_{t=0} = 0$

3. $u_t = u_{xx} - 2u_x + x + 2t$, $0 < x < 1$, $u|_{x=0} = u|_{x=l} = t$, $u|_{t=0} = e^x \sin \pi x$.
4. $u_t = u_{xx} + u - x + 2 \sin 2x \cos x$, $0 < x < \frac{\pi}{2}$, $u|_{x=0} = 0$, $u_x|_{x=\frac{\pi}{2}} = 1$, $u|_{t=0} = x$.
5. $u_t = u_{xx} + 4u + x^2 - 2t - 4x^2 t + 2\cos^2 x$, $0 < x < \pi$, $u_x|_{x=0} = 0$, $u_x|_{x=\pi} = 2\pi t$, $u|_{t=0} = 0$.
6. $u_t - u_{xx} + 2u_x - u = e^x \sin x - t$, $0 < x < \pi$, $u|_{x=0} = 1 + t$, $u|_{x=\pi} = 1 + t$, $u|_{t=0} = 1 + e^x \sin 2x$.

20.46. Solve the following mixed problems:
1. $u_t - u_{xx} - u = xt(2 - t) + 2\cos t$, $0 < x < \pi$, $u_x|_{x=0} = t^2$, $u_x|_{x=\pi} = t^2$, $u|_{t=0} = \cos 2x$.
2. $u_t - u_{xx} - 9u = 4 \sin^2 t \cos 3x - 9x^2 - 2$, $0 < x < \pi$, $u_x|_{x=0} = 0$, $u_x|_{x=\pi} = 2\pi$, $u|_{t=0} = x^2 + 2$.
3. $u_t = u_{xx} + 6u + 2t(1 - 3t) - 6x + 2 \cos x \cos 2x$, $0 < x < \pi/2$, $u_x|_{x=0} = 1$, $u|_{x=\pi/2} = t^2 + \pi/2$, $u|_{t=0} = x$.
4. $u_t = u_{xx} + 6u + x^2(1 - 6t) - 2(t + 3x) + \sin 2x$, $0 < x < \pi$, $u_x|_{x=0} = 1$, $u_x|_{x=\pi} = 2\pi t + 1$, $u|_{t=0} = x$.
5. $u_t = u_{xx} + 4u_x + x - 4t + 1 + e^{-2x} \cos^2 \pi x$, $0 < x < 1$, $u|_{x=0} = t$, $u|_{x=1} = 2t$, $u|_{t=0} = 0$.

The problem of heat propagation in a homogeneous ball of radius R centered at the origin of coordinates in the case where the temperature of any point in the ball depends only on the distance from this point to the ball's center can be reduced to solving the heat conduction equation

$$\frac{\partial u}{\partial t} = a^2 \left(\frac{\partial^2 u}{\partial r^2} + \frac{2}{r} \frac{\partial u}{\partial r} \right) \tag{20.55}$$

with the initial condition

$$u|_{t=0} = u_0(r). \tag{20.56}$$

If the ball's surface participates in heat exchange with the surrounding medium whose temperature is zero, the boundary condition is of the form

$$(u_r + hu)|_{r=R} = 0. \tag{20.57}$$

Putting $v = ru$, we obtain

$$\frac{\partial v}{\partial t} = a^2 \frac{\partial^2 v}{\partial r^2}, \tag{20.58}$$

$$v|_{r=0} = 0, \quad \left[v_r + \left(h - \frac{1}{R} \right) v \right] \Big|_{r=R} = 0 \tag{20.59}$$

$$v|_{t=0} = r u_0(r). \tag{20.60}$$

Thus, the problem (20.55)-(20.57) has been reduced to the problem (20.58)-(20.60) of the propagation of heat along a rod whose one end ($r = 0$) is kept at zero temperature and the other ($r = R$) is involved in heat exchange with the surrounding medium (see Problem 20.43).

20.47. Given a homogeneous ball of radius R centered at the origin of coordinates, find the temperature inside the ball under the following conditions:

1. The surface of the ball is kept at zero temperature, and the initial temperature depends only on the distance from the ball's center, that is, $u|_{t=0} = u_0(r)$.
2. The surface of the ball is involved in convective heat exchange that obeys Newton's law with a medium whose temperature is zero, and $u|_{t=0} = u_0(r)$.
3. The ball's surface is involved in convective heat exchange with a medium that has a temperature $u_1 = \text{const}$, and $u|_{t=0} = u_0 = \text{const}$.
4. Starting at time $t = 0$, there is an influx of heat across its surface with a flux density $q = \text{const}$, and the initial temperature is $u|_{t=0} = u_0 = \text{const}$.

20.48. Given a thin square plate $(0 < x < l,\ 0 < y < l)$ for which the initial temperature distribution $u|_{t=0} = u_0(x, y)$ is known. The lateral sides $x = 0$ and $x = l$ and the base sides $y = 0$ and $y = l$ are kept at zero temperature over the entire time of observation. Find the temperature of a point of the plate at time $t > 0$.

Solution of Problems 20.48-20.52 requires using Bessel functions (see the theory between Problems 20.19 and 20.20).

For instance, the problem of radial heat propagation in an infinite circular cylinder of radius R whose lateral surface is kept at zero temperature can be reduced to solving the equation

$$\frac{\partial u}{\partial t} = a^2\left(\frac{\partial^2 u}{\partial r^2} + \frac{1}{r}\frac{\partial u}{\partial r}\right) \tag{20.61}$$

with the boundary condition

$$u|_{r=R} = 0 \tag{20.62}$$

and the initial condition

$$u|_{t=0} = u_0(r). \tag{20.63}$$

Employing Fourier's method, we find the solution of problem (20.61)-(20.63) in the form of a series:

$$u(r, t) = \sum_{n=1}^{\infty} a_n J_0\left(\frac{\mu_n r}{R}\right) e^{-(a\mu_n/R)^2 t},$$

where μ_n are the positive roots of the equation $J_0(\mu) = 0$, and the coefficients a_n can be found via the initial condition (20.63).

20.49. Given an infinitely long circular cylinder of radius R, find the temperature distribution inside it at time t under the following conditions:

1. The surface of the cylinder is constantly kept at zero temperature, and the temperature inside the cylinder at the initial moment

is $u|_{t=0} = AJ_0\ (\mu_k r/R)$, where μ_k is a positive root of the equation $J_0(\mu) = 0$.

2. The surface of the cylinder is kept at a constant temperature u_0, while the initial temperature inside the cylinder is zero.

3. Through the lateral surface of the cylinder there is heat emission into the surrounding medium kept at zero temperature, and the initial temperature is $u|_{t=0} = u_0(r)$.

20.50. Find the solution of the mixed problem

$$u_t = u_{xx} + \frac{1}{x} u_x - \frac{1}{x^2} u + f(t) J_1(\mu_k x),$$

where μ_k is a positive root of the equation $J_1(\mu) = 0$, $0 < x < 1$,

$$|u|_{x=0}| < \infty, \quad u|_{x=1} = 0, \quad u|_{t=0} = 0$$

with

1. $f(t) = \sin t$. 2. $f(t) = e^{-t}$.

20.51. Find the solution of the mixed problem

$$u_t = u_{rr} + \frac{1}{r} u_r + tJ_0(\mu_1 r),$$

where μ_1 is a positive root of the equation $J_0(\mu) = 0$, $0 < r < 1$.

$$u|_{r=0}| < \infty, \quad u|_{r=1} = 0, \quad u|_{t=0} = 0.$$

20.52. Solve the following mixed problems:

1. $u_t = xu_{xx} + u_x - \frac{1}{4x} u + tJ_1(\mu_k \sqrt{x})$,

where μ_k is a positive root of the equation $J_1(\mu) = 0$, $0 < x < 1$

$$|u|_{x=0}| < \infty, \quad u|_{x=1} = 0, \quad u|_{t=0} = 0.$$

2. $u_t = xu_{xx} + u_x - \frac{9}{4x} u, \quad 0 < x < 1,$

$$|u|_{x=0}| < \infty, \quad u|_{x=1} = 0, \quad u|_{t=0} = J_3(\mu_k \sqrt{x}),$$

where μ_k is a positive root of the equation $J_3(\mu) = 0$.

Answers to Problems of Sec. 20

20.1. 1. $A \sin \frac{\pi n x}{l} \cos \frac{\pi n a t}{l}$. 2. $\frac{32h}{\pi^3} \sum_{k=0}^{\infty} \frac{1}{(2k+1)^3} \sin \frac{(2k+1)\pi x}{l} \times$ $\times \cos \frac{(2k+1)\pi a t}{l}$. *Hint.* $u_0(x) = \frac{4h}{l^2} x(l-x)$. 3. $\frac{2hl^2}{\pi^2 c(l-c)} \sum_{k=1}^{\infty} \times$ $\times \frac{1}{k^2} \sin \frac{k\pi c}{l} \sin \frac{k\pi x}{l} \cos \frac{k\pi a t}{l}$; $\frac{8h}{\pi^2} \sum_{k=0}^{\infty} \frac{(-1)^k}{(2k+1)^2} \sin \frac{(2k+1)\pi x}{l} \times$ $\cos \frac{(2k+1)\pi a t}{l}$ $\left(\text{at } c = \frac{l}{2}\right)$. *Hint.* $u_0(x) = \begin{cases} \frac{hx}{c} & \text{if } 0 \leqslant x \leqslant c, \\ \frac{h(l-x)}{l-c} & \text{if } c \leqslant x \leqslant l. \end{cases}$

20.2. 1. $\frac{4lv_0}{\pi^2 a}\sum_{k=0}^{\infty}\frac{1}{(2k+1)^2}\sin\frac{(2k+1)\pi x}{l}\sin\frac{(2k+1)\pi at}{l}$.

2. $\frac{2lv_0}{\pi^2 a}\sum_{k=1}^{\infty}\frac{\cos\frac{k\pi\alpha}{l}-\cos\frac{k\pi\beta}{l}}{k^2}\sin\frac{k\pi x}{l}\sin\frac{k\pi at}{l}$.

3. $\frac{8A\alpha}{\pi^2 a}\sum_{k=1}^{\infty}\frac{\cos\frac{\pi k\alpha}{l}\sin\frac{\pi kx_0}{l}}{k\left[1-\frac{(2\alpha k)^2}{l^2}\right]}\sin\frac{\pi kx}{l}\sin\frac{\pi kat}{l}$.

20.3. 1. $\sum_{k=0}^{\infty}\left[a_k\cos\frac{(2k+1)\pi at}{2l}+b_k\sin\frac{(2k+1)\pi at}{2l}\right]\sin\frac{(2k+1)\pi x}{2l}$, where $a_k=\frac{2}{l}\int_0^l u_0(x)\sin\frac{(2k+1)\pi x}{2l}\,dx$ and $b_k=\frac{4}{\pi a(2k+1)}\times$ $\times\int_0^l u_1(x)\sin\frac{(2k+1)\pi x}{2l}\,dx$. *Hint.* $u|_{x=0}=0$, $u_x|_{x=l}=0$, $u|_{t=0}=u_0(x)$, $u_t|_{t=0}=u_1(x)$. 2. $\frac{1}{l}\int_0^l[u_0(\xi)+tu_1(\xi)]\,d\xi+\sum_{k=1}^{\infty}\Big(a_k\times$ $\times\cos\frac{k\pi at}{l}+b_k\sin\frac{k\pi at}{l}\Big)\cos\frac{k\pi x}{l}$, where $a_k=\frac{2}{l}\int_0^l u_0(x)\cos\frac{k\pi x}{l}\,dx$, $b_k=\frac{2}{\pi ak}\int_0^l u_1(x)\cos\frac{k\pi x}{l}\,dx$. *Hint.* $u_x|_{x=0}=u_x|_{x=l}=0$, $u|_{t=0}=$ $=u_0(x)$, $u_t|_{t=0}=u_1(x)$. 3. $\sum_{n=1}^{\infty}(a_n\cos\lambda_n at+b_n\sin\lambda_n at)\cos\lambda_n x$, where λ_n $(n=1, 2, \ldots)$ are the eigenvalues and $X_n(x)=\cos\lambda_n x$ the eigenfunctions of the boundary value problem $X''(x)+\lambda^2 X=0$, $X'(0)=0$, $X'(l)+hX(l)=0$ (λ_n are the positive roots of the equation $\tan\lambda l=h/\lambda$), $a_n=\frac{1}{\|X_n\|^2}\int_0^l u_0(x)\cos\lambda_n x\,dx$, $b_n=$ $=\frac{1}{\|X_n\|^2 a\lambda_n}\int_0^l u_1(x)\cos\lambda_n x\,dx$, $\|X_n\|^2=\frac{l}{2}\left[1+\frac{h}{l(\lambda_n^2+h^2)}\right]$. *Hint.* $u_x|_{x=0}=0$, $u_x|_{x=l}=-hu|_{x=l}$, $h=\frac{k}{E\sigma}$, where E is Young's modulus, σ the cross-sectional area of the rod, and k the coefficient that characterizes the rigidity of the fixture; $u|_{t=0}=u_0(x)$, $u_t|_{t=0}=u_1(x)$.

20.4. $u(x, t)=\frac{8Pl}{E\sigma\pi^2}\sum_{k=1}^{\infty}(-1)^k\frac{\sin\frac{(2k+1)\pi x}{2l}\cos\frac{(2k+1)\pi at}{2l}}{(2k+1)^2}$, where

σ is the cross-sectional area of the rod, and E is Young's modulus. *Hint.* $u|_{x=0}=0$, $u_x|_{x=l}=0$, $u|_{t=0}=\frac{Px}{E\sigma}$, $\left.\frac{\partial u}{\partial t}\right|_{t=0}=0$.

20.5. $i(x, t)=-E_0\sqrt{\frac{C}{L}}\cos\frac{\pi x}{2l}\sin\frac{\pi at}{2l}$, $a=\frac{1}{\sqrt{LC}}$. *Hint.* The current $i(x, t)$ satisfies the equation $LCi_{tt}=i_{xx}$, with L the self-inductance, and C the capacitance per unit length of the conductor. The initial conditions are $i|_{t=0}=0$, $i_t|_{t=0}=-\frac{E_0\pi}{2lL}\cos\frac{\pi x}{2l}$, while the boundary conditions are $i_x|_{x=0}=0$, $i|_{x=l}=0$.

20.6. 1. $bx(l-x)+\frac{4l^2b}{\pi^2}\sum_{k=1}^{\infty}\frac{(-1)^k\sin\frac{k\pi x}{l}\cos\frac{k\pi}{l}t}{k^2}$. *Hint.* Look for the solution in the form $u=v+w$, where $v=bx(l-x)$ satisfies the nonhomogeneous equation and zero boundary conditions, while w satisfies the homogeneous equation, zero boundary conditions, and the following initial conditions:

$v|_{t=0}=bx(x-l)$, $v_t|_{t=0}=0$. 2. $\frac{2}{\pi}t\sin t\sin x+\sum_{k=2}^{\infty}\frac{4}{k\pi(1-k^2)}\times$ $\times(\cos t-\cos kt)\sin kx$.

20.7. $\frac{4A}{\pi}\sum_{k=0}^{\infty}\frac{(2k+1)\sin\omega t-\frac{l\omega}{a\pi}\sin\frac{(2k+1)\pi at}{l}}{(2k+1)^2(\mu_k^2-\omega^2)}\sin\frac{(2k+1)\pi x}{l}$, where $\omega_k=\frac{(2k+1)\pi a}{l}$.

20.8. $u(x, t)=\frac{gx(2l-x)}{2a^2}-\frac{16gl^2}{\pi^3a^2}\sum_{k=0}^{\infty}\frac{\cos\frac{(2k+1)\pi at}{2l}\sin\frac{(2k+1)\pi x}{2l}}{(2k+1)^3}$.

Hint. The problem can be reduced to solving the equation $u_{tt}=a^2u_{xx}+g$, where g is the acceleration gravity, with $u|_{x=0}=\left.\frac{\partial u}{\partial x}\right|_{x=l}=0$, $u|_{t=0}=\left.\frac{\partial u}{\partial t}\right|_{t=0}=0$. Look for the solution of this problem in the form $u=v+w$, where $u=Ax^2+Bx+C$ (select A, B, and C in such a way that u satisfies the nonhomogeneous equation and the given boundary conditions).

20.9. 1. $\frac{xt}{l}+\sum_{k=1}^{\infty}\frac{(-1)^k2l}{(k\pi)^2}\sin\frac{k\pi x}{l}\sin\frac{k\pi t}{l}$. 2. $t+1+x(t^3-t+1)+$

$$+\sum_{k=1}^{\infty}\left\{\frac{2}{(\pi k)^2}\left[\frac{6(-1)^{k+1}}{(\pi k)^2}-1\right]\sin\pi kt+\frac{(-1)^k12t}{\pi^3k^3}\right\}\sin\pi kx.$$

20.10. $A \dfrac{\sin \frac{\omega x}{a}}{\sin \frac{\omega l}{a}} \sin \omega t + \dfrac{2A\omega a}{l} \sum_{k=1}^{\infty} \dfrac{(-1)^{k-1} \sin \frac{k\pi at}{l} \sin \frac{k\pi x}{l}}{\omega^2 - (k\pi a/l)^2}$. *Hint.* The problem can be reduced to solving the equation $u_{tt} = a^2 u_{xx}$ with zero initial conditions and the following boundary conditions: $u|_{x=0} = 0$, $u|_{x=l} = A \sin \omega t$. Look for the solution in the form $u = v + w$, with $v = X(x) \sin \omega t$. Select v in such a way that it satisfies the equation and the given boundary conditions.

20.11. $u(x, t) = \dfrac{Q}{E\sigma} x - \dfrac{8Ql}{\pi^2 E\sigma} \sum_{k=0}^{\infty} \dfrac{(-1)^k}{(2k+1)^2} \cos \dfrac{(2k+1)\pi at}{2l} \times$ $\times \sin \dfrac{(2k+1)\pi x}{2l}$, where E is the elastic modulus, and σ the cross-sectional area of the rod. *Hint.* The problem can be reduced to solving the equation $u_{tt} = a^2 u_{xx}$ with zero initial conditions and the following boundary conditions: $u|_{x=0} = 0$, $u_x|_{x=l} = Q/E\sigma$. Put $u = v + w$, with $v = Ax$ (select A in such a way that v satisfies the given boundary conditions).

20.12. $u(x, t) = \dfrac{Aa}{E\sigma\omega} \dfrac{\sin \frac{\omega}{a} x \sin \omega t}{\cos \frac{\omega l}{a}} + \dfrac{2Aa\omega}{E\sigma l} \sum_{k=0}^{\infty} \dfrac{(-1)^{k-1} 2l}{(2k+1)\pi} \times$ $\times \dfrac{\sin \frac{(2k+1)\pi}{2l} x}{\omega^2 - \left[\frac{(2k+1)\pi a}{2l}\right]^2} \sin a \dfrac{(2k+1)\pi t}{2l}$. *Hint.* The problem can be reduced to solving the equation $u_{tt} = a^2 u_{xx}$ with zero initial conditions and the following boundary conditions: $u|_{x=0} = 0$, $u|_{x=l} = (A/E\sigma) \sin \omega t$. Look for the solution of this problem in the form $u = v + w$, where $v = f(x) \sin \omega t$; select $f(x)$ in such a manner that v satisfies the equation and the given boundary conditions.

20.13. $u(x, t) = e^{-\alpha t} \sum_{k=1}^{\infty} (a_k \cos \mu_k t + b_k \sin \mu_k t) \sin \dfrac{k\pi x}{l}$, where $\mu_k = \sqrt{\dfrac{a^2\pi^2 k^2}{l^2} - \alpha^2}$, $a_k = \dfrac{2}{l} \int_0^l u_0(x) \sin \dfrac{\pi k x}{l}\, dx$, $b_k = \dfrac{\alpha}{\mu_k} a_k + \dfrac{2}{l\mu_k} \int_0^l u_1(x) \sin \dfrac{k\pi x}{l}\, dx$. *Hint.* The problem can be reduced to solving the equation $u_{tt} + 2\alpha u_t = a^2 u_{xx}$ (α is a small positive number) under the following conditions: $u|_{x=0} = u|_{x=l} = 0$, $u|_{t=0} = u_0(x)$, $u_t|_{t=0} = u_1(x)$.

20.14. 1. $-\dfrac{8}{\pi^3} \sum_{k=0}^{\infty} \dfrac{\sin (2k+1)\pi x}{(2k+1)^3} \cos \left(\sqrt{(2k+1)^2 \pi^2 + 4t}\right)$.

2. $-\frac{8e^{-t}}{\pi}\sum_{k=0}^{\infty}\frac{1}{(2k+1)^3}\left[\cos(2k+1)t+\frac{1}{2k+1}\sin(2k+1)t\right]\times$ $\times\sin(2k+1)x$.

3. $8e^{-t}\sum_{k=0}^{\infty}\frac{1}{(2k+1)^2}\left[(-1)^k-\frac{2}{\pi(2k+1)}\right]\sin\frac{2k+1}{2}t\cos\frac{2k+1}{2}x$.

4. $t(1-x)+\sum_{k=1}^{\infty}e^{-\frac{t}{2}}\frac{1}{(k\pi)^3}\left[2\cos\lambda_k t+\frac{1}{\lambda_k}\sin\lambda_k t-2\right]\sin\pi kx$, $\lambda_k=\sqrt{(k\pi)^2-\frac{1}{4}}$.

5. $(2-x)t+\sum_{k=1}^{\infty}\left(\frac{4t}{k\pi\lambda_k^2}-\frac{k\pi}{\lambda_k^3}\sin\lambda_k t\right)\sin\frac{k\pi x}{2}$, $\lambda_k=\sqrt{\left(\frac{k\pi}{2}\right)^2-1}$.

6. $\frac{xt}{l}+\sum_{k=1}^{\infty}\frac{2(-1)^{k+1}}{\pi k\lambda_k^2}\left(t-\frac{\sin\lambda_k t}{\lambda_k}\right)\sin\frac{\pi kx}{l}$, $\lambda_k=\sqrt{(\pi k/l)^2-1}$.

20.15. 1. $\sin 2x\cos 2t+\sum_{k=1}^{\infty}(-1)^k\frac{2}{k^3}(1-\cos kt)\sin kx$.

2. $-\sum_{k=0}^{\infty}c_k\left[-1+e^{-\frac{t}{2}}\left(\cos\mu_k t+\frac{1}{2\mu_k}\sin\mu_k t\right)\right]\sin(2k+1)\pi x$, $c_k=\frac{4}{(2k+1)^3\pi^3}$, $\mu_k=\sqrt{(2k+1)^2\pi^2-\frac{1}{4}}$. *Hint.* Look for the solution in the form of the series $u(x,t)=\sum_{k=1}^{\infty}T_k(t)\sin k\pi x$. *Remark.* The solution can also be sought in the form $u=v+w$, where $v=\frac{1}{2}x(1-x)$ satisfies the equation and the given boundary conditions. Then $u(x,t)=\frac{x(1-x)}{2}-\sum_{k=0}^{\infty}\left(\cos\mu_k t+\frac{1}{2\mu_k}\times\right.$ $\left.\times\sin\mu_k t\right)e^{-t/2}\sin(2k+1)\pi x$.

20.16. 1. $2xt+(2e^t-e^{-t}-3te^{-t})\cos x$. 2. $3+x(t+t^2)+(5te^t-$ $-8e^t+4t+8)\sin x$. 3. $x(t+1)+\left(\frac{1}{5}e^{\frac{5}{2}t}-e^{\frac{t}{2}}+\frac{4}{5}\right)\cos\frac{3}{2}x$. 4. $xt+\left(\frac{1}{10}-\frac{1}{6}e^{2t}+\frac{1}{15}e^{5t}\right)e^{-x}\sin 3x$. 5. $xt+(1-e^{-t}-$ $-te^{-t})\cos 3x$. 6. $\frac{1}{8}(e^{2t}+e^{-2t})-\frac{1}{4}-\frac{t^2}{2}\cos 2x$. 7. $\frac{1}{9}\sin x\times$ $\times(\cosh 3t-1)+\sin 3x(\cosh t-1)$. 8. $xt+(2e^t-e^{2t})e^{-x}\sin x$.

20.17. $A\cos\frac{a\pi\sqrt{2}}{p}t\sin\frac{\pi x}{p}\sin\frac{\pi y}{p}$.

20.18. $3\cos\sqrt{5}\,t\sin x\sin 2y+\sin 5t\sin 3x\sin 4y$.

20.19. $\frac{16Ap^2q^2}{\pi^6}\sum_{k,\,l=0}^{\infty}\frac{\sin\frac{(2k+1)\pi x}{p}\sin\frac{(2l+1)\pi y}{q}}{(2k+1)^3(2l+1)^3}\cos\pi a\mu_{k,\,l}t$, where

$\mu_{k,\,l}=\sqrt{\frac{(2k+1)^2}{p^2}+\frac{(2l+1)^2}{q^2}}$.

20.20. 1. $A\cos\frac{a\mu_k t}{R}J_0\left(\frac{\mu_k r}{R}\right)$. 2. $\sum_{n=1}^{\infty}\left(a_n\cos\frac{\mu_n}{R}at+b_n\times\right.$ $\left.\times\sin\frac{\mu_n}{R}at\right)J_0\left(\frac{\mu_n r}{R}\right)$, where $a_n=\frac{2}{R^2J_1^2(\mu_n)}\int_0^R rf(r)J_0\left(\frac{\mu_n r}{R}\right)dr$,

$b_n=\frac{2}{a\mu_nRJ_1^2(\mu_n)}\int_0^R rF(r)J_0\left(\frac{\mu_n r}{R}\right)dr$ (μ_n are the positive roots of the equation $J_0(\mu)=0$).

3.
$$u(r,t)=8A\sum_{n=1}^{\infty}\frac{J_0\left(\mu_n\frac{r}{R}\right)}{\mu_n^3J_1(\mu_n)}\cos\frac{a\mu_n t}{R},\tag{20.64}$$
where μ_n $(n=1,2,\ldots)$ are the positive roots of the equation $J_0(\mu)=0$. *Hint*. The problem can be reduced to solving the equation $u_{rr}+\frac{1}{r}u_r=\frac{1}{a^2}u_{tt}$ with $u|_{r=R}=0$, $|u|_{r=0}<\infty$; $u|_{t=0}=A\left(1-\frac{r^2}{R^2}\right)$, $A=\text{const}$, $u_t|_{t=0}=0$. For calculating the expansion coefficients in (20.64) employ the following formulas:
$$\int_0^x \xi J_0(\xi)\,d\xi=xJ_1(x),$$
$$\int_0^x \xi^3J_0(\xi)\,d\xi=2x^2J_0(x)+(x^3-4x)J_1(x).$$

20.21. 1. $u(x,t)=[\mu_k^{-4}(2-\mu_k^2)\cos\mu_kt+\mu_k^{-2}t^2+\mu_k^{-4}(\mu_k^2-2)]J_0(\mu_kx)$. *Hint*. Look for the solution in the form $u=v+w$, where $v=(at^2+c)J_0(\mu_kx)$ is a particular solution of the nonhomogeneous equation, w is a solution of the homogeneous equation, $w|_{t=0}=-v|_{t=0}$, $w_t|_{t=0}=-v_t|_{t=0}$. 2. $u(x,t)=(\mu_k^2-1)^{-1}(\cos t+\sin t-\cos\mu_kt-\mu_k^{-1}\sin\mu_kt)J_0(\mu_kx)$. *Hint*. Look for the solution in the form $u=v+w$, where $v=(a\sin t+b\cos t)J_0(\mu_kx)$ is a particular solution of the nonhomogeneous equation, $w=(A\cos\mu_kt+B\sin\mu_kt)J_0(\mu_kx)$ is a solution of the homogeneous solution, $w|_{t=0}=-v|_{t=0}$, and $w_t|_{t=0}=-v_t|_{t=0}$.

20.22. 1. $\frac{1}{2}\left[1-\frac{J_0(2x)}{J_0(2)}\cos 2t\right]$. 2. $\frac{J_0(2x)}{J_0(2)}\cos 2t$. 3. $t-1+J_0(\mu_1x)\cos\mu_1t$.

20.23. 1. $\left[1-\frac{J_0(x)}{J_0(1)}\right]\cos t$. 2. $1+\frac{1}{9}\sin 3t\left[1-\frac{J_0(3x)}{J_0(3)}\right]$. 3. $\frac{1}{2}\times\left[\frac{J(2x)}{J_0(2)}-1\right]\cos 2t+J_0(\mu_1 x)\cos\mu_1 t$.

20.24. $\frac{J_0(x\sqrt{3})}{J_0(\sqrt{3})}\cos 2t+\frac{J_0(2x\sqrt{2})}{J_0(2\sqrt{2})}\sin 3t$.

20.25. $\frac{p_0}{\omega^2\rho}\left[\frac{J_0\left(\frac{\omega}{a}r\right)}{J_0\left(\frac{\omega}{a}R\right)}-1\right]\sin\omega t+\frac{2p_0\omega R^3}{a\rho}\sum_{n=1}^{\infty}\frac{J_0\left(\frac{\mu_n r}{R}\right)\sin\frac{\alpha\mu_n t}{R}}{\mu_n^2(\omega^2R^2-a^2\mu_n^2)J_1(\mu_n)}$, where ρ is the surface density of the membrane. *Hint.* The problem can be reduced to solving the equation $\frac{1}{a^2}u_{tt}=u_{rr}+r^{-1}u_r+$ $+\rho^{-1}\sin\omega t$, $0<r<R$, $|u|_{r=0}|<\infty$, $u|_{r=R}=0$, $u|_{t=0}=u_t|_{t=0}=0$.

20.26. 1. $J_1(\mu_k x)\cos\mu_k t+J_1(\mu_m x)\cos\mu_m t$. 2. $J_1(\mu_k x)\cos\mu_k t+$ $+\mu_m^{-1}J_1(\mu_m x)\sin\mu_m t$.

20.27. $(1+\mu_k^2)^{-1}(e^t-\cos\mu_k t-\mu_k^{-1}\sin\mu_k t)J_1(\mu_k x)$.

20.28. $\frac{1}{2J_1(1)}J_1(x)\sin t+\frac{1}{2J_1(3)}J_1(3x)\sin 3t$.

20.29. 1. $(\cos\mu_k t+\mu_k^{-1}\sin\mu_k x)J_2(\mu_k x)$. 2. $\left(\frac{1}{2}\cos\mu_k t+\right.$ $\left.+\frac{3}{2}\mu_k^{-1}\times\sin\mu_k t\right)J_2(\mu_k x)$.

20.30. 1. $(\mu_1^{-2}t-\mu_1^{-3}\sin\mu_1 t)J_2(\mu_1 x)$. 2. $(\mu_1^2-1)^{-1}(\cos t-$ $-\cos\mu_1 t)\times J_2(\mu_1 x)$.

20.31. 1. $\mu_1^{-1}J_3(\mu_1 x)\sin\mu_1 t$. 2. $(\cos\mu_1 t+\mu_1^{-1}\sin\mu_1 t)J_3(\mu_1 x)$.

20.32. 1. $[\mu_k(1+\mu_k^2)]^{-1}(\sin\mu_k t-\mu_k\cos\mu_k t+\mu_k e^{-t})J_3(\mu_k x)$. 2. $\mu_k^{-2}(2\mu_k^{-2}+t-t^2-\mu_k^{-1}\sin\mu_k t-2\mu_k^{-2}\cos\mu_k t)J_3(\mu_k x)$.

20.33. $J_0(2\mu_1\sqrt{x})\cos\mu_1 t$. *Hint.* If we put $u=X(x)T(t)$, we arrive at two equations:

$$X''+\frac{X'}{x}+\frac{\lambda^2}{x}X=0,\qquad T''+\lambda^2T=0. \tag{20.65}$$

Substitution of $\eta=2\lambda\sqrt{x}$ into Eq. (20.65) yields Bessel's equation $X''(\eta)+\frac{1}{\eta}X'(\eta)+X(\eta)=0$, whose general solution is $X(\eta)=$ $=aJ_0(\eta)+bY_0(\eta)$.

20.34. $\frac{1}{l}\sum_{n=1}^{\infty}A_n\frac{J_0\left(\mu_n\sqrt{\frac{x}{l}}\right)}{J_1^2(\mu_n)}\cos\frac{\mu_n at}{2\sqrt{l}}$, where $A_n=\int_0^l u_0(x)J_0\times$ $\times\left(\mu_n\sqrt{\frac{x}{l}}\right)dx$, and μ_n $(n=1, 2, \ldots)$ are the positive roots of the equation $J_0(\mu)=0$. *Hint.* The problem can be reduced to solving the equation $u_{tt}=a^2(xu_x)_x$, $0<x<l$, $a=\sqrt{g}$, with $|u|_{x=0}|<\infty$, $u|_{x=l}=0$, $u|_{t=0}=u_0(x)$, $u_t|_{t=0}=0$.

20.35. $\sum_{n=1}^{\infty}(A_n\cos a\lambda_n t+B_n\sin a\lambda_n t)\,J_0\left(\mu_n\sqrt{\frac{x}{l}}\right)$, where $\lambda_n=\sqrt{\frac{\mu_n^2}{4l}-\left(\frac{\omega}{a}\right)^2}$, $A_n=\frac{1}{lJ_1^2(\mu_n)}\int_0^l u_0(x)\,J_0\left(\mu_n\sqrt{\frac{x}{l}}\right)dx$, $B_n=\frac{1}{a\lambda_n lJ_1^2(\lambda_n)}\int_0^l u_1(x)\,J_0\left(\mu_n\sqrt{\frac{x}{l}}\right)dx$, with μ_n $(n=1, 2, \ldots)$ the positive roots of the equation $J_0(\mu)=0$. *Hint.* The problem can be reduced to solving the equation $u_{tt}=a^2(xu_x)_x+\omega^2 u$, $0<x<l$, $a=\sqrt{g}$, with $|u|_{x=0}|<\infty$, $u|_{x=l}=0$, $u|_{t=0}=u_0(x)$, $u_t|_{t=0}=u_1(x)$.

20.36. $\frac{2}{\mu_k}J_0(\mu_k\sqrt{x})\sin\frac{\mu_k}{2}t$.

20.37. 1. $\left(4\mu_1^{-2}t-8\mu_1^{-3}\sin\frac{\mu_1}{2}t\right)J_0(\mu_1\sqrt{x})$. 2. $4(\mu_1^2-4)^{-1}\left(\sin t-2\mu_1^{-1}\sin\frac{\mu_1 t}{2}\right)J_0(\mu_1\sqrt{x})$.

20.38. $\frac{2}{\mu_k}\sin\frac{\mu_k}{2}tJ_2(\mu_k\sqrt{x})$.

20.39. $\frac{2}{\mu_1}\sin\frac{\mu_1}{2}tJ_3(\mu_1\sqrt{x})$.

20.40. 1. $\sum_{n=1}^{\infty}a_n e^{-(n\pi a/l)^2 t}\sin\frac{\pi n x}{l}$, where $a_n=\frac{2}{l}\int_0^l u_0(x)\sin\frac{\pi n x}{l}dx$. If $u_0(x)=A=\text{const}$, then $u(x,t)=\frac{4A}{\pi}\sum_{k=0}^{\infty}\frac{1}{2k+1}e^{-(2k+1)^2\pi^2a^2t/l^2}\times\sin\frac{(2k+1)\pi x}{l}$. If $u_0(x)=Ax(x-l)$, then $u(x,t)=\frac{8Al^2}{\pi^3}\times\sum_{k=0}^{\infty}\frac{1}{(2k+1)^3}e^{-(2k+1)^2\pi^2a^2t/l^2}\sin\frac{(2k+1)\pi x}{l}$. 2. $\frac{2}{l}\sum_{n=1}^{\infty}a_n\times\frac{\sigma^2+\mu_n^2}{\sigma(\sigma+1)+\mu_n^2}e^{-\mu_n^2a^2t/l^2}\sin\frac{\mu_n x}{l}$, where $a_n=\int_0^l u_0(x)\sin\frac{\mu_n x}{l}dx$, μ_n $(n=1, 2, \ldots)$ are the positive roots of the equation $\tan\mu=-\mu/\sigma$, $\sigma=hl>0$. *Hint.* The boundary conditions have the following form: $u|_{x=0}=0$, $(u_x+hu)|_{x=l}=0$. 3. $\frac{2}{l}\sum_{n=1}^{\infty}b_n e^{-\frac{\mu_n^2a^2}{l^2}t}\times$

$\times \dfrac{\mu_n \cos\frac{\mu_n x}{l} + \sigma \sin\frac{\mu_n x}{l}}{\sigma(\sigma+2)+\mu_n^2}$, where $b_n = \int_0^l u_0(x) \left(\mu_n \cos\frac{\mu_n x}{l} + \sigma \sin \times \frac{\mu_n x}{l}\right) dx$, μ_n $(n = 1, 2, \ldots)$ are the positive roots of the equation $\cot\mu = \frac{1}{2}\left(\frac{\mu}{\sigma} - \frac{\sigma}{\mu}\right)$, $\sigma = hl$, $u_0(x) = u|_{t=0}$. *Hint*. The boundary conditions have the form $(u_x - hu)|_{x=0} = (u_x + hu)|_{x=l} = 0$. 4. u_0. *Hint*. The boundary conditions have the form $u_x|_{x=0} = u_x|_{x=l} = 0$. 5. $\frac{u_0}{2} + \frac{2u_0}{\pi}\sum_{k=0}^{\infty}(-1)^k \frac{1}{2k+1} e^{-(2k+1)^2\pi^2a^2t/l^2} \times \cos\frac{(2k+1)\pi x}{l}$, $\lim_{t\to\infty} u(x, t) = \frac{u_0}{2}$. 6. $\frac{u_0}{2} - \frac{4u_0}{\pi^2}\sum_{k=0}^{\infty}\frac{1}{(2k+1)^2} \times e^{-4(2k+1)^2\pi^2a^2t/l^2} \cos\frac{2(2k+1)\pi x}{l}$, $\lim_{t\to\infty} u(x, t) = u_0/2$.

20.41. 1. $\frac{32}{\pi^3}\sum_{n=0}^{\infty}\frac{(-1)^n}{(2n+1)^3} e^{-\left(\frac{2n+1}{2}\pi\right)^2 t} \cos\frac{2n+1}{2}\pi x.$

2. $\frac{4}{\pi}\sum_{k=0}^{\infty}\frac{1}{2k+1} e^{-\left(\frac{\pi^2(2k+1)^2}{l^2}+1\right)t} \sin\frac{(2k+1)\pi x}{l}.$

3. $-\frac{8}{\pi}\sum_{k=0}^{\infty}\frac{1}{(2k+1)^2} e^{-(2k+1)^2} \sin(2k+1)x.$

20.42. 1. $u_1 + \frac{u_2 - u_1}{l}x + \frac{2}{\pi}\sum_{n=1}^{\infty}\frac{1}{n}\{(u_0 - u_1)[1 - (-1)^n] + (-1)^n(u_2 - u_0)\} e^{-n^2\pi^2a^2t/l^2} \sin\frac{\pi n x}{l}$, $\lim_{t\to\infty} u(x, t) = u_1 + (u_2 - u_1)\frac{x}{l}$. 2. $u_1 + \frac{8Al^2}{\pi^3}\sum_{k=0}^{\infty}\frac{1}{(2k+1)^3} e^{-(2k+1)^2\pi^2a^2t/l^2} \sin\frac{(2k+1)\pi x}{l} - 4u_1\sum_{k=0}^{\infty} e^{-(2k+1)^2\pi^2a^2t/l^2} \sin\frac{(2k+1)\pi x}{l}$, $\lim_{t\to\infty} u(x, t) = u_1$. 3. $u_2 + \frac{4(A - u_2)}{\pi}\sum_{k=0}^{\infty}(-1)^k\frac{1}{2k+1} e^{-\frac{(2k+1)^2\pi^2a^2t}{4l^2}} \cos\frac{(2k+1)\pi x}{2l} - \frac{8A}{\pi^2}\sum_{k=0}^{\infty}\frac{1}{(2k+1)^2} e^{-\frac{(2k+1)^2\pi^2a^2t}{4l^2}} \cos\frac{(2k+1)\pi x}{2l}$. 4. $\frac{qx}{k} + u_1 + \sum_{n=0}^{\infty}\left[a_n - \frac{4}{\pi^2}\frac{(2n+1)\pi u_1 + \frac{lq}{k}}{(2n+1)^2}\right] e^{-\frac{(2n+1)^2\pi^2a^2t}{4l^2}} \sin\frac{(2n+1)\pi x}{2l}$, where $a_n =$

$= \frac{2}{l} \int_0^l u_0(x) \sin \frac{(2n+1)\pi x}{2l} dx$. *Hint.* The boundary conditions have the form $u|_{x=0} = u_1$, $ku_x|_{x=l} = q$.

20.43. 1. $\dfrac{u_2 \sinh \frac{h}{a} x - u_1 \sinh \frac{h}{a}(x-l)}{\sinh \frac{h}{a} l} + \frac{2}{l} \sum_{n=1}^{\infty} \left(\frac{\pi n}{l} \cdot \frac{(-1)^n u_2 - u_1}{\lambda_n^2} + a_n \right) e^{-a^2\lambda_n^2 t} \sin \frac{\pi n x}{l}$, where $\lambda_n^2 = \left(\frac{\pi n}{l}\right)^2 + \left(\frac{h}{a}\right)^2$, $a_n = \int_0^l u_0 \times (x) \sin \frac{\pi n x}{l} dx$. *Hint.* The problem can be reduced to solving the equation

$$u_t = a^2 u_{xx} - h^2 u \tag{20.66}$$

with the boundary conditions $u|_{x=0} = u_1$ and $u|_{x=l} = u_2$ and the initial condition $u|_{t=0} = u_0(x)$. Look for the solution of this problem in the form $u(x, t) = v(x) + w(x, t)$, where v is the solution of Eq. (20.66) $(a^2 v''(x) - h^2 v = 0)$ that satisfies the given boundary conditions, while $w(x, t)$ is a solution of Eq. (20.66) with zero boundary conditions and the initial condition $w|_{t=0} = u_0(x) - v(x)$. 2. $u_1 \dfrac{h \cosh \frac{h}{a}(l-x) + h_1 a \sinh \frac{h}{a}(l-x)}{h \cosh \frac{hl}{a} + h_1 a \sinh \frac{hl}{a}} - 2u_1 a^2 \sum_{n=1}^{\infty} \frac{\mu_n (\mu_n^2 + h_1^2)}{(a^2\mu_n^2 + h^2)} \times \frac{1}{[l(\mu_n^2 + h_1^2) + h_1]} e^{-(a^2\mu_n^2 + h^2)t} \sin \mu_n x$, where μ_n $(n = 1, 2, \ldots)$ are the positive roots of the equation $\tan l\mu = -\mu/h_1$. *Hint.* The boundary conditions have the form $u|_{x=0} = u_1$, $(u_x + h_1 u)|_{x=l} = 0$. Look the solution of this problem in the form $u(x, t) = v(x) + w(x, t)$, where $v(x)$ is the solution of the equation $a^2 v''(x) - h^2 v = 0$ that satisfies the boundary conditions $v|_{x=0} = u_1$ and $(v_x + h_1 v)|_{x=l} = 0$. while $w(x, t)$ is a solution of Eq. (20.66) (see Problem 20.43.1) with $w|_{x=0} = 0$, $(w_x + hw)|_{x=l} = 0$, and $w|_{t=0} = -v(x)$.

20.44. $At \dfrac{l-x}{l} - \dfrac{2Al^2}{\pi^3 a^2} \sum_{n=1}^{\infty} \dfrac{1}{n^3} (1 - e^{-n^2\pi^2 a^2 t/l^2}) \sin \dfrac{n\pi x}{l}$.

20.45. 1. $x - l + \dfrac{8l}{\pi^2} \sum_{k=0}^{\infty} \dfrac{e^{-\lambda_k^2 t}}{(2k+1)^2} \cos \lambda_k x$; $\lambda_k = \dfrac{\pi(2k+1)}{2l}$. 2. $t \cos x + \frac{1}{8}(e^{-8t} - 1) \cos 3x$. 3. $xt + \sin \pi x \, e^{x-t-\pi^2 t}$. 4. $x + t \sin x + \frac{1}{8}(1 - e^{-8t}) \sin 3x$. 5. $tx^2 + \frac{1}{4}(e^{4t} - 1) + t \cos 2x$. 6. $t + 1 + (1 - e^{-t}) e^x \sin x + e^{x-4t} \sin 2x$.

20.46. 1. $xt^2+e^t+\sin t-\cos t+e^{-3t}\cos 2x$. 2. $x^2+2e^{9t}+(2t-\sin 2t)\cos 3x$. 3. $x+t^2+\frac{1}{5}(e^{5t}-1)\cos x+\frac{1}{3}(1-e^{-3t})\cos 3x$.

4. $x^2t+x+\sum_{k=1}^{\infty}\frac{C_{2k-1}}{(2k-1)^2-6}\{1-e^{-6(2k-1)^2t}\}\sin(2k-1)x$, $C_{2k-1}=\frac{2}{\pi}\left(\frac{1}{2k+1}-\frac{1}{2k-3}\right)$. 5. $t(x+1)+e^{-2x}\sum_{k=1}^{\infty}\frac{C_k}{k^2\pi^2+4}[1-e^{-(k^2\pi^2+4)t}]\sin k\pi x$,

$$C_k=\begin{cases}0 & \text{if } k=2m,\\ \frac{1}{\pi}\left(\frac{2}{2m-1}+\frac{1}{2m+1}+\frac{1}{2m-3}\right) & \text{if } k=2m-1.\end{cases}$$

20.47. 1. $\frac{2}{Rr}\sum_{n=1}^{\infty}a_n e^{-(\pi na/R)^2t}\sin\frac{\pi nr}{R}$, $a_n=\int_0^R ru_0(r)\sin\frac{\pi nr}{R}\,dr$.

2. $\frac{2}{Rr}\sum_{n=1}^{\infty}a_n\frac{\sigma^2+\mu_n^2}{\sigma(\sigma+1)+\mu_n^2}e^{-(\mu_n a/R)^2t}\sin\frac{\mu_n r}{R}$, $a_n=\int_0^R tu_0(r)\sin\times\frac{\mu_n r}{R}\,dr$, μ_n $(n=1, 2, \ldots)$ are the positive roots of the equation $\tan\mu=-\mu/\sigma$, $\sigma=hR-1$ $(\sigma>-1)$. 3. $u_1+2(u_1-u_0)\times\frac{hR^2}{r}\sum_{n=1}^{\infty}(-1)^n a_n e^{-(a\mu_n/R)^2t}\sin\frac{\mu_n}{R}r$, where μ_n are the positive roots of the equation $\tan\mu=-\mu/\sigma$, $\sigma=hR-1$ $(\sigma>-1)$, h is the heat exchange coefficient in the boundary condition $[u_r+h(u-u_1)]|_{r=R}=0$, and

$$a_n=\frac{\sqrt{\mu_n^2+\sigma^2}}{\mu_n[\mu_n^2+\sigma(\sigma+1)]}.$$

4. $u_0+\frac{q}{k}\left(\frac{3a^2}{R}t+\frac{5r^2-3R^2}{10R}-\frac{2R^2}{r}\sum_{n=1}^{\infty}\frac{e-(a\mu_n/R)^2t}{\mu_n^3\cos\mu_n}\sin\frac{\mu_n r}{R}\right.$,

where μ_n $(n=1, 2, 3, \ldots)$ are the positive roots of the equation $\tan\mu=\mu$. *Hint.* The problem can be reduced to solving Eq. (20.55) with $|u|_{r=0}|<\infty$ and $u_r|_{r=R}=q/k$.

20.48. $\sum_{j=1}^{\infty}\sum_{k=1}^{\infty}a_{jk}e^{-(a\pi/l)^2(j^2+k^2)t}\sin\frac{j\pi x}{l}\sin\frac{k\pi y}{l}$,

$$a_{jk}=\frac{4}{l^2}\int_0^l\int_0^l u_0(x, y)\sin\frac{j\pi x}{l}\sin\frac{k\pi y}{l}\,dx\,dy.$$

Hint. Apply Fourier's method to the equation $u_t=a^2\nabla^2u$ with $u|_{x=0}=u|_{x=l}=u|_{y=0}=u|_{y=l}=0$ and $u|_{t=0}=u_0(x, y)$.

20.49. 1. $Ae^{-(a\mu_k/R)^2t}J_0(\mu_k r/R)$. 2. $u_0\left[1+2\sum_{n=1}^{\infty}\frac{J_0(\mu_n r/R)}{\mu_n J_0'(\mu_n)}\times\right.$ $\left.\times e^{-(a\mu_n/R)^2t}\right]$, where μ_n $(n=1, 2, \ldots)$ are the positive roots of the equation $J_0(\mu)=0$. 3. $\frac{2}{R^2}\sum_{n=1}^{\infty}\frac{a_n\mu_n^2}{\mu_n^2+h^2R^2}e^{-(a\mu_n/R)^2t}J_0\left(\mu_n\frac{r}{R}\right)$, where $a_n=\frac{1}{J_0^2(\mu_n)}\int_0^R ru_0(r)J_0\left(\mu_n\frac{r}{R}\right)dr$ and μ_n $(n=1, 2, \ldots)$ are the positive roots of the equation $\mu J_0'(\mu)+hRJ_0(\mu)=0$. *Hint.* The boundary conditions are $|u|_{r=0}|<\infty$, $(u_r+hu)|_{r=R}=0$.

20.50. 1. $(1+\mu_k^4)^{-1}(e^{-\mu_k^2t}+\mu_k^2\sin t-\cos t)J_1(\mu_k x)$. 2. $(\mu_k^2-1)^{-1}\times$ $\times(e^{-t}-e^{-\mu_k^2t})J_1(\mu_k x)$.

20.51. $[\mu_1^{-2}t+\mu_1^{-4}(e^{-\mu_1^2t}-1)]J_0(\mu_1 r)$.

20.52. 1. $(16\mu_k^{-4}e^{-\mu_k^2t/4}+4\mu_k^{-2}t-16\mu_k^{-4})J_1(\mu_k\sqrt{x})$. 2. $e^{-\mu_k^2t/4}J_3(\mu_k\sqrt{x})$.

21 Other Methods

21.1. Prove that the problem

$$u_{tt}=a^2u_{xx},\quad t>0,\quad x>0;$$
$$u|_{t=0}=0,\quad u_t|_{t=0}=0,\quad u|_{x=0}=g(t)$$

has a unique solution,

$$u(x, t)=\begin{cases}0, & x\geqslant at,\\ g\left(t-\frac{x}{a}\right), & x<at,\end{cases}$$

if $g\in C^2(t\geqslant 0)$ and $g(0)=g'(0)=g''(0)=0$.

21.2. Prove that the problem

$$u_{tt}=a^2u_{xx},\quad t>0,\quad x>0;$$
$$u|_{t=0}=u_0(x),\quad u_t|_{t=0}=u_1(x),\quad u|_{x=0}=0$$

has a unique solution,

$$u(x, t)=\begin{cases}\frac{1}{2}[u_0(x+at)+u_0(x-at)]+\frac{1}{2a}\int_{x-at}^{x+at}u_1(\xi)\,d\xi, & x\geqslant at,\\ \frac{1}{2}[u_0(x+at)-u_0(at-x)]+\frac{1}{2a}\int_{a-x}^{at+x}u_1(\xi)\,d\xi, & x<at\end{cases}$$

if $u_0 \in C^2 (x \geqslant 0)$, $u_1 \in C^1 (x \geqslant 0)$, and $u_0 (0) = u_0'' (0) = u_1 (0) = 0$. Show that this solution follows from D'Alembert's formula (12.6) if both $u_0 (x)$ and $u_1 (x)$ are continued at $x < 0$ so that they become odd functions.

21.3. Prove that the problem

$$u_{tt} = a^2 u_{xx}, \quad t > 0, \quad x > 0;$$
$$u|_{t=0} = 0, \quad u_t|_{t=0} = 0, \quad u_x|_{x=0} = g(t)$$

has a unique solution,

$$u(x, t) = \begin{cases} 0, & x \geqslant at, \\ -a \int\limits_0^{t-\frac{x}{a}} g(\tau)\, d\tau, & x < at, \end{cases}$$

if $g \in C^1 (t \geqslant 0)$ and $g(0) = g'(0) = 0$.

21.4. Prove that the problem

$$u_{tt} = a^2 u_{xx}, \quad t > 0, \quad x > 0;$$
$$u|_{t=0} = u_0(x), \quad u_t|_{t=0} = u_1(x), \quad u_x|_{x=0} = 0$$

has a unique solution,

$$u(x, t) = \begin{cases} \frac{1}{2} [u_0(x+at) + u_0(x-at)] + \frac{1}{2a} \int\limits_{x-at}^{x+at} u_1(\xi)\, d\xi, & x \geqslant at, \\ \frac{1}{2} [u_0(x+at) + u_0(at-x)] + \frac{1}{2a} \left[\int\limits_0^{x+at} u_1(\xi)\, d\xi + \int\limits_0^{at-x} u_1(\xi)\, d\xi \right], & x < at, \end{cases}$$

if $u_0 \in C^2 (x \geqslant 0)$, $u_1 \in C^1 (x \geqslant 0)$, and $u_0' (0) = u_1' (0) = 0$. Show that this solution follows from D'Alembert's formula if both $u_0(x)$ and $u_1 (x)$ are continued for $x < 0$ so that they become even functions.

21.5. Prove that the problem

$$u_{tt} = a^2 u_{xx}, \quad t > 0, \quad 0 < x < l;$$
$$u|_{t=0} = 0, \quad u_t|_{t=0} = 0, \quad u|_{x=0} = g(t), \quad u|_{x=l} = 0$$

has a unique solution,

$$u(x, t) = \sum_{n=0}^{\infty} \left[\tilde{g}\left(t - \frac{x}{a} - \frac{2nl}{a}\right) - \tilde{g}\left(t + \frac{x}{a} - \frac{2(n+1)l}{a}\right) \right],$$

$$\tilde{g}(t) = \begin{cases} g(t), & t \geqslant 0, \\ 0, & t < 0, \end{cases}$$

if $g \in C^2 (t \geqslant 0)$ and $g(0) = g'(0) = g''(0) = 0$.

21.6. Prove that the problem

$$u_{tt} = a^2 u_{xx}, \quad t > 0, \quad 0 < x < l;$$

$$u|_{t=0} = u_0(x), \quad u_t|_{t=0} = u_1(x), \quad u|_{x=0} = 0, \quad u|_{x=l} = 0$$

has a unique solution,

$$u(x, t) = \frac{1}{2}[\tilde{u}(x+at) + \tilde{u}_0(x-at)] + \frac{1}{2a}\int_{x-at}^{x+at} \tilde{u}_1(\xi)\, d\xi,$$

where $\tilde{u}_0(x)$ and $\tilde{u}_1(x)$ are odd, $2l$-periodic functions that coincide with $u_0(x)$ and $u_1(x)$ for $0 \leqslant x \leqslant l$, if $u_0 \in C^2[0, l]$, $u_1 \in C^1[0, l]$, and $u_0(0) = u_0(l) = u_1(0) = u_1(l) = u_0''(0) = u_0''(l) = 0$.

In Problems 21.7-21.23 prove that each problem has a unique solution and find this solution.

21.7. $u_{tt} = a^2 u_{xx}, \quad t > 0, \quad x > 0;$

$$u|_{t=0} = 0, \quad u_t|_{t=0} = 0, \quad (u_x - \beta u)|_{x=0} = g(t)$$

$$g \in C^1(t \geqslant 0), \quad g(0) = g'(0) = 0.$$

21.8. $u_{tt} = a^2 u_{xx}, \quad t > 0, \quad x > 0;$

$$u|_{t=0} = u_0(x), \quad u_t|_{t=0} = 0, \quad (u_x - \beta u)|_{x=0} = 0,$$

$$u_0 \in C^2 \quad (x \geqslant 0), \quad u_0'(0) - \beta u_0(0) = 0.$$

21.9. $u_{tt} = a^2 u_{xx}, \quad t > 0, \quad 0 < x < l;$

$$u|_{t=0} = 0, \quad u_t|_{t=0} = 0, \quad u_x|_{x=0} = g(t), \quad u_x|_{x=l} = 0.$$

$$g \in C^1(t \geqslant 0), \quad g(0) = g'(0) = 0.$$

21.10. $u_{tt} = a^2 u_{xx}, \quad t > 0, \quad 0 < x < l;$

$$u|_{t=0} = u_0(x), \quad u_t|_{t=0} = u_1(x), \quad u_x|_{x=0} = 0, \quad u_x|_{x=l} = 0,$$

$$u_0 \in C^2([0, l]), \quad u_1 \in C^1([0, l]),$$

$$u_0'(0) = u_1'(0) = u_0'(l) = u_1'(l) = 0.$$

21.11. $u_{tt} = a^2 u_{xx}, \quad t > 0, \quad 0 < x < l;$

$$u|_{t=0} = 0, \quad u_t|_{t=0} = 0, \quad u|_{x=0} = g(t), \quad u_x|_{x=l} = 0,$$

$$g \in C^2(t \geqslant 0), \quad g(0) = g'(0) = g''(0) = 0.$$

21.12. $u_{tt} = a^2 u_{xx}, \quad t > 0, \quad 0 < x < l;$

$$u|_{t=0} = u_0(x), \quad u_t|_{t=0} = u_1(x), \quad u|_{x=0} = 0, \quad u_x|_{x=l} = 0,$$

$$u_0 \in C^2([0, l]), \quad u_1 \in C^1([0, l]),$$

$$u_0(0) = u_0''(0) = u_1(0) = u_0'(l) = u_1'(l) = 0.$$

21.13. $u_{tt} = u_{xx}, \quad t > 0, \quad x > 0;$

$$u|_{t=0} = x^2, \quad u_t|_{t=0} = x, \quad u|_{x=0} = t^2.$$

21.14. $u_{tt} = 4u_{xx} + 16t^2, \quad t > 0, \quad x > 0;$

$$u|_{t=0} = \frac{1}{6}x^4, \quad u_t|_{t=0} = 2\sin x, \quad u|_{x=0} = 4t^4.$$

21.15. $9u_{tt}=u_{xx}, \quad t>0, \quad x>0;$

$$u|_{t=0}=27x^3, \quad u_t|_{t=0}=0, \quad u|_{x=0}=t^3.$$

21.16. $u_{tt}=u_{xx}+2, \quad t>0, \quad x>0;$

$$u|_{t=0}=x+\cos x, \quad u_t|_{t=0}=1, \quad u_x|_{x=0}=1.$$

21.17. $u_{tt}=u_{xx}, \quad t>0, \quad x>0;$

$$u|_{t=0}=x, \quad u_t|_{t=0}=1, \quad u_x|_{x=0}=\cos t.$$

21.18. $u_{tt}=9u_{xx}+e^t, \quad t>0, \quad x>0;$

$$u|_{t=0}=1+x, \quad u_t|_{t=0}=4-3\cos\frac{x}{3}, \quad u_x|_{x=0}=2-\cos t.$$

21.19. $u_{tt}=3u_{xx}+2(1-6t^2)e^{-2x}, \quad t>0, \quad x>0;$

$$u|_{t=0}=1, \quad u_t|_{t=0}=x, \quad (u_x-2u)|_{x=0}=-2+t-4t^2.$$

21.20. $u_{tt}=u_{xx}, \quad t>0, \quad x>0;$

$$u|_{t=0}=0; \quad u_t|_{t=0}=0, \quad (u_x+u)|_{x=0}=1-\cos t.$$

21.21. $u_{tt}=u_{xx}+4, \quad t>0, \quad x>0;$

$$u|_{t=0}=1-x, \quad u_t|_{t=0}=0, \quad (u_x+u)|_{x=0}=\frac{3}{2}t^2.$$

21.22. $u_{tt}=u_{xx}, \quad t>0, \quad x>0;$

$$u|_{t=0}=x^2, \quad u_t|_{t=0}=0, \quad (u_t-u)|_{x=0}=2t-t^2.$$

21.23. 1. $u_{tt}=u_{xx}-6, \quad t>0, \quad x>0;$

$$u|_{t=0}=x^2, \quad u_t|_{t=0}=0, \quad (u_t+2u_x)|_{x=0}=-4t.$$

2. $u_{tt}=4u_{xx}+2, \quad t>0, \quad x>0;$

$$u|_{t=0}=2-x, \quad u_t|_{t=0}=2, \quad (u_t+3u_x)|_{x=0}=3t-e^t.$$

21.24. Find the largest region in which each problem given below has a unique solution, and find this solution.

1. $u_{tt}=u_{xx};$

$$u|_{t=0}=x^3, \quad u_t|_{t=0}=0, \quad 0\leqslant x\leqslant 2, \quad u|_{x=0}=t^3, \quad 0\leqslant t\leqslant 2.$$

2. $u_{tt}=u_{xx};$

$$u|_{t=0}=2x^3, \quad u_t|_{t=0}=0, \quad 0\leqslant x\leqslant 4, \quad u|_{t=3x}=0, \quad 0\leqslant x\leqslant 1.$$

21.25. Prove that the problem

$$u_{tt}=a^2\nabla^2 u, \quad t>0, \quad |x|>1, \quad x\in R^3,$$
$$u|_{t=0}=0, \quad u_t|_{t=0}=0, \quad u|_{|x|=1}=g(t)$$

has a unique solution,

$$u(x, t)=\begin{cases} 0, & |x|\geqslant 1+at, \\ \dfrac{1}{|x|}g\left(t+\dfrac{1-|x|}{a}\right), & 1<|x|<1+at, \end{cases}$$

if $g \in C^2$ $(t \geqslant 0)$ and $g(0) = g'(0) = g''(0) = 0$. Show that if $g(t)$ is a finite function, then $u(x, t) = 0$ for every fixed x, $|x| \geqslant 1$, and sufficiently large t's. In the case where $g(t) \neq 0$ for $0 < t < T$ and $g(t) = 0$ for $t \geqslant T$, find the time t_x at which the back wave-front of the wave passes through point x, $|x| > 1$.

21.26. Find the solution of the problem

$$u_{tt} = a^2 \nabla^2 u, \quad t > 0, \quad |x| > 1, \quad x \in R^3;$$
$$u|_{t=0} = \alpha(|x|), \quad u_t|_{t=0} = \beta(|x|), \quad u|_{|x|=1} = 0,$$

where $\alpha(r) \in C^2$ $(r \geqslant 1)$, $\beta(r) \in C^1$ $(r \geqslant 1)$, $\alpha(1) = 0$, $\alpha''(1) + 2\alpha'(1) = 0$, and $\beta(1) = 0$. Prove that if both $\alpha(r)$ and $\beta(r)$ are finite functions, then $u(x, t) = 0$ for every fixed x, $|x| \geqslant 1$, and sufficiently large t's.

21.27. Find the solution of the problem

$$u_{tt} = \nabla^2 u, \quad t > 0, \quad |x| > 1, \quad x \in R^3:$$
$$u|_{t=0} = 0, \quad u_t|_{t=0} = 0, \quad \left.\frac{\partial u}{\partial \mathbf{n}}\right|_{|x|=1} = g(t),$$

where $g \in C^1$ $(t \geqslant 0)$ and $g(0) = g'(0) = 0$. Prove that if $g(t)$ is a finite function, then there exists a function $c(x)$ such that $|u(x, t)| \leqslant c(x) e^{-t}$, while for $u(x, t)$ to vanish for each fixed x, $|x| \geqslant 1$, and for sufficiently large t's it is necessary and sufficient that $\int_0^\infty e^t g(t)\, dt = 0$.

21.28. Find the solution of the problem

$$u_{tt} = \nabla^2 u, \quad t > 0, \quad |x| > 1, \quad x \in R^3;$$
$$u|_{t=0} = \alpha(|x|), \quad u_t|_{t=0} = \beta(|x|), \quad \left.\frac{\partial u}{\partial \mathbf{n}}\right|_{|x|=1} = 0,$$

where $\alpha \in C^2$ $(r \geqslant 1)$, $\beta \in C^1$ $(r \geqslant 1)$, and $\alpha'(1) = \beta'(1) = 0$. Prove that if both $\alpha(r)$ and $\beta(r)$ are finite functions, then there exists a function $c(x)$ such that $|u(x, t)| \leqslant c(x) e^{-t}$, while for $u(x, t)$ to vanish for each fixed x, $|x| \geqslant 1$, and for sufficiently large t's it is necessary and sufficient that $\int_0^\infty r e^r [\alpha(r) - \beta(r)]\, dr = 0$.

Solve Problems 21.29-21.37.

21.29. $u_{tt} = \nabla^2 u, \quad t > 0, \quad |x| > 1, \quad x \in R^3;$

$$u|_{t=0} = 0, \quad u_t|_{t=0} = 0, \quad \left(ku + \frac{\partial u}{\partial n}\right)\Big|_{|x|=1} = g(t), \quad k = \text{const}.$$

21.30. $u_t = a^2 u_{xx} + f(x, t), \quad t > 0, \quad x > 0;$

$$u|_{t=0} = u_0(x), \quad u|_{x=0} = 0.$$

21.31. $u_t = a^2 u_{xx}, \quad t > 0, \quad x > 0;$

$$u|_{t=0} = 0, \quad u|_{x=0} = g(t).$$

21.32. $u_t = a^2 u_{xx}, \quad t>0, \quad x>0;$

$$u|_{t=0} = u_0(x), \quad u_x|_{x=0} = 0.$$

21.33. $u_t = a^2 u_{xx}, \quad t>0, \quad x>0;$

$$u|_{t=0} = 0, \quad u_x|_{x=0} = g_1(t).$$

21.34. $u_t = u_{xx}, \quad t>0, \quad x>0;$

$$u|_{t=0} = 0, \quad (u-u_x)|_{x=0} = g(t).$$

21.35. $u_t = a^2 u_{xx}, \quad t>0, \quad x>0;$

$$u|_{t=0} = u_0(x), \quad (u_x - hu)|_{x=0} = 0, \quad h \geqslant 0.$$

21.36. $\frac{\partial^2 u}{\partial t^2} + \frac{\partial^4 u}{\partial x^4} = 0, \quad t>0, \quad x>0;$

$$u|_{t=0} = u_0(x), \quad \left.\frac{\partial u}{\partial t}\right|_{t=0} = 0, \quad u|_{x=0} = g(t), \quad \left.\frac{\partial^2 u}{\partial x^2}\right|_{x=0} = 0.$$

21.37. $u_t = \alpha^2(x) u_{xx}$, $t>0$, $x \neq 0$, where $\alpha(x) = a$ for $x<0$ and $\alpha(x) = b$ for $x>0$;

$$u|_{t=0} = \theta(x), \quad u|_{x=-0} = u|_{x=+0}, \quad u_x|_{x=-0} = k u_x|_{x=+0}.$$

Answers to Problems of Sec. 21

21.7. 0 at $x \geqslant at$ and $-a e^{\beta(x-at)} \int_0^{t-\frac{x}{a}} e^{a\beta\tau} g(\tau)\, d\tau$ at $x<at$.

21.8. $\frac{1}{2}[u_0(x+at)+u_0(x-at)]$ at $x \geqslant at$ and $\frac{1}{2}[u_0(x+at)+$ $+u_0(at-x)] - \beta e^{\beta(x-at)} \int_0^{at-x} u_0(\xi) e^{\beta\xi}\, d\xi$ at $x<at$.

21.9. $\sum_{n=0}^{\infty} \left[f\left(t - \frac{x}{a} - \frac{2nl}{a}\right) - f\left(t + \frac{x}{a} - \frac{2(n+1)l}{a}\right)\right]$, where $f(x) = 0$ at $x \leqslant 0$ and $f(x) = -a \int_0^x g(\tau)\, d\tau$ at $x>0$.

21.10. $\frac{1}{2}[\tilde{u}_0(x+at)+\tilde{u}_0(x-at)] + \frac{1}{2a}\int_{x-at}^{x+at} \tilde{u}_1(\xi)\, d\xi$, where $\tilde{u}_0(x)$ and $\tilde{u}_1(x)$ are even, $2l$-periodic functions that coincide with $u_0(x)$ and $u_1(x)$ at $0 \leqslant x \leqslant l$.

21.11. $\sum_{n=0}^{\infty} (-1)^n \left[\tilde{g}\left(t - \frac{x}{a} - \frac{2ln}{a}\right) + \tilde{g}\left(t + \frac{x}{a} - \frac{2l(n+1)}{a}\right)\right]$, $\tilde{g}(t) = 0$ at $t<0$ and $\tilde{g}(t) = g(t)$ at $t \geqslant 0$.

21.12. $\frac{1}{2}[\tilde{u}_0(x+at)+\tilde{u}_0(x-at)]+\frac{1}{2a}\int\limits_{x-at}^{x+at}\tilde{u}_1(\xi)\,d\xi$, where $\tilde{u}_0(x)$ and $\tilde{u}_1(x)$ are odd functions that coincide with $u_0(x)$ and $u_1(x)$ at $0\leqslant x\leqslant l$, while $\tilde{u}_0(x-l)$ and $\tilde{u}_1(x-l)$ are even functions.

21.13. x^2+xt+t^2. **21.14.** $4t^4+4t^2x^2+\frac{1}{6}x^4+\sin 2t\sin x$.

21.15. $9xt^2+27x^3$ at $x\geqslant\frac{1}{3}t$; t^3+27x^2 at $x<\frac{1}{3}t$.

21.16. $x+t+t^2+\cos x\cos t$.

21.17. $x+t$ at $x\geqslant t$ and $2t+\sin(x-t)$ at $x<t$.

21.18. $x+3t+e^t-3\sin t\cos\frac{x}{3}$ at $x\geqslant 3t$ and $2x+e^t-3\cos t\sin\frac{x}{3}$ at $x<3t$. **21.19.** $1+xt+t^2e^{-2x}$.

21.20. 0 at $x\geqslant t$ and $1-\frac{1}{2}e^{t-x}-\frac{1}{2}[\sin(x-t)+\cos(x-t)$ at $x<t$.

21.21. $1-x+2t^2$ at $x\geqslant t$ and $2t^2-t-\frac{1}{2}(x-t)^2+e^{t-x}$ at $x<t$.

21.22. x^2+t^2. **21.23.** 1. x^2-2t^2. 2. $2+2t-x+t^2$ at $x\geqslant 2t$ and $xt-\frac{1}{4}x^2+2e^{t-x/2}$ at $x<2t$.

21.24. 1. x^3+3xt^2 at $0\leqslant x+t\leqslant 2$, $0\leqslant x-t\leqslant 2$, and $3x^2t+t^3$ at $0\leqslant x+t\leqslant 2$, $-2\leqslant x-t\leqslant 0$; 2. $2x^3+6xt^2$ at $0\leqslant x+t\leqslant 4$, $0\leqslant x-t\leqslant 4$, and $(x+t)^3+8(x-t)^3$ at $0\leqslant x+t\leqslant 4$, $-2\leqslant x-t\leqslant 0$. **21.25.** $t_x=T+\frac{|x|-1}{a}$.

21.26. $\frac{2}{2|x|}[(|x|+at)\alpha(|x|+at)+(|x|-at)\alpha(|x|-at)]+\frac{1}{2a|x|}\int\limits_{|x|-at}^{|x|+at}\xi\beta(\xi)\,d\xi$ at $|x|\geqslant 1+at$ and $\frac{1}{2|x|}[(|x|+at)\alpha\times(|x|+at)-(2-|x|+at)\alpha(2-|x|+at)]+\frac{1}{2a|x|}\times\int\limits_{2-|x|+at}^{|x|+at}\xi\beta(\xi)\,d\xi$ at $1<|x|<1+at$. **21.27.** 0 at $|x|\geqslant 1+t$ and $\frac{1}{|x|}e^{|x|-t-1}\int\limits_0^{t+1-|x|}e^{\tau}g(\tau)\,d\tau$ at $1<|x|<1+t$.

21.28. $\frac{1}{2|x|}[(|x|+t)\alpha(|x|+t)+(|x|-t)\alpha(|x|-t)]+\frac{1}{2|x|}\int\limits_{|x|-t}^{|x|+t}\xi\beta(\xi)\,d\xi$ at $|x|\geqslant 1+t$ and $\frac{1}{2|x|}[(|x|+t)\alpha(|x|+t)+(2-|x|+t)\alpha(2-|x|+t)]+$

$$+\frac{1}{2|x|}\int_{2-|x|+t}^{|x|+t}\xi\beta(\xi)\,d\xi-\frac{1}{|x|}e^{|x|-t-2}\int_{1}^{2-|x|+t}\xi e^{\xi}\,[\alpha(\xi)-$$
$-\beta(\xi)]\,d\xi$ at $1<|x|<1+t$.

21.29. 0 at $|x|\geqslant 1+t$ and $\frac{1}{|x|}e^{(k+1)(|x|-t-1)}\int_{0}^{t+1-|x|}e^{(k+1)\tau}g(\tau)\,d\tau$ at $1<|x|<1+t$.

21.30. $$\frac{1}{2a\sqrt{\pi t}}\int_{0}^{\infty}u_0(\xi)\,[e^{-(x-\xi)^2/(4a^2t)}-e^{-(x+\xi)^2/(4a^2t)}]\,d\xi+\frac{1}{2a\sqrt{\pi}}\times$$
$$\times\int_{0}^{t}\int_{0}^{\infty}\frac{f(\xi,\tau)}{\sqrt{t-\tau}}\left[e^{-\frac{(x-\xi)^2}{4a^2(t-\tau)}}-e^{-\frac{(x+\xi)^2}{4a^2(t-\tau)}}\right]d\xi\,d\tau.$$

21.31. $$\frac{x}{2a\sqrt{\pi}}\int_{0}^{t}\frac{g(t-\tau)}{\tau^{3/2}}e^{-x^2/(4a^2\tau)}\,d\tau.$$

21.32. $$\frac{1}{2a\sqrt{\pi t}}\int_{0}^{\infty}u_0(\xi)\,[e^{-(x-\xi)^2/(4a^2t)}+e^{-(x+\xi)^2/(4a^2t)}]\,d\xi.$$

21.33. $$\frac{-a}{\sqrt{\pi}}\int_{0}^{t}\frac{g(t-\tau)}{\sqrt{\tau}}e^{-x^2/(4a^2\tau)}\,d\tau.$$

21.34. $$\frac{-1}{\sqrt{\pi}}\int_{0}^{t}\frac{g(t-\tau)}{\sqrt{\tau}}e^{-x^2/(4\tau)}\,d\tau+\frac{2}{\sqrt{\pi}}e^{x}\int_{0}^{t}g(t-\tau)e^{\tau}\times$$
$$\times\int_{\sqrt{\tau}+\frac{x}{2\sqrt{\tau}}}e^{-\alpha^2}\,d\alpha\,d\tau.$$

21.35. $$\frac{1}{2a\sqrt{\pi t}}\int_{0}^{\infty}u_0(\xi)\left\{e^{-\frac{(x-\xi)^2}{4a^2t}}+e^{-\frac{(x+\xi)^2}{4a^2t}}-2h\int_{0}^{\infty}e^{-\frac{(x+\xi+\eta)^2}{4a^2t}-h\eta}\times\right.$$
$\left.\times\,d\eta\right\}d\xi.$ **21.36.** $$\frac{1}{2\sqrt{\pi t}}\int_{0}^{\infty}u_0(\xi)\left\{\cos\left[\frac{(x-\xi)^2}{4t}-\frac{\pi}{4}\right]-\right.$$
$$\left.-\cos\left[\frac{(x+\xi)^2}{4t}-\frac{\pi}{4}\right]\right\}d\xi+\frac{x}{2\sqrt{\pi}}\int_{0}^{t}\frac{g(t-\tau)}{\tau^{3/2}}\cos\left(\frac{x^2}{4\tau}-\frac{\pi}{4}\right)d\tau.$$

21.37. $$\frac{2ka}{(b-ka)\sqrt{\pi}}\int_{-x/(2a\sqrt{t})}^{\infty}e^{-t^2}\,dt$$ at $x<0$ and
$$\frac{ka}{b+ka}\left\{1\;\frac{2b}{ka\sqrt{\pi}}\int_{0}^{x/(2b\sqrt{t})}e^{-t^2}\,dt\right\}$$ at $x>0$.

Appendix

Examples of Solution Techniques for Some Typical Problems

A1 Method of Characteristics

Example 1. To find the solution to the Cauchy problem for the equation

$$y^2 u_{xy} + u_{yy} - \frac{2}{y} u_y = 0 \tag{A1.1}$$

in the half-plane $y > 0$ that satisfies the initial conditions

$$u|_{y=1} = 1 - x, \quad u_y|_{y=1} = 3. \tag{A1.2}$$

Solution. First we find the general solution to Eq. (A1.1) in the half-plane $y > 0$. To this end we reduce the equation to canonical form. The characteristic equation $-y^2\, dx\, dy + (dx)^2 = 0$ splits into two equations, $dx = 0$ and $-y^2\, dy + dx = 0$, for which $x = C$ and $3x - y^3 = C$ are the general solutions. Hence, in Eq. (A1.1) we must introduce new variables, $\xi = x$ and $\eta = 3x - y^3$. Then $u_y = -3y^2 u_\eta$, $u_{xy} = -3y^2 u_{\xi\eta} - 9y^2 u_{\eta\eta}$, and $u_{yy} = 9y^4 u_{\eta\eta} - 6yu_\eta$, and the canonical form of Eq. (A1.1) is $u_{\xi\eta} = 0$. Integrating this equation, we find that $u = f(\xi) + g(\eta) = f(x) + g(3x - y^3)$.

Now we use the initial conditions (A1.2):

$$\left.\begin{aligned} f(x) + g(3x-1) &= 1 - x, \\ -3g'(3x-1) &= 3. \end{aligned}\right\}$$

Solving this system yields $f(x) = 2x + C$ and $g(x) = -x - C$. Hence, the solution to the problem (A1.1), (A1.2) is the function

$$u(x, y) = 2x + C + (-3x + y^3 - C), \text{ or } u(x, y) = y^3 - x.$$

Example 2. To find the solution to Goursat's problem for the equation

$$u_{xx} + 3u_{xy} - 4u_{yy} - u_x + u_y = 0 \tag{A1.3}$$

in the entire plane that satisfies the conditions

$$u|_{y=4x} = 5x + e^x, \quad u|_{y=-x} = 1. \tag{A1.4}$$

Solution. Let us find the general solution to Eq. (A1.3). The characteristic equation $(dy)^2 - 3dx\, dy - 4(dx)^2 = 0$ splits into two equations, $dy + dx = 0$ and $dy - 4dx = 0$, for which $y + x = C$ and $y - 4x = C$ are the general solutions. Substituting $\xi = y + x$ and $\eta = y - 4x$ into Eq. (A1.3), we reduce Eq. (A1.3) to the canonical form $u_{\xi\eta} - (1/5)\, u_\eta = 0$. Integration yields $u = f(\eta)\, e^{-\xi/5} + g(\xi) = f(y - 4x)\, e^{-(y+x)/5} + g(y + x)$.

Now we use the initial conditions (A1.4):

$$\left.\begin{aligned} f(0)\, e^{-x} + g(5x) &= 5x + e^x, \\ f(-5x) + g(0) &= 1. \end{aligned}\right\} \qquad (A1.5)$$

Solving this system yields $f(x) = 1 - g(0)$ and $g(x) = x + e^{x/5} - f(0)\, e^{-x/5}$. Hence, $u(x, y) = [1 - g(0)]\, e^{-(x+y)/5} + x + y + e^{(x+y)/5} - f(0)\, e^{-(x+y)/5}$. Bearing in mind that (A1.5) at $x = 0$ implies $f(0) + g(0) = 1$, we arrive at the solution to the problem (A1.3), (A1.4), $u(x, y) = x + y + e^{(x+y)/5}$.

Example 3. To find the solution to the mixed problem for the equation

$$u_{tt} - 4u_{xx} = 6xt \qquad (A1.6)$$

in the region $x > 0$, $t > 0$ that satisfies the following conditions:

$$u\,|_{t=0} = x^3, \quad u_t\,|_{t=0} = 0, \quad u\,|_{x=0} = t^3. \qquad (A1.7)$$

Solution. The general solution to Eq. (A1.6) has the form $u(x, t) = f(x + 2t) + g(x - 2t) + xt^3$. The conditions (A1.7) then yield

$$\left.\begin{aligned} f(x) + g(x) &= x^3, & x \geqslant 0, \\ f'(x) - g'(x) &= 0, & x \geqslant 0, \\ f(2t) + g(-2t) &= t^3, & t \geqslant 0. \end{aligned}\right\} \qquad (A1.8)$$

The first two equations of this system yield $f(x) = (1/2)\, x^3 + C$ and $g(x) = (1/2)\, x^3 - C$, $x \geqslant 0$. Substituting $f(x)$ into the third equation yields $g(x) = (3/8)\, x^3 - C$, $x \leqslant 0$. Hence, the following function is the solution to the problem (A1.6), (A1.7):

$$u(x, t) = \begin{cases} \frac{1}{2}(x + 2t)^3 + \frac{1}{2}(x - 2t)^3 + xt^3, & x \geqslant 2t, \\ \frac{1}{2}(x + 2t)^3 + \frac{3}{8}(x - 2t)^3 + xt^3, & x < 2t. \end{cases}$$

Example 4. To find the solution to the mixed problem for the equation

$$u_{tt} - 9u_{xx} = 2 \qquad (A1.9)$$

in the region $x > 0$, $t > 0$ that satisfies the following conditions:

$$u|_{t=0} = x + x^3, \; u_t|_{t=0} = -9x^2, \; (u - u_x)|_{x=0} = t^2 - 1. \qquad (A1.10)$$

Solution. The general solution to Eq. (A1.9) has the form $u(x, t) = f(x + 3t) + g(x - 3t) + t^2$. Conditions (A1.10) yield

$$\left.\begin{aligned} f(x) + g(x) &= x + x^3, & x \geqslant 0, \\ 3f'(x) - 3g'(x) &= -9x^2, & x \geqslant 0, \\ f(3t) + g(-3t) - f'(3t) - g'(-3t) &= -1, & t \geqslant 0. \end{aligned}\right\} \qquad (A1.11)$$

The first two equations of this system yield $f(x) = (1/2)\, x + C$ and $g(x) = (1/2)\, x + x^3 - C$, $x \geqslant 0$. Substituting $f(x)$ into the third equation yields $g'(x) - g(x) = C + 1/2 - (1/2)\, x$, whence $g(x) = C_1 e^x + (1/2)\, x - C$, $x \leqslant 0$. The fact that $g(x)$ is continuous at $x = 0$ yields $C_1 = 0$, that is, $g(x) = (1/2)x - C$, $x \leqslant 0$. Hence, the following function is the solution to the problem (A1.9), (A1.10):

$$u(x, t) = \begin{cases} (x-3t)^2 + x + t^2, & x \geqslant 3t, \\ x + t^2, & x < 3t. \end{cases}$$

A2 Fourier's Method

Example 5. To solve the mixed problem for the nonhomogeneous equation of the hyperbolic type

$$u_{tt} - u_{xx} = 2t, \quad 0 < x < 1, \quad t > 0 \tag{A2.1}$$

with the initial conditions

$$u|_{t=0} = 0, \quad u_t|_{t=0} = x \tag{A2.2}$$

and the boundary conditions

$$u|_{x=0} = 0, \quad u_x|_{x=1} = t. \tag{A2.3}$$

Solution. We start by selecting a function w such that it satisfies the boundary conditions (A2.3). For instance, take $w = xt$. Then $w_{tt} - w_{xx} = 0$, $w|_{t=0} = 0$, and $w_t|_{t=0} = x$. Hence, the function

$$v(x, t) = u(x, t) - xt \tag{A2.4}$$

satisfies the equation

$$v_{tt} - v_{xx} = 2t, \tag{A2.5}$$

the homogeneous boundary conditions

$$v|_{x=0} = 0, \quad v_x|_{x=1} = 0, \tag{A2.6}$$

and zero initial conditions,

$$v|_{t=0} = 0, \quad v_t|_{t=0} = 0. \tag{A2.7}$$

To apply Fourier's method to the homogeneous equation $v_{tt} - v_{xx} = 0$ under conditions (A2.6) and (A2.7) we put $v(x, t) = X(x)T(t)$. We arrive at the following Sturm-Liouville problem:

$$X''(x) + \lambda^2 X = 0, \quad X(0) = 0, \quad X'(1) = 0.$$

Solving the problem, we find its eigenvalues $\lambda_n = \pi/2 + \pi n$, $n = 0, 1, 2, \ldots$, and the corresponding eigenfunctions

$$X_n(x) = \sin \lambda_n x. \tag{A2.8}$$

We look for the solution to the problem (A2.5)-(A2.7) in the form

$$v(x, t) = \sum_{n=0}^{\infty} T_n(t) \sin \lambda_n x, \tag{A2.9}$$

where

$$T_n(0) = 0, \quad T_n'(0) = 0. \tag{A2.10}$$

Substituting (A2.9) into Eq. (A2.5) yields

$$\sum_{n=0}^{\infty} (T_n''(t) + \lambda_n^2 T_n(t)) \sin \lambda_n x = 2t. \tag{A2.11}$$

To find the $T_n(t)$, we write the following expansion in a Fourier series in the system of functions (A2.8) in (0, 1):

$$1=\sum_{n=0}^{\infty} a_n \sin \lambda_n x. \tag{A2.12}$$

Since $\int_0^1 \sin^2 \lambda_n x\, dx = \frac{1}{2}$, we can write $a_n = 2\int_0^1 \sin \lambda_n x\, dx = \frac{2}{\lambda_n}$, and from (A2.11) and (A2.12) we obtain

$$T_n''(t)+\lambda_n^2 T_n(t)=\frac{4t}{\lambda_n}. \tag{A2.13}$$

The general solution to Eq. (A2.13) has the form $T_n(t)= =\frac{4t}{\lambda_n^3}+A\sin\lambda_n t+B\cos\lambda_n t$. Employing condition (A2.10), we find that $B=0$ and $A=-4/\lambda_n^4$. Substituting $T_n(t)=\frac{4t}{\lambda_n^3}-\frac{4}{\lambda_n^4}\sin\lambda_n t$ into (A2.9) and employing (A2.4), we arrive at the solution to the problem (A2.1)-(A2.3):

$$u=xt+4\sum_{n=0}^{\infty}\frac{1}{\lambda_n^4}(\lambda_n t-\sin\lambda_n t)\sin\lambda_n x,$$

where $\lambda_n = \pi/2 + \pi n$.

Example 6. To solve the mixed problem for the nonhomogeneous equation of the parabolic type

$$u_t - u_{xx} = t(x+1), \quad 0<x<1, \quad t>0 \tag{A2.14}$$

with the initial condition

$$u|_{t=0} = 0 \tag{A2.15}$$

and the boundary conditions

$$u_x|_{x=0} = t^2, \quad u|_{x=1} = t^2. \tag{A2.16}$$

Solution. The function $w = xt^2$ satisfies the boundary condition (A2.16), the equation $w_t - w_{xx} = 2xt$, and the initial condition $w|_{t=0} = 0$. Therefore, the function

$$v = u - xt^2 \tag{A2.17}$$

satisfies the equation

$$v_t - v_{xx} = (1-x)t \tag{A2.18}$$

and the conditions

$$v|_{t=0} = 0, \quad v_x|_{x=0} = 0, \quad v|_{x=1} = 0. \tag{A2.19}$$

To apply Fourier's method to the homogeneous equation $v_t - v_{xx} = 0$ under conditions (A2.19) we put $v = X(x)T(t)$. We arrive at the Sturm-Liouville problem

$$X''(x)+\lambda^2 X(x)=0, \quad X'(0)=0, \quad X(1)=0,$$

whose eigenvalues are the numbers $\lambda_n = \pi/2 + \pi n$, $n = 0, 1, 2, \ldots$, and whose eigenfunctions are

$$X_n(x) = \cos\lambda_n x. \tag{A2.20}$$

We look for the solution to the problem (A2.18), (A2.19) in the form

$$v(x, t)=\sum_{n=0}^{\infty} T_n(t)\cos\lambda_n x. \tag{A2.21}$$

Substituting this into Eq. (A2.18), we obtain

$$\sum_{n=0}^{\infty} (T'_n(t)+\lambda_n^2 T_n(t))\cos\lambda_n x=(1-x)\,t. \tag{A2.22}$$

Next we expand $1-x$ in a Fourier series in the system of functions (A2.20) in $(0, 1)$:

$$1-x=\sum_{n=0}^{\infty} a_n\cos\lambda_n x. \tag{A2.23}$$

Since $a_n=2\int_0^1 (1-x)\cos\lambda_n x\,dx=\frac{2}{\lambda_n^2}$, from (A2.22) and (A2.23) it follows that

$$T'_n(t)+\lambda_n^2 T(t)=\frac{2t}{\lambda_n^2}. \tag{A2.24}$$

The following function is the solution to Eq. (A2.24) at $T_n(0)=0$:

$$T_n(t)=2\lambda_n^{-6}(e^{-\lambda_n^2 t}+\lambda_n^2 t-1). \tag{A2.25}$$

From (A2.17), (A2.21), and (A2.25) we then find the solution to the problem (A2.14)-(A2.16):

$$u=xt^2+2\sum_{n=0}^{\infty}\lambda_n^{-6}(e^{-\lambda_n^2}+\lambda_n^2 t-1)\cos\lambda_n x,$$

where $\lambda_n=\pi/2+\pi n$.

A3 Integral Equations with a Degenerate Kernel

Example 7. To solve the integral equation

$$\varphi(x)=\lambda\int_{-\pi}^{\pi}(x\sin y+y\cos x)\,\varphi(y)\,dy+a\sin x+bx \tag{A3.1}$$

for all admissible values of a, b, and λ.

Solution. We introduce the notation

$$C_1=\int_{-\pi}^{\pi}\sin y\cdot\varphi(y)\,dy,\qquad C_2=\int_{-\pi}^{\pi}y\varphi(y)\,dy; \tag{A3.2}$$

then Eq. (A3.1) takes the form

$$\varphi(x)=\lambda C_1 x+\lambda C_2\cos x+a\sin x+bx. \tag{A3.3}$$

Then from (A3.2) and (A3.3) we obtain

$$\left.\begin{aligned} C_1 &= \int_{-\pi}^{\pi} \sin y\,(\lambda C_1 y + \lambda C_2 \cos y + a \sin y + by)\,dy, \\ C_2 &= \int_{-\pi}^{\pi} y\,(\lambda C_1 y + \lambda C_2 \cos y + a \sin y + by)\,dy, \end{aligned}\right\}$$

from which we find that

$$\left.\begin{aligned} C_1 &= \lambda C_1 \cdot 2\pi + a\pi + 2\pi b, \\ C_2 &= \lambda C_1 \frac{2\pi^3}{3} + a2\pi + b\,\frac{2\pi^3}{3}. \end{aligned}\right\} \tag{A3.4}$$

We write system (A3.4) as follows:

$$\left.\begin{aligned} C_1(1-2\pi\lambda) &= a\pi + 2\pi b, \\ -\lambda \frac{2\pi^3}{3} C_1 + C_2 &= 2a\pi + \frac{2\pi^3 b}{3}. \end{aligned}\right\} \tag{A3.5}$$

The system determinant $\Delta(\lambda)$ is equal to $1 - 2\pi\lambda$. If $\Delta(\lambda) \neq 0$, that is, $\lambda \neq 1/2\pi$, then the system has a unique solution for all a and b:

$$C_1 = \frac{a\pi + 2\pi b}{1-2\pi\lambda}, \quad C_2 = \frac{2\pi^3\lambda\,(a\pi + 2\pi b)}{3\,(1-2\pi\lambda)} + 2a\pi + \frac{2\pi^3 b}{3}. \tag{A3.6}$$

Substituting (A3.6) into (A3.3), we find for $\lambda \neq 1/2\pi$ the unique solution to the integral equation (A3.1).

But suppose $\lambda = 1/2\pi$. Then (A3.5) takes the form

$$\left.\begin{aligned} C_1 \cdot 0 &= (a+2b)\,\pi, \\ -\frac{\pi^2}{3} C_1 + C_2 &= 2a\pi + \frac{2\pi^3 b}{3}. \end{aligned}\right\} \tag{A3.7}$$

This system has a solution if and only if

$$a + 2b = 0, \tag{A3.8}$$

which is a necessary and sufficient condition for Eq. (A3.1) to have a solution at $\lambda = 1/2\pi$ $\Big(1/2\pi$ is the characteristic number of the integral solution

$$\varphi(x) = \lambda \int_{-\pi}^{\pi} (x \sin y + y \cos x)\,\varphi(y)\,dy\Big).$$

The general solution of the linear homogeneous system

$$\left.\begin{aligned} C_1 \cdot 0 &= 0, \\ -\frac{\pi^2}{3} C_1 + C_2 &= 0 \end{aligned}\right\}$$

corresponding to (A3.7) has the form

$$\tilde{C}_1 = C, \quad \tilde{C}_2 = \frac{\pi^2}{3} C,$$

where C is an arbitrary constant. For a particular solution to (A3.7) we can take

$$C_1^0 = 0, \quad C_2^0 = 2a\pi - \frac{a\pi^3}{3}.$$

For this reason the general solution to (A3.7) has the form

$$C_1 = C, \quad C_2 = \frac{\pi^2}{3} C + a\pi \left(2 - \frac{\pi^2}{3}\right). \tag{A3.9}$$

Substituting (A3.9) in (A3.3), we can find all the solutions of Eq. (A3.1) at $\lambda = 1/2\pi$ under condition (A3.8). These solutions can be given by a single formula,

$$\varphi(x) = \left(A - \frac{a}{2}\right) x + \left[\frac{A\pi^2}{3} + a\left(1 - \frac{\pi^2}{6}\right)\right] \cos x + a \sin x,$$

where A is an arbitrary constant.

A4 Variational Problems

Example 8. Find the minimum of the functional

$$I(v) = \int_G \left[|\operatorname{grad} v|^2 + \frac{4v}{\sqrt{x_1^2 + x_2^2}} \right] dx_1\, dx_2 \tag{A4.1}$$

among functions that belong to the class $\mathring{C}^1(G)$, where $G = \{1 < |x| < 3\}$, $x = (x_1, x_2)$.

Solution. It is known that there is a function $v_0(x_1, x_2)$ that belongs to $\mathring{C}^1(G)$ and minimizes (A4.1). This function is the solution to the boundary value problem

$$\nabla^2 u = \frac{2}{r}, \quad u|_{|x|=1} = u|_{|x|=3} = 0.$$

Writing ∇^2 in polar coordinates, we obtain

$$(r u_r)' = 2, \quad u|_{|x|=1} = u|_{|x|=3} = 0. \tag{A4.2}$$

The solution to the boundary value problem (A4.2) is the function $v_0 = 2(r-1) - (4/\ln 3)\ln r$. Since v_0 is independent of φ, we can write $|\operatorname{grad} v_0|^2 = \left|\frac{\partial v_0}{\partial r}\right|^2 = \left(2 - \frac{4}{\ln 3}\frac{1}{r}\right)^2$. Then

$$I(v_0) = \int_0^{2\pi}\int_1^3 \left\{\left(2 - \frac{4}{\ln 3}\frac{1}{r}\right)^2 + \left[8(r-1) - \frac{16}{\ln 3}\ln r\right]\frac{1}{r}\right\} r\, dr\, d\varphi$$

$$= 2\pi \int_1^3 \left(4r - \frac{16}{\ln 3} + \frac{16}{\ln^2 3}\frac{1}{r} + 8r - 8 - \frac{16}{\ln 3}\ln r\right) dr$$

$$= 2\pi \int_1^3 \left(12r - \frac{16}{\ln 3} - 8 + \frac{16}{\ln^2 3}\frac{1}{r} - \frac{16}{\ln 3}\ln r\right) dr = 32\pi\left(\frac{1}{\ln 3} - 1\right).$$

Hence, the minimum of (A4.1) is equal to $32\pi\left(\frac{1}{\ln 3} - 1\right)$.

References

Antosik, I., J. Mikusinski, and R. Sikorski
[1] *Theory of Distributions: The Sequential Approach*, Elsevier, Amsterdam (1973).

Arsenin, V. Ya.
[1] *Basic Equations and Special Functions of Mathematical Physics*, Nauka, Moscow (1974) [English transl.: Iliffe, London (1968)].

Beklemishev, D. V.
[1] *A Course of Analytic Geometry and Linear Algebra*, Nauka, Moscow (1980) [in Russian].

Brychkov, Yu. A., and A. P. Prudnikov
[1] *Integral Transformations of Generalized Functions*, Nauka, Moscow (1977) [in Russian].

Budak, B. M., A. A. Samarskii, and A. N. Tikhonov
[1] *A Collection of Problems on Mathematical Physics*, Nauka, Moscow (1980) [English transl.: Pergamon Press, Oxford (1964)].

Courant, R., and D. Hilbert
[1] *Methods of Mathematical Physics*, 2 vols., Interscience, New York (1953, 1962).

Gel'fand, I. M., and G. E. Shilov
[1] *Generalized Functions*, vols. 1-3, Fizmatgiz, Moscow (1959, 1958, 1958) [English transl.: Academic Press, New York (1964, 1968, 1967)].

Godunov, S. K.
[1] *Equations of Mathematical Physics*, 2nd ed., Nauka, Moscow (1979) [in Russian].

Gradshtein, I. S., and I. M. Ryzhik
[1] *Tables of Integrals, Series and Products*, Nauka, Moscow (1971) [English transl.: Academic Press, New York (1966)].

Hörmander, L.
[1] *Linear Partial Differential Operators*, Springer, Berlin (1963).

Jahnke, E., and F. Emde
[1] *Tables of Functions with Formulae and Curves*, Dover, New York (1945).

Jeffreys, H., and B. Swirles
[1] *Methods of Mathematical Physics*, 3rd ed., Cambridge University Press, London (1962).

Kolmogorov, A. N., and S. V. Fomin
[1] *Introductory Real Analysis*, Nauka, Moscow (1976) [English transl.: Dover, New York (1975)].

Korn, G. A., and T. M. Korn
[1] *Mathematical Handbook for Scientists and Engineers*, 2nd ed., McGraw-Hill, New York (1968).

Koshlyakov, N. S., E. B. Gliner, and M. M. Smirnov
[1] *Differential Equations of Mathematical Physics*, Vysshaya shkola, Moscow (1970) [North-Holland, Amsterdam (1964)].

Kudryavtsev, L. D.
[1] *Fundamentals of Mathematical Analysis*, 2 vols., Vysshaya shkola, Moscow (1981) [in Russian].

Ladyzhenskaya, O. A.
[1] *Boundary Value Problems of Mathematical Physics*, Nauka, Moscow (1973) [in Russian].

Lavrent'ev, M. A., and B. V. Shabat
[1] *Methods of the Theory of Functions of a Complex Variable*, Nauka, Moscow (1973) [in Russian].

Mikhailov, V. P.
[1] *Partial Differential Equations*, Nauka, Moscow (1976) [English transl.: Mir Publishers, Moscow (1978)].

Miranda, C.
[1] *Partial Differential Equations of Elliptic Type*, 2nd ed., Springer, New York (1970).

Nikiforov, A. F., and V. B. Uvarov
[1] *Special Functions of Mathematical Physics*, Nauka, Moscow (1978) [in Russian].

Nikol'skii, S. M.
[1] *A Course of Mathematical Analysis*, 2 vols., Nauka, Moscow (1975) [English transl.: Mir Publishers, Moscow (1977; reprinted 1981)].

Petrovskii, I. G.
[1] *Lectures on Partial Differential Equations*, Nauka, Moscow (1970) [English transl.: Wiley-Interscience, New York (1955)].
[2] *Lectures on the Theory of Integral Equations*, Nauka, Moscow (1965) [English transl.: Mir Publishers, Moscow (1971; reprinted 1975)].

Polozhii, G. N.
[1] *Equations of Mathematical Physics*, Vysshaya shkola, Moscow (1964) [in Russian].

Pontryagin, L. S.
[1] *Ordinary Differential Equations*, Nauka, Moscow (1974) [English transl.: Addison-Wesley, Reading, Mass. (1962)].

Schwartz, L.
[1] *Méthodes mathématiques pour les sciences physiques*, Hermann, Paris (1961).

Sidorov, Yu. V., M. V. Fedoryuk, and M. I. Shabunin
[1] *Lectures on the Theory of Functions of a Complex Variable*, Nauka, Moscow (1982) [English transl.: Mir Publishers, Moscow (1985)].

Smirnov, M. M.
[1] *Problems on the Equations of Mathematical Physics*, Nauka, Moscow (1975) [English transl.: Noordhoff, Groningen (1966)].

Smirnov, V. I.
[1] *A Course of Higher Mathematics*, vol. 2: *Advanced Calculus*, Nauka, Moscow (1967) [English transl.: Addison-Wesley, Reading, Mass. (1964)].
[2] *A Course of Higher Mathematics*, vol. 4: *Boundary Value Problems, Integral Equations and Partial Differential Equations*, Nauka, Moscow (1974) [English transl.: Addison-Wesley, Reading, Mass. (1964)].

Sneddon, I. N.
[1] *Fourier Transforms*, McGraw-Hill, New York (1951).
[2] *The Use of Integral Transforms*, McGraw-Hill, New York (1972).

Sobolev, S. L.
[1] *Partial Differential Equations of Mathematical Physics*, Nauka, Moscow (1966) [English transl.: Addison-Wesley, Reading, Mass. (1964)].

Stepanov, V. V.
[1] *A Course of Differential Equations*, Fizmatgiz, Moscow (1959) [in Russian].

Tikhonov, A. N., and A. A. Samarskii
[1] *Equations of Mathematical Physics*, Nauka, Moscow (1977) [English transl.: Pergamon Press, Oxford (1963)].

Tricomi, F. G.
[1] *Integral Equations*, Interscience, New York (1957).

Vladimirov, V. S.
[1] *Generalized Functions in Mathematical Physics*, Nauka, Moscow (1979) [English transl.: Mir Publishers, Moscow (1979)].
[2] *Equations of Mathematical Physics*, Nauka, Moscow (1981) [English transl.: Mir Publishers, Moscow (1984)].

Yosida, K.
[1] *Lectures on Differential and Integral Equations*, Interscience, New York (1960).

Zemanian, A. H.
[1] *Distribution Theory and Transform Analysis*, McGraw-Hill, New York (1965).

Subject Index

Made in the USA
Las Vegas, NV
12 November 2024

11563098R00162